SMARTBOOK

2nd Ed. w/Change 1

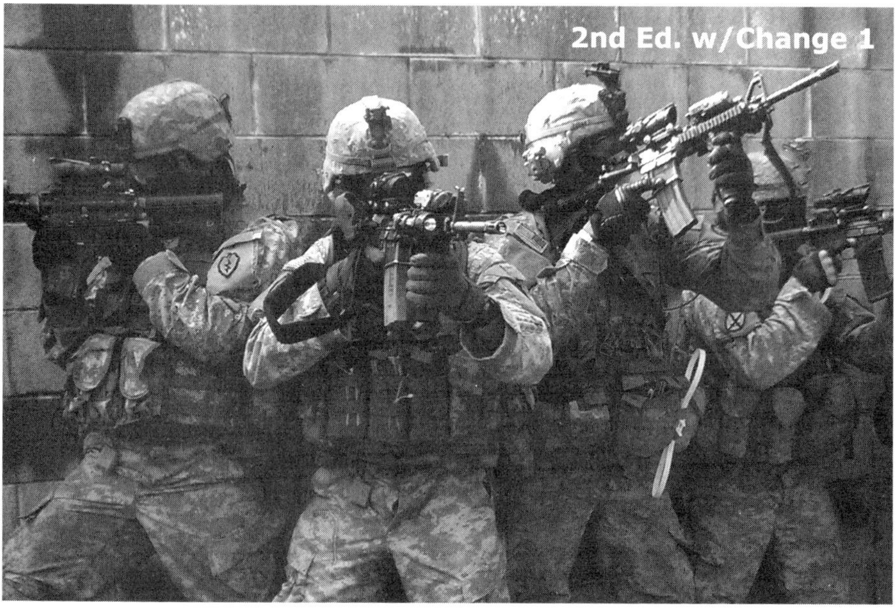

Small Unit TACTICS

Leading, Planning & Conducting Tactical Operations

Second Edition with Change 1 (Mar 2017)

The Lightning Press

Christopher Larsen
Norman M. Wade

The Lightning Press

2227 Arrowhead Blvd.
Lakeland, FL 33813
24-hour Voicemail/Fax/Order: 1-800-997-8827
E-mail: SMARTbooks@TheLightningPress.com

www.TheLightningPress.com

(SUTS2) The Small Unit Tactics SMARTbook, 2nd Rev. Ed. (w/Chg 1)

Leading, Planning & Conducting Tactical Operations

* Change 1 to SUTS2 (Mar 2017) incorporates minor text edits from ADRP 3-0 (Nov 2016), FM 6-0 (Chg 2, Apr 2016), and Train, Advise, Assist (chap 4). An asterisk marks changed pages.

Copyright © 2017, Christopher Larsen & Norman M. Wade

ISBN: 978-1-935886-64-8

All Rights Reserved

No part of this book may be reproduced or utilized in any form or other means, electronic or mechanical, including photocopying, recording or by any information storage and retrieval systems, without permission in writing by the publisher. Inquiries should be addressed to The Lightning Press.

Notice of Liability

The information in this SMARTbook and quick reference guide is distributed on an "As Is" basis, without warranty. While every precaution has been taken to ensure the reliability and accuracy of all data and contents, neither the author nor The Lightning Press shall have any liability to any person or entity with respect to liability, loss, or damage caused directly or indirectly by the contents of this book. If there is a discrepancy, refer to the source document. This SMARTbook does not contain classified or sensitive information restricted from public release. "The views presented in this publication are those of the author and do not necessarily represent the views of the Department of Defense or its components."

SMARTbook is a trademark of The Lightning Press.

About our cover photo: Prepare to breach. Soldiers from Borzoi Company, 1st Battalion, 27th Infantry Regiment, 2nd Stryker Brigade Combat Team, 25th Infantry Division, stack outside the objective compound while Sappers from the 34th Engineer Company prepare to breach the target building during training in South Korea. (Dept of Army photo).

Credits: Photos courtesy of HJ Images; photos by Jeong, Hae-jung. All other photos courtesy Dept. of the Army and/or Dept. of Defense and credited individually where applicable.

Printed and bound in the United States of America.

(SUTS2) Notes to Reader

Leading, Planning & Conducting Tactical Operations

Tactics is the employment and ordered arrangement of forces in relation to each other. Through tactics, commanders use combat power to accomplish missions. The tactical-level commander uses combat power in battles, engagements, and small-unit actions.

Establishing a common frame of reference, **doctrine** provides a menu of practical options based on experience. It provides an authoritative guide for leaders and Soldiers but requires original applications that adapt it to circumstances.

The Small Unit Tactics SMARTbook **translates and bridges** operational-level doctrine into tactical application -- in the form of tactics, techniques and procedures -- and provides the **"how to"** at the small-unit level, providing a ready reference at the battalion, company, platoon, squad and fire team level.

* Change 1 to SUTS2 (Mar 2017) incorporates minor text edits from ADRP 3-0 (Nov 2016), FM 6-0 (Chg 2, Apr 2016) and Train, Advise, Assist (chap 4). An asterisk marks changed pages.

Tactics, Techniques and Procedures (TTPs)

Principles alone do not guide operations. Tactics, techniques, and procedures provide additional levels of detail and more specific guidance, based on evolving knowledge and experience.

- **Tactics**. Tactics is the employment and ordered arrangement of forces in relation to each other. Effective tactics translate combat power into decisive results. Primarily descriptive, tactics vary with terrain and other circumstances; they change frequently as the enemy reacts and friendly forces explore new approaches.

- **Techniques**. Employing a tactic usually requires using and integrating several techniques and procedures. Techniques are non prescriptive ways or methods used to perform missions, functions, or tasks. They are the primary means of conveying the lessons learned that units gain in operations.

- **Procedures**. Procedures are standard, detailed steps that prescribe how to perform specific tasks. They normally consist of a series of steps in a set order. Procedures are prescriptive; regardless of circumstances, they are executed in the same manner. Techniques and procedures are the lowest level of doctrine.

SMARTbooks - DIME is our DOMAIN!

SMARTbooks: Reference Essentials for the Instruments of National Power (D-I-M-E: Diplomatic, Informational, Military, Economic)! Recognized as a "whole of government" doctrinal reference standard by military, national security and government professionals around the world, SMARTbooks comprise a comprehensive professional library.

SMARTbooks can be used as quick reference guides during actual operations, as study guides at education and professional development courses, and as lesson plans and checklists in support of training. Visit **www.TheLightningPress.com**!

Offense and Defense (Decisive Operations)
Ref: ADP 3-90, Offense and Defense (Aug '12).

Tactics is the employment and ordered arrangement of forces in relation to each other (CJCSM 5120.01). Through tactics, commanders use combat power to accomplish missions. The tactical-level commander employs combat power in the conduct of engagements and battles. This section addresses the tactical level of war, the art and science of tactics, and hasty versus deliberate operations.

The Tactical Level of War
ADP 3-90 is the primary manual for offensive and defensive tasks at the tactical level. It does not provide doctrine for stability or defense support of civil authorities tasks. It is authoritative and provides guidance in the form of combat tested concepts and ideas for the employment of available means to win in combat. These tactics are not prescriptive in nature, and they require judgment in application.

The tactical level of war is the level of war at which battles and engagements are planned and executed to achieve military objectives assigned to tactical units or task forces (JP 3-0). Activities at this level focus on the ordered arrangement and maneuver of combat elements in relation to each other and to the enemy to achieve combat objectives. It is important to understand tactics within the context of the levels of war. The strategic and operational levels provide the context for tactical operations. Without this context, tactical operations are just a series of disconnected and unfocused actions. Strategic and operational success is a measure of how one or more battles link to winning a major operation or campaign. In turn, tactical success is a measure of how one or more engagements link to winning a battle.

The Offense
The offense is the decisive form of war. While strategic, operational, or tactical considerations may require defending for a period of time, defeat of the enemy eventually requires shifting to the offense. Army forces strike the enemy using offensive action in times, places, or manners for which the enemy is not prepared to seize, retain, and exploit the operational initiative. Operational initiative is setting or dictating the terms of action throughout an operation (ADRP 3-0).

The main purpose of the offense is to defeat, destroy, or neutralize the enemy force. Additionally, commanders conduct offensive tasks to secure decisive terrain, to deprive the enemy of resources, to gain information, to deceive and divert the enemy, to hold the enemy in position, to disrupt his attack, and to set the conditions for future successful operations.

The Defense
While the offense is the most decisive type of combat operation, the defense is the stronger type. Army forces conduct defensive tasks as part of major operations and joint campaigns, while simultaneously conducting offensive and stability tasks as part of decisive action outside the United States.

Commanders choose to defend to create conditions for a counteroffensive that allows Army forces to regain the initiative. Other reasons for conducting a defense include to retain decisive terrain or deny a vital area to the enemy, to attrit or fix the enemy as a prelude to the offense, in response to surprise action by the enemy, or to increase the enemy's vulnerability by forcing the enemy to concentrate forces.

Tactical Enabling Tasks
Commanders direct tactical enabling tasks to support the conduct of decisive action. Tactical enabling tasks are usually shaping or sustaining. They may be decisive in the conduct of stability tasks. Tactical enabling tasks discussed in ADRP 3-90 include reconnaissance, security, troop movement, relief in place, passage of lines, encirclement operations, and urban operations. Stability ultimately aims to create a condition so the local populace regards the situation as legitimate, acceptable, and predictable.

Offense and Defense (Unifying Logic Chart)

Unified Land Operations

Seize, retain, and exploit the initiative to gain and maintain a position of relative advantage in sustained land operations in order to create the conditions for favorable conflict resolution.

Executed through...

Decisive Action
offensive | defensive | stability | DSCA

Offensive tasks

- **Movement to contact**
  - Search and attack
  - Cordon and search
- **Attack**
  - Ambush
  - Counterattack
  - Demonstration
  - Spoiling attack
  - Feint
  - Raid
- **Exploitation**
- **Pursuit**

Forms of maneuver
- Envelopment
- Flank attack
- Frontal attack
- Infiltration
- Penetration
- Turning movement

Defensive tasks

- **Area defense**
- **Mobile defense**
- **Retrograde operations**
  - Delay
  - Withdrawal
  - Retirement

Forms of the defense
- Defense of a linear obstacle
- Perimeter defense
- Reverse slope defense

Tactical enabling tasks
Tactical mission tasks

Ref: ADP 3-90, Offense and Defense, fig. 1, p. iv.

Refer to AODS5:*The Army Operations & Doctrine SMARTbook (Guide to Army Operations and the Six Warfighting Functions) for discussion of the fundamentals, principles and tenets of Army operations, plus chapters on each of the six warfighting functions: mission command, movement and maneuver, intelligence, fires, sustainment, and protection.*

(SUTS2) References

The following references were used in part to compile The Small Unit Tactics SMARTbook. Additionally listed are related resources useful to the reader. All references are available to the general public and designated as "approved for public release; distribution is unlimited." The Small Unit Tactics SMARTbook does not contain classified or sensitive information restricted from public release.

* Change 1 to SUTS2 (Feb 2017) incorporates minor text edits from ADRP 3-0 (Nov 2016), FM 6-0 (Chg 2, Apr 2016) and FM 7-0 (Oct 2016). An asterisk marks changed pages.

Army Doctrinal Publications (ADPs) and Army Doctrinal Reference Publications (ADRPs)

ADP/ADRP 1-02	Feb 2015	Operational Terms and Military Symbols
ADP/ADRP 3-0*	Nov 2016	Operations
ADP/ADRP 3-07	Aug 2012	Stability
ADP/ARDRP 3-90	Aug 2012	Offense and Defense
ADP/ADRP 5-0	May 2012	The Operations Process
ADP/ADRP 6-0	May 2012	Mission Command (with Chg 1, Sept 2012)

Army Tactics, Techniques and Procedures (ATTPs)

ATP 5-19	Apr 2014	Risk Management
ATTP 3-06.11	Jun 2011	Combined Arms Operations in Urban Terrain
ATTP 3-97.11	Jan 2011	Cold Region Operations
FM 3-19.4	Mar 2002	Military Police Leader's Handbook (change 1)
FM 3-21.8	Mar 2007	The Infantry Rifle Platoon and Squad
FM 3-21.10	Jul 2006	The Infantry Rifle Company
FM 3-24	Dec 2006	Counterinsurgency
FM 3-90-1*	Mar 2013	Offense and Defense, Volume 1
FM 3-90-2*	Mar 2013	Reconnaissance, Security, and Tactical Enabling Tasks, Volume 1
FM 3-97.6	Nov 2000	Mountain Operations
FM 6-0*	May 2014	Commander and Staff Organization and Operations (w/Change 2, Apr 2016)
FM 7-0*	Oct 2016	Train to Win in a Complex World
FM 6-01.1	Jul 2012	Knowledge Management Operations
FM 7-85	Jun 1987	Ranger Operations
FM 7-92	Dec 2001	The Infantry Reconnaissance Platoon and Squad (Airborne, Air Assault, Light Infantry) w/change 1
FM 7-93	Oct 1995	Long-Range Surveillance Unit Operations
FM 90-3	Aug 1993	Desert Operations
FM 90-5	Aug 1993	Jungle Operations

Joint Publications

JP 3-0	Aug 2011	Joint Operations

Additional Resources and Publications

SH 21-76	Jul 2006	The Ranger Handbook

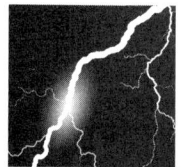

(SUTS2) Table of Contents

Chap 1: Tactical Mission Fundamentals

I. The Art of Tactics..1-1
    I. The Tactical Level of War...1-1
        Individuals, Crews, and Small Units..1-1
        Battles, Engagements and Small-Unit Actions...............................1-2
    II. The Science and Art of Tactics 1-2
    III. Hasty vs. Deliberate Operations...1-4

II. The Army's Operational Concept.. 1-5*
    I. The Goal of Unified Land Operations 1-5*
    II. Decisive Action 1-5*
        Operations (Unified Logic Chart)..1-7*
        Tasks of Decisive Action 1-8*
    III. Unified Action... 1-10*
        A. Joint Operations.. 1-10*
        B. Multinational Operations 1-10a*
    IV. Train to Win in a Complex World ....1-10b"
    V. Homeland Defense and Decisive Action 1-10d*

III. Tactical Mission Tasks 1-11
    A. Mission Symbols 1-11
    B. Effects on Enemy Forces 1-12
    C. Actions by Friendly Forces..1-13
    Tactical Doctrinal Taxonomy 1-14

IV. Understand, Visualize, Describe, Direct, Lead, Assess........1-15
    I. Understand 1-15
    II. Visualize...1-16
        The Army Operational Framework 1-18*
    III. Describe..1-16
        A. Commander's Intent 1-20
        B. Planning Guidance..1-20
        C. Commander's Critical Information Requirements (CCIR).......1-20
        D. Essential Elements of Friendly Information (EEFI).................1-21
    IV. Direct 1-17
        Elements of Combat Power... 1-22*
        The Six Warfighting Functions 1-23
    V. Lead...1-17
    VI. Assess 1-17
    Activities of the Operations Process 1-24

V. Troop Leading Procedures 1-25
    I. Troop Leading Procedure Steps...1-26
    II. METT-TC (Mission Variables)..1-31
    III. OCOKA - Military Aspects of the Terrain....................................1-32
    IV. Risk Management (RM)..1-36

VI. Combat Orders 1-37
- I. Warning Order (WARNO) 1-38
- II. Operations Order (OPORD) 1-39
- III. Fragmentary Order (FRAGO) 1-40
- Techniques for Issuing Orders 1-41
- The Operations Order (OPORD) - A Small Unit Perspective 1-42
- On Point 1-44

VII. Preparation and Pre-Combat Inspection (PCI) 1-45
- I. Preparation 1-45
- II. The Pre-Combat Inspection (PCI) 1-45
- On Point 1-50

VIII. Rehearsals 1-51
- I. Rehearsal Techniques 1-52
- II. Rehearsals - Company Level & Smaller 1-54

IX. The After Action Review (AAR) 1-55
- I. Types of After Action Reviews 1-56
- II. Steps in the After Action Review 1-57
- III. AARs - A Small Unit Perspective 1-58
- On Point 1-60

Chap 2 — The Offense

The Offense 2-1
- I. Primary Offensive Tasks 2-2
- II. Purposes of Offensive Tasks 2-3
- III. Forms of Maneuver 2-4
    - A. Envelopment 2-4
    - B. Turning Movement 2-4
    - C. Infiltration 2-5
    - D. Penetration 2-5
    - E. Frontal Attack 2-5
- IV. Common Offensive Control Measures 2-5a
- V. Transition 2-5c
- VI. Characteristics of Offensive Operations 2-6

I. Movement to Contact 2-7
- Meeting Engagement 2-7
- I. Organization 2-8
    - Search and Attack 2-8
    - Approach-March Technique 2-8
- II. Planning & Preparation 2-9
- III. Conducting the MTC - A Small Unit Perspective 2-10
- On Point 2-12

II. Attack 2-13
- I. Organization 2-14
- II. Planning & Preparation 2-15
- III. Conducting the Attack - A Small Unit Perspective 2-16
- On Point 2-18
    - Maneuver Control Measures 2-18
    - Fire Control Measures 2-18

III. Exploitation 2-19
  I. Organization 2-20
  II. Planning & Preparation 2-20
  III. Conducting the Exploitation - A Small Unit Perspective 2-20
  On Point 2-22
    Transition 2-22
IV. Pursuit 2-23
  I. Organization 2-24
    A. Frontal Pursuit 2-24
    B. Combination Pursuit 2-25
  II. Planning & Preparation 2-25
  III. Conducting the Pursuit - A Small Unit Perspective 2-26
  On Point 2-28
    Transition 2-28
V. Small Unit Offensive Tactical Tasks 2-29
  I. Seize 2-30
  II. Suppress 2-30
  III. Support by Fire 2-31
  IV. Clear 2-32
  V. Attack by Fire 2-34

Chap 3: The Defense

The Defense 3-1
  I. Purposes of Defensive Operations 3-2
  II. Defensive Tasks 3-2
  III. Characteristics of the Defense 3-3
  IV. Common Defensive Control Measures 3-3a
  V. Transition 3-4
I. Mobile Defense 3-5
  I. Organization 3-6
    A. The Fixing Force 3-6
    B. The Striking Force 3-6
  II. Planning & Preparation 3-7
  III. Conducting the Mobile Defense - A Small Unit Perspective 3-8
  On Point 3-10
II. Area Defense 3-11
  I. Organization 3-12
    Primary Positions 3-12
    Alternate Positions 3-12
    Supplementary Positions 3-12
    Subsequent Positions 3-12
  II. Planning & Preparation 3-14
    A. Range Card 3-14
    B. Sector Sketch 3-14
    C. Sectors of Fire 3-15
    D. Engagement Areas 3-16
  III. Conducting the Area Defense - A Small Unit Perspective 3-17
  IV. Priorities of Work in the Defense 3-18

III. Retrograde 3-19
    I. Delay 3-20
        A. Delay Within a Sector 3-20
        B. Delay Forward of a Specified Line for a Specified Time 3-20
    II. Withdrawal 3-21
        A. Assisted 3-21
        B. Unassisted 3-21
    III. Retirement 3-21
    Conducting the Retrograde - A Small Unit Perspective 3-20
    On Point 3-22
        Reconstitution 3-22

IV. Small Unit Defensive Techniques 3-23
    I. Defend an Area 3-23
    II. Defend a Battle Position 3-24
    III. Defend a Strongpoint 3-25
    IV. Defend a Perimeter 3-27
    V. Defend a Reverse Slope 3-27

Chap 4: Train, Advise & Assist

Train, Advise & Assist 4-1*
    Security Force Assistance (SFA) 4-2*

I. Military Engagement, Security Cooperation & Deterrance 4-5*

II. Stability Operations 4-7*
    I. Nature of Stability Operations 4-7*
        End State Conditions for Stability 4-8*
    II. Stability Operations Across the Range of Military Operations 4-10*
    III. Small Unit Stability Tasks 4-11*
        A. Establish and Occupy a Lodgement Area or a 4-11*
        Forward Operating Base (FOB)
        B. Monitor Compliance with an Agreement 4-11*
        C. Establishing Observation Posts and Checkpoints 4-12*
        D. Search 4-13*
        E. Patrol 4-13*
        F. Escort a Convoy 4-13*
        G. Open and Secure Routes 4-13*
        H. Conduct Reserve Operations 4-13*
        I. Control Crowds 4-13*

III. Peace Operations (PO) 4-15*
    1. Peacekeeping Operations (PKO) 4-16*
    2. Peace Enforcement Operations (PEO) 4-16*
    3. Peace Buliding (PB) 4-16*
    4. Peacemaking (PM) 4-16*
    5. Conflict Prevention 4-16*

IV. Counterinsurgency Operations (COIN) 4-17*
    I. Insurgency 4-17*
    II. Governance and Legitimacy 4-18*

Chap 5: Tactical Enabling Tasks

I. Security Operations 5-1
    I. Forms of Security Operations 5-1
        Security Fundamentals 5-2
        A. Screen 5-1
        B. Guard 5-4
        C. Cover 5-5
        D. Area Security 5-6
        E. Local Security 5-7
        * Combat Outposts 5-7
    II. Fundamentals of Security Operations 5-3
    On Point 5-8

II. Reconnaissance Operations 5-9
    Reconnaissance Objective 5-9
    I. Reconnaissance Fundamentals 5-10
    II. Organization 5-11
    III. Planning & Preparation 5-11
    IV. Forms of the Reconnaissance 5-11
        A. The Route Reconnaissance 5-12
            Recon Push 5-13
            Recon Pull 5-13
        B. The Zone Reconnaissance 5-14
        C. The Area Reconnaissance 5-16
            Single-Team Method 5-16
            Double-Team Method 5-17
            Dissemination of Information 5-17
        D. Reconnaissance in Force (RIF) 5-18
    On Point 5-18

III. Relief in Place 5-19
    I. Organization 5-20
    II. Planning & Preparation 5-20
        Hasty or Deliberate 5-20
    III. Conducting the Relief in Place - A Small Unit Perspective 5-21
        Techniques: Sequential, Simultaneous or Staggered 5-21
    On Point 5-22

IV. Passage of Lines 5-23
    I. Conducting the Relief in Place - A Small Unit Perspective 5-24
        Departing the Forward Line of Troops (FLOT) 5-24
        Reentering the Forward Line of Troops (FLOT) 5-25
    II. Organization 5-26
    On Point 5-26

V. Encirclement Operations 5-27
    I. Offensive Encirclement Operations 5-27
    II. Defending Encircled 5-27

VI. Troop Movement 5-29
    I. Methods of Troop Movement 5-29
    II. Movement Techniques 5-30

Chap 6: Special Purpose Attacks

Special Purpose Attacks 6-1
I. Ambush 6-3
    Near Ambush 6-4
    Far Ambush 6-4
    I. Organization 6-5
        A. Near Ambush 6-6
        B. Far Ambush 6-8
    II. Planning & Preparation 6-9
        A. Near Ambush 6-10
        B. Far Ambush 6-10
    III. Conducting the Ambush - A Small Unit Perspective 6-11
        A. Near Ambush 6-12
        B. Far Ambush 6-14
    IV. Ambush Categories 6-16
        Deliberate 6-16
        Hasty 6-16
        Area Ambush 6-16
        Point Ambush 6-16
II. Raid 6-17
    I. Organization 6-18
        A. Security Team 6-18
        B. Support Team 6-18
        C. Assault Team 6-18
    II. Planning & Preparation 6-19
    III. Conducting the Raid - A Small Unit Perspective 6-20
        A. Infiltrate to the Objective 6-20
        B. Actions on the Objective 6-21
    On Point 6-22

Chap 7: Urban & Regional Environments

Urban & Regional Environments (Overview) 7-1
I. Urban Operations 7-3
    - Urban Operations (UO) 7-4
    - Understanding the Urban Environment 7-6
    - Threat 7-8
    I. Find 7-3
    II. Isolate the Building 7-3
        Cordon 7-10

         III. Assault a Building 7-10
            A. Entering a Building...7-11
            B. Clearing Rooms 7-13
            C. Moving in the Building...7-14
         IV. Follow Through 7-16

II. Fortified Areas 7-17
      Characteristics..7-17
      I. Find...7-18
      II. Fix..7-18
      III. Finish (Fighting Enemies in Fortifications)..7-18
         A. Securing the Near and Far Side--Breaching Protective Obstacles..........7-18
         B. Knocking out Bunkers 7-19
         C. Assaulting Trench Systems...7-20
      IV. Follow Through 7-22

III. Desert Operations 7-23
      I. Desert Environments...7-23
      II. Desert Effects on Personnel 7-25
      III. Mission Command Considerations 7-27
      IV. Tactical Considerations...7-28

IV. Cold Region Operations 7-31
      I. Cold Regions...7-31
      II. Cold Weather Effects on Personnel...7-34
      III. Mission Command Considerations 7-35
      IV. Tactical Considerations...7-36

V. Mountain Operations 7-39
      I. Mountain Environments 7-39
      II. Effects on Personnel..7-40
      III. Tactical Considerations...7-44
      IV. Mission Command Considerations 7-46

VI. Jungle Operations..7-47
      I. Jungle Environments 7-47
      II. Effects on Personnel..7-50
      III. Mission Command Considerations 7-51
      IV. Tactical Considerations...7-52

Chap 8: Patrols & Patrolling

Patrols and Patrolling 8-1
      Combat Patrols..8-1
      Reconnaissance Patrols 8-1
      I. Organization of Patrols..8-2
      II. Planning & Conducting a Patrol 8-4
      III. Elements of a Combat Patrol...8-6
         A. Assault Element 8-6
         B. Support Element 8-6
         C. Security Element..8-6

Table of Contents-7

I. Traveling Techniques ... 8-7
- I. Traveling ... 8-8
- II. Traveling Overwatch ... 8-8
- III. Bounding Overwatch ... 8-9
- On Point ... 8-10

II. Attack Formations ... 8-11
- Fire Team Formations ... 8-11
- Attack Formation Considerations ... 8-12
- I. The Line ... 8-13
- II. The File ... 8-14
  - Variation: The Staggered Column ... 8-15
- III. The Wedge ... 8-16
  - Variation: The Diamond ... 8-17
- On Point ... 8-18
  - Security Checks While on Patrol ... 8-18
  - 5 and 20 Meter Checks ... 8-18

III. Crossing a Danger Area ... 8-19
- Types of Danger Areas ... 8-19
- I. Patch-to-the-Road Method ... 8-20
- II. Heart-Shaped Method ... 8-22
- III. Bypass Method ... 8-24
- IV. Box Method ... 8-25
- V. Crossing Large Open Areas ... 8-25
- On Point ... 8-26
  - Enemy Contact ... 8-26

IV. Establishing a Security Halt ... 8-27
- I. Cigar-Shaped Method ... 8-29
- II. Wagon Wheel Method ... 8-30
- III. Priorities of Work at the Objective Rally Point (ORP) ... 8-31
- On Point ... 8-32
  - En Route Rally Point (ERP) ... 8-32
  - Objective Rally Point (ORP) ... 8-32

V. Establishing a Hide Position ... 8-33
- Considerations ... 8-33
- I. Back-to-Back Method ... 8-34
- II. Star Method ... 8-35
- On Point ... 8-36
  - Site Selection ... 8-36
  - Site Sterilization ... 8-36

VI. Establishing a Patrol Base ... 8-37
- Site Selection ... 8-37
- The Triangle Method ... 8-38
- Planning Considerations ... 8-39
- Security Measures ... 8-39
- Priorities of Work ... 8-40

Index

Index ... Index-1 to Index-4

8-Table of Contents

Chap 1: Tactical Mission Fundamentals

I. The Art of Tactics

Ref: ADRP 3-90, Offense & Defense (Aug '12), chap. 1.

Tactics is the employment of units in combat. It includes the ordered arrangement and maneuver of units in relation to each other, the terrain and the enemy to translate potential combat power into victorious battles and engagements. (Dept. of Army photo by Staff Sgt. Russell Bassett).

I. The Tactical Level of War

Through tactics, commanders use combat power to accomplish missions. The tactical-level commander employs combat power to accomplish assigned missions.

The tactical level of war is the level of war at which battles and engagements are planned and executed to achieve military objectives assigned to tactical units or task forces (JP 3-0). Activities at this level focus on the ordered arrangement and maneuver of combat elements in relation to each other and to the enemy to achieve combat objectives. It is important to understand tactics within the context of the levels of war. The strategic and operational levels provide the context for tactical operations. Without this context, tactical operations are reduced to a series of disconnected and unfocused actions.

Tactical operations always require judgment and adaptation to the unique circumstances of a specific situation. Techniques and procedures are established patterns that can be applied repeatedly with little or no judgment in a variety of circumstances. Tactics, techniques, and procedures (TTP) provide commanders and staffs with a set of tools to use in developing the solution to a tactical problem.

Individuals, crews, and small units

Individuals, crews, and small units act at the tactical level. At times, their actions may produce strategic or operational effects. However, this does not mean these elements are acting at the strategic or operational level. Actions are not strategic unless they contribute directly to achieving the strategic end state. Similarly, actions are considered operational only if they are directly related to operational movement or the sequencing of battles and engagements. The level at which an action occurs is determined by the perspective of the echelon in terms of planning, preparation, and execution.

Battles, Engagements and Small-Unit Actions

Tactics is the employment and ordered arrangement of forces in relation to each other. Through tactics, commanders use combat power to accomplish missions. The tactical-level commander uses combat power in battles, engagements, and small-unit actions. A battle consists of a set of related engagements that lasts longer and involves larger forces than an engagement. Battles can affect the course of a campaign or major operation. An engagement is a tactical conflict, usually between opposing, lower echelons maneuver forces (JP 1-02). Engagements are typically conducted at brigade level and below. They are usually short, executed in terms of minutes, hours, or days.

II. The Science and Art of Tactics

The tactician must understand and master the science and the art of tactics, two distinctly different yet inseparable concepts. Commanders and leaders at all echelons and supporting commissioned, warrant, and noncommissioned staff officers must be tacticians to lead their soldiers in the conduct of full spectrum operations.

A. The Science

The science of tactics encompasses the understanding of those military aspects of tactics—capabilities, techniques, and procedures—that can be measured and codified. The science of tactics includes the physical capabilities of friendly and enemy organizations and systems, such as determining how long it takes a division to move a certain distance. It also includes techniques and procedures used to accomplish specific tasks, such as the tactical terms and control graphics that comprise the language of tactics. While not easy, the science of tactics is fairly straightforward. Much of what is contained in this manual is the science of tactics—techniques and procedures for employing the various elements of the combined arms team to achieve greater effects.

Mastery of the science of tactics is necessary for the tactician to understand the physical and procedural constraints under which he must work. These constraints include the effects of terrain, time, space, and weather on friendly and enemy forces. However—because combat is an intensely human activity—the solution to tactical problems cannot be reduced to a formula. This realization necessitates the study of the art of tactics.

B. The Art

The art of tactics consists of three interrelated aspects: the creative and flexible array of means to accomplish assigned missions, decision making under conditions of uncertainty when faced with an intelligent enemy, and understanding the human dimension—the effects of combat on soldiers. An art, as opposed to a science, requires exercising intuitive faculties that cannot be learned solely by study. The tactician must temper his study and evolve his skill through a variety of relevant, practical experiences. The more experience the tactician gains from practice under a variety of circumstances, the greater his mastery of the art of tactics.

Military professionals invoke the art of tactics to solve tactical problems within his commander's intent by choosing from interrelated options, including—

- Types and forms of operations, forms of maneuver, and tactical mission tasks
- Task organization of available forces, to include allocating scarce resources
- Arrangement and choice of control measures
- Tempo of the operation
- Risks the commander is willing to take

Close Combat

Ref: ADP 3-0, Operations (Nov '16), p. 1-11.

Close combat is indispensable and unique to land operations. Only on land do combatants routinely and in large numbers come face-to-face with one another. When other means fail to drive enemy forces from their positions, Army forces close with and destroy or capture them. The outcome of battles and engagements depends on Army forces' ability to prevail in close combat.

The complexity of urban terrain and density of noncombatants reduce the effectiveness of advanced sensors and long-range and air-delivered weapons. Thus, a weaker enemy often attempts to negate Army advantages by engaging Army forces in urban environments. Operations in large, densely populated areas require special considerations. From a planning perspective, commanders view cities as both topographic features and a dynamic system of varying operational entities containing hostile forces, local populations, and infrastructure.

Regardless of the importance of technological capabilities, success in operations requires Soldiers to accomplish the mission. Today's operational environment requires professional Soldiers and leaders whose character, commitment, and competence represent the foundation of a values-based, trained, and ready Army. Today's Soldiers and leaders adapt and learn while training to perform tasks both individually and collectively. Soldiers and leaders develop the ability to exercise judgment and disciplined initiative under stress. Army leaders and their subordinates must remain—

- Honorable servants of the Nation.
- Competent and committed professionals.
- Dedicated to living by and upholding the Army Ethic.
- Able to articulate mission orders to operate within their commander's intent.
- Committed to developing their subordinates and creating shared understanding while building mutual trust and cohesion.
- Courageous enough to accept prudent risk and exercise disciplined initiative while seeking to exploit opportunities in a dynamic and complex operational environment.
- Trained to operate across the range of military operations.
- Able to operate in combined arms teams within unified action and leverage other capabilities in achieving their objectives.
- Able to apply cultural understanding to make the right decisions and take the right actions.
- Opportunistic and offensively minded.

Effective close combat relies on lethality with a high degree of situational understanding. The capacity for physical destruction is a foundation of all other military capabilities, and it is the most basic building block of military operations. Army leaders organize, equip, train, and employ their formations for unmatched lethality over a wide range of conditions. Lethality is a persistent requirement for Army organizations, even in conditions where only the implicit threat of violence suffices to accomplish the mission through nonlethal engagements and activities.

An inherent, complementary relationship exists between using lethal force and applying military capabilities for nonlethal purposes. Though each situation requires a different mix of violence and constraint, lethal and nonlethal actions used together complement each other and create multiple dilemmas for opponents. Lethal actions are critical to accomplishing offensive and defensive tasks. However, nonlethal actions are also important contributors to combined arms operations, regardless of which element of decisive action dominates.

III. Hasty vs. Deliberate Operations

A hasty operation is an operation in which a commander directs his immediately available forces, using fragmentary orders (FRAGOs), to perform activities with minimal preparation, trading planning and preparation time for speed of execution. A deliberate operation is an operation in which a commander's detailed intelligence concerning the situation allows him to develop and coordinate detailed plans, including multiple branches and sequels. He task organizes his forces specifically for the operation to provide a fully synchronized combined arms team. He conducts extensive rehearsals while conducting shaping operations to set the conditions for the conduct of his decisive operation. Most operations lie somewhere along a continuum between these two extremes.

Choices and Trade-offs

The leader must choose the right point along the continuum to operate. His choice involves balancing several competing factors. He bases his decision to conduct a hasty or deliberate operation on his current knowledge of the enemy situation, and his assessment of whether the assets available (to include time), and the means to coordinate and synchronize those assets, are adequate to accomplish the mission. If they are not he takes additional time to plan and prepare for the operation or bring additional forces to bear on the problem. The commander makes that choice in an environment of uncertainty, which always entails some risk.

Risk Reduction

Uncertainty and risk are inherent in tactical operations and cannot be eliminated. A commander cannot be successful without the capability of acting under conditions of uncertainty while balancing various risks and taking advantage of opportunities. Although the commander strives to maximize his knowledge about his forces, the terrain and weather, civil considerations, and the enemy, he cannot let a lack of information paralyze him. The more intelligence on the enemy, the better able the commander is to make his assessment. Less information means that the commander has a greater risk of making a poor decision for the specific situation.

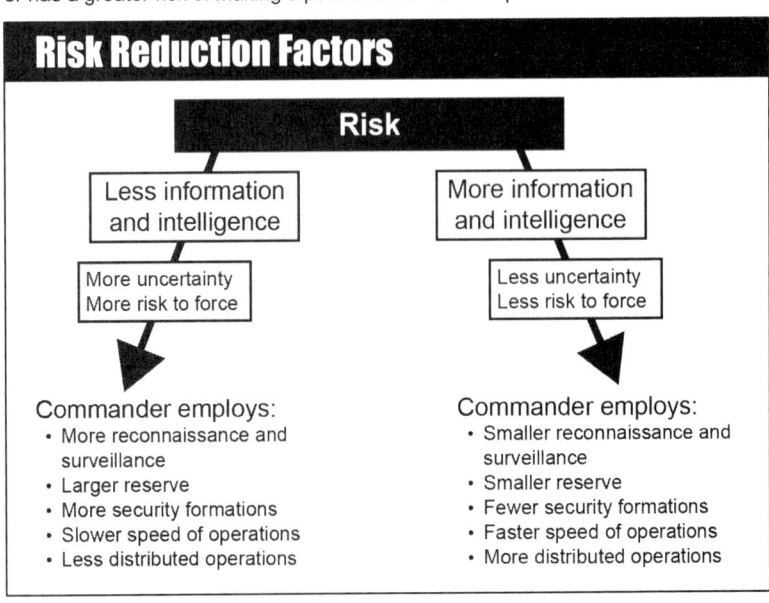

Ref: ADRP 3-90, Offense and Defense, fig. 1-1, p. 1-5.

II. The Army's Operational Concept

Ref: ADRP 3-0, Operations (Nov '16), chap. 1 and 3.

An operation is a sequence of tactical actions with a common purpose or unifying theme (JP 1). Army forces, as part of the joint force, contribute to the joint fight through the conduct of unified land operations. Unified land operations are simultaneous offensive, defensive, and stability or defense support of civil authorities tasks to seize, retain, and exploit the initiative and consolidate gains to prevent conflict, shape the operational environment, and win our Nation's wars as part of unified action (ADRP 3-0). ADP 3-0 is the Army's basic warfighting doctrine and is the Army's contribution to unified action.

I. The Goal of Unified Land Operations

Unified land operations is the Army's operational concept and the Army's contribution to unified action. Unified land operations are simultaneous offensive, defensive, and stability or defense support of civil authorities tasks to seize, retain, and exploit the initiative and consolidate gains to prevent conflict, shape the operational environment, and win our Nation's wars as part of unified action. The goal of unified land operations is to apply landpower as part of unified action to defeat the enemy on land and establish conditions that achieve the joint force commander's end state. Unified land operations is how the Army applies combat power through 1) simultaneous offensive, defensive, and stability, or defense support of civil authorities tasks, to 2) seize, retain, and exploit the initiative, and 3) consolidate gains. Where possible, military forces working with unified action partners seek to prevent or deter threats. However, if necessary, military forces possess the capability in unified land operations to prevail over aggression.

II. Decisive Action

Decisive action is the continuous, simultaneous combinations of offensive, defensive, and stability or defense support of civil authorities tasks. In unified land operations, commanders seek to seize, retain, and exploit the initiative while synchronizing their actions to achieve the best effects possible. Operations conducted outside the United States and its territories simultaneously combine three elements—offense, defense, and stability. Within the United States and its territories, decisive action combines the elements of defense support of civil authorities and, as required, offense and defense to support homeland defense.

Decisive action begins with the commander's intent and concept of operations. As a single, unifying idea, decisive action provides direction for an entire operation. Based on a specific idea of how to accomplish the mission, commanders and staffs refine the concept of operations during planning and determine the proper allocation of resources and tasks. They adjust the allocation of resources and tasks to specific units throughout the operation, as subordinates develop the situation or conditions change.

Refer to AODS5: The Army Operations & Doctrine SMARTbook for complete discussion of the fundamentals, principles and tenets of Army operations and organization (ADP/ADRP 3-0 Operations, 2016); chapters on each of the six warfighting functions: mission command (ADP/ADRP 6-0), movement and maneuver (ADPs 3-90, 3-07, 3-28, 3-05), intelligence (ADP/ADRP 2-0), fires (ADP/ADRP 3-09), sustainment (ADP/ADRP 4-0), and protection (ADP/ADRP 3-37); Doctrine 2015 guide and glossary of terms.

Unified land operations addresses more than combat between armed opponents. Army forces conduct operations amid populations. This requires Army forces to defeat the enemy and simultaneously shape civil conditions. Offensive and defensive tasks defeat enemy forces, whereas stability tasks shape civil conditions. Winning battles and engagements is important, but that alone may not be the most significant task. Shaping civil conditions (in concert with civilian organizations, civil authorities, and multinational forces) often proves just as important to campaign success. In many joint operations, stability or defense support of civil authorities tasks often prove more important than offensive and defensive tasks.

The emphasis on different tasks of decisive action changes with echelon, time, and location. In an operation dominated by stability, part of the force might conduct simultaneous offensive and defensive tasks in support of establishing stability. Within the United States, defense support of civil authorities may be the only activity actually conducted. Simultaneous combinations of the tasks, which commanders constantly adapt to conditions, are the key to successful land operations in achieving the end state.

The Tasks of Decisive Action
Decisive action requires simultaneous combinations of offense, defense, and stability or defense support of civil authorities tasks.
See following pages (pp. 1-8 to 1-9) for discussion of the tasks of decisive action.

The Purpose of Simultaneity
Simultaneously conducting offensive, defensive, and stability or defense support of civil authorities tasks requires the synchronized application of combat power. Simultaneity means doing multiple things at the same time. It requires the ability to conduct operations in depth and to integrate them so that their timing multiplies their effectiveness throughout an area of operations and across the multiple domains.

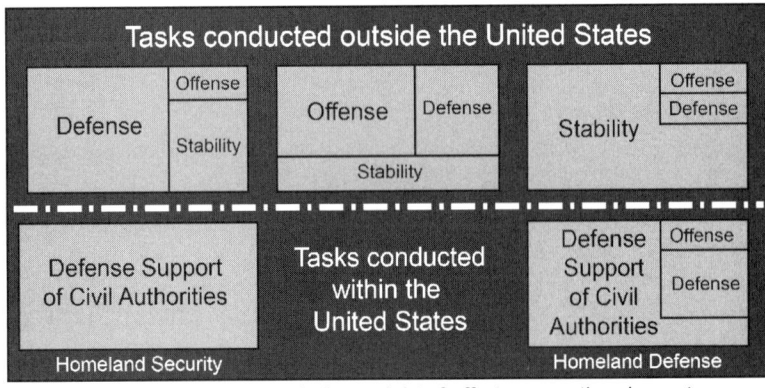

The mission determines the relative weight of effort among the elements.

Ref: ADP 3-0 (Nov '16), fig. 3-1. Decisive action.

Transitioning in Decisive Action
Conducting decisive action involves more than simultaneous execution of all its tasks. It requires commanders and staffs to consider their units' capabilities and capacities relative to each task. Commanders consider their missions; decide which tactics, techniques, and procedures to use; and balance the tasks of decisive action while preparing their commander's intent and concept of operations. They determine which tasks the force can accomplish simultaneously, if phasing is required, what additional resources it may need, and how to transition from one task to another.

Operations (Unified Logic Chart)

Ref: ADP 3-0, Operations (Nov '16), figure 1. ADP 3-0 unified logic chart.

An operation is a sequence of tactical actions with a common purpose or unifying theme (JP 1). Army forces, as part of the joint force, contribute to the joint fight through the conduct of unified land operations. Unified land operations are simultaneous offensive, defensive, and stability or defense support of civil authorities tasks to seize, retain, and exploit the initiative and consolidate gains to prevent conflict, shape the operational environment, and win our Nation's wars as part of unified action (ADRP 3-0). ADP 3-0 is the Army's basic warfighting doctrine and is the Army's contribution to unified action.

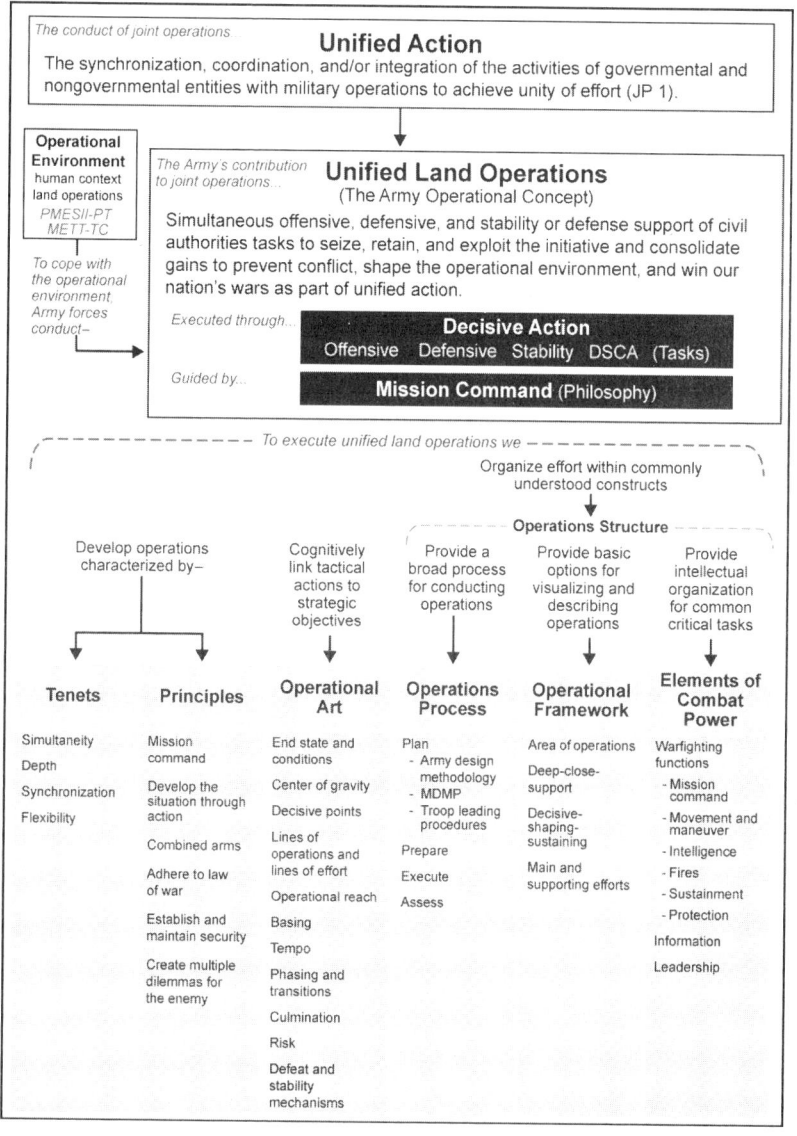

(Tactical Mission Fundamentals) II. The Army's Operational Concept 1-7*

Tasks of Decisive Action

Ref: ADRP 3-0, Operations (Nov '16), pp. 3-4 to 3-5 and table 3-1, p. 3-2.

Decisive action requires simultaneous combinations of offense, defense, and stability or defense support of civil authorities tasks.

1. Offensive Tasks

An offensive task is a task conducted to defeat and destroy enemy forces and seize terrain, resources, and population centers. Offensive tasks impose the commander's will on the enemy. Against a capable, adaptive enemy, the offense is the most direct and sure means of seizing, retaining, and exploiting the initiative to gain physical and psychological advantages and achieve definitive results. In the offense, the decisive operation is a sudden, shattering action against an enemy weakness that capitalizes on speed, surprise, and shock. If that operation does not destroy the enemy, operations continue until enemy forces disintegrate or retreat to where they no longer pose a threat. Executing offensive tasks compels the enemy to react, creating or revealing additional weaknesses that the attacking force can exploit.

See chap. 2.

Offensive Tasks

Primary Tasks
- Movement to contact
- Attack
- Exploitation
- Pursuit

Purposes
- Dislocate, isolate, disrupt and destroy enemy forces
- Seize key terrain
- Deprive the enemy of resources
- Refine intelligence
- Deceive and divert the enemy
- Provide a secure environment for stability operations

2. Defensive Tasks

A defensive task is a task conducted to defeat an enemy attack, gain time, economize forces, and develop conditions favorable for offensive or stability tasks. Normally the defense alone cannot achieve a decisive victory. However, it can set conditions for a counteroffensive or counterattack that enables Army forces to regain the initiative. Defensive tasks are a counter to the enemy offense. They defeat attacks, destroying as much of the attacking enemy as possible. They also preserve and maintain control over land, resources, and populations. The purpose of defensive tasks is to retain key terrain, guard populations, protect lines of communications, and protect critical capabilities against enemy attacks and counterattacks. Commanders can conduct defensive tasks to gain time and economize forces so offensive tasks can be executed elsewhere.

See chap. 3.

Defensive Tasks

Primary Tasks
- Mobile defense
- Area defense
- Retrograde

Purposes
- Deter or defeat enemy offensive operations
- Gain time
- Achieve economy of force
- Retain key terrain
- Protect the populace, critical assets and infrastructure
- Refine intelligence

3. Stability Tasks

Stability tasks are tasks conducted as part of operations outside the United States in coordination with other instruments of national power to maintain or reestablish a safe and secure environment and provide essential governmental services, emergency infrastructure reconstruction, and humanitarian relief (ADP 3-07). These tasks support governance, whether it is imposed by a host nation, an interim government, or military government. Stability tasks involve both coercive and constructive actions. They help to establish or maintain a safe and secure environment and facilitate reconciliation among local or regional adversaries. Stability tasks assist in building relationships among unified action partners, and promote specific U.S security interests. Stability tasks can also help establish political, legal, social, and economic institutions while supporting the transition to legitimate host-nation governance. Stability tasks cannot succeed if they only react to enemy initiatives. Stability tasks must maintain the initiative by pursuing objectives that resolve the causes of instability.

Stability Tasks

Primary Tasks
- Establish civil security
- Establish civil control
- Restore essential services
- Support to governance
- Support to economic and infrastructure development
- Conduct security cooperation

Purposes
- Provide a secure environment
- Secure land areas
- Meet the critical needs of the populace
- Gain support for host-nation government
- Shape the environment for interagency and host-nation success
- Promote security, build partner capacity, and provide access
- Refine intelligence

Refer to TAA2: The Military Engagement, Security Cooperation & Stability SMARTbook, 2nd Ed. (ADRP 3-07).

4. Defense Support of Civil Authority Tasks

Defense support of civil authorities is support provided by United States Federal military forces, Department of Defense civilians, Department of Defense contract personnel, Department of Defense component assets, and National Guard forces (when the Secretary of Defense, in coordination with the Governors of the affected States, elects and requests to use those forces in Title 32, United States Code, status) in response to requests for assistance from civil authorities for domestic emergencies, law enforcement support, and other domestic activities, or from qualifying entities for special events (DODD 3025.18). For Army forces, defense support of civil authorities is a task that takes place only in the homeland and U.S. territories. Defense support of civil authorities is conducted in support of another primary or lead federal agency, or in some cases, local authorities.

Defense Support of Civil Authorities Tasks

Primary Tasks
- Provide support for domestic disasters
- Provide support for domestic CBRN incidents
- Provide support for domestic civilian law enforcement agencies
- Provide other designated support

Purposes
- Save lives
- Restore essential services
- Maintain or restore law and order
- Protect infrastructure and property
- Support maintenance or restoration of local government
- Shape the environment for interagency success

Refer to HDS1: Homeland Defense & DSCA SMARTbook. (JP 3-28)

III. Unified Action

Unified action is the synchronization, coordination, and/or integration of the activities of governmental and nongovernmental entities with military operations to achieve unity of effort (JP 1). Unity of effort is coordination and cooperation toward common objectives, even if the participants are not necessarily part of the same command or organization, which is the product of successful unified action (JP 1). As military forces synchronize actions, they achieve unity of effort. Unified action includes actions of Army, joint, and multinational forces synchronized or coordinated with activities of other government agencies, nongovernmental and intergovernmental organizations, and the private sector. Through engagement, military forces play a key role in unified action before, during, and after operations. The Army's contribution to unified action is unified land operations. Army forces are uniquely suited to shape operational environments through their forward presence and sustained engagements with unified action partners and local civilian population.

Army forces remain the preeminent fighting force in the land domain. However, Army forces both depend on and support joint forces across multiple domains (land, air, maritime, space, and cyberspace). This integration across multiple domains, as well as both the contributions that Army forces provide and the benefits that Army forces derive from operating in multiple domains, is multi-domain battle. The Army depends on the other Services for strategic and operational mobility, joint fires, and other key enabling capabilities. The Army supports other Services, combatant commands, and unified action partners with foundational capabilities such as ground-based indirect fires and ballistic missile defense, defensive cyberspace operations, electronic protection, communications, intelligence, rotary-wing aircraft, logistics, and engineering. Unified action partners are those military forces, governmental and nongovernmental organizations, and elements of the private sector with whom Army forces plan, coordinate, synchronize, and integrate during the conduct of operations.

The Army's ability to set and sustain the theater of operations is essential to allowing the joint force to seize the initiative while restricting the enemy's options. The Army possesses capacities to establish, maintain, and defend vital infrastructure. It also provides to the joint force commander unique capabilities, such as port and airfield opening; logistics; chemical defense; and reception, staging, and onward movement, and integration.

Interagency coordination is inherent in unified action. Interagency coordination is, within the context of Department of Defense involvement, the coordination that occurs between elements of Department of Defense, and engaged United States Government agencies and departments for the purpose of achieving an objective (JP 3-0). Army forces conduct and participate in interagency coordination using established liaison, Soldier and leader engagement, and planning processes.

A. Joint Operations

Single Services may accomplish tasks and missions in support of Department of Defense objectives. However, the Department of Defense primarily employs two or more Services (from two military departments) in a single operation, particularly in combat, through joint operations. Joint operations is a general term to describe military actions conducted by joint forces and those Service forces employed in specified command relationships with each other, which of themselves, do not establish joint forces (JP 3-0). A joint force is a general term applied to a force composed of significant elements, assigned or attached, of two or more Military Departments operating under a single joint force commander (JP 3-0). Joint operations exploit the advantages of interdependent Service capabilities through unified action, and joint planning integrates military power with other instruments of national power (diplomatic, informational, and economic) to achieve a desired military end state.

B. Multinational Operations

Ref: ADP 3-0, Operations (Nov '16), figure 1. ADP 3-0 unified logic chart.

Multinational operations is a collective term to describe military actions conducted by forces of two or more nations, usually undertaken within the structure of a coalition or alliance (JP 3-16). While each nation has its own interests and often participates within limitations of national caveats, all nations bring value to an operation. Each nation's force has unique capabilities, and each usually contributes to the operation's legitimacy in terms of international or local acceptability. Army forces should anticipate participating in multinational operations and plan accordingly.

Refer to FM 3-16 for more information on multinational operations.

Alliance

An alliance is the relationship that results from a formal agreement between two or more nations for broad, long-term objectives that further the common interests of the members (JP 3-0). Military alliances, such as the North Atlantic Treaty Organization (commonly known as NATO), allow partners to establish formal, standard agreements.

Coalition

A coalition is an arrangement between two or more nations for common action (JP 5-0). Nations usually form coalitions for focused, short-term purposes. A coalition action is an action outside the bounds of established alliances, usually for single occasions or longer cooperation in a narrow sector of common interest. Army forces may conduct coalition actions under the authority of a United Nations resolution.

Soldiers assigned to a multinational force face many demands. These include dealing with cultural issues, different languages, interoperability challenges, national caveats on the use of respective forces and sharing of information and intelligence, rules of engagement, and underdeveloped methods and systems for commanding and controlling. Commanders analyze the mission's particular requirements to exploit the multinational force's advantages and compensate for its limitations. Establishing effective liaison with multinational partners is an important means for increasing the commander's understanding.

Multinational sustainment requires detailed planning and coordination. Normally, each nation provides a national support element to sustain its deployed forces. However, integrated multinational sustainment may improve efficiency and effectiveness. When authorized and directed, an Army theater sustainment command can provide logistics and other support to multinational forces. Integrating support requirements of several nations' forces, often spread over considerable distances and across international boundaries, is challenging. Commanders consider multinational force capabilities, such as mine clearance, that may exceed U.S. forces' capabilities.

Refer to TAA2: Military Engagement, Security Cooperation & Stability SMARTbook (Foreign Train, Advise, & Assist) for further discussion. Topics include Security Cooperation & Security Assistance (Train, Advise, & Assist), Stability Operations (ADRP 3-07), Peace Operations (JP 3-07.3), Counterinsurgency Operations (JP & FM 3-24), Civil-Military Operations (JP 3-57), Multinational Operations (JP 3-16), Interorganizational Coordination (JP 3-08), and more.

IV. Train to Win in a Complex World

Ref: FM 7-0, Train to Win in a Complex World (Oct '16).

The Army trains to win in a complex world. To fight and win in a chaotic, ambiguous, and complex environment, the Army trains to provide forces ready to conduct unified land operations. The Army does this by conducting tough, realistic, and challenging training. Unit and individual training occurs all the time—at home station, at combat training centers, and while deployed.

Army forces face threats that will manifest themselves in combinations of conventional and irregular forces, including insurgents, terrorists, and criminals. Some threats will have access to sophisticated technologies such as night vision systems, unmanned systems (aerial and ground), and weapons of mass destruction. Some threats will merge cyberspace and electronic warfare capabilities to operate from disparate locations. Additionally, they may hide among the people or in complex terrain to thwart the Army's conventional combat overmatch. Adding to this complexity is continued urbanization and the threat's access to social media. This complex environment will therefore require future Soldiers to train to perform at the highest levels possible.

Principles of Training

- Train As You Fight
- Training Is Commander Driven
- Training Is Led By Trained Officers and Noncommissioned Officers (NCOs)
- Train To Standard
- Train Using Appropriate Standard
- Training Is Protected
- Training Is Resourced
- Train To Sustain
- Train To Maintain
- Training Is Multiechelon and Combined Arms

Ref: FM 7-0 (Oct '16), p. 1-4.

Training is the most important thing the Army does to prepare for operations. Training is the cornerstone of readiness. Readiness determines our Nation's ability to fight and win in a complex global environment. To achieve a high degree of readiness, the Army trains in the most efficient and effective manner possible. Realistic training with limited time and resources demands that commanders focus their unit training efforts to maximize training proficiency.

Unit training and leader development are the Army's life-blood. Army leaders train units to be versatile. They develop subordinate leaders—military and Army civilians—to be competent, confident, agile, and adaptive using the Army leader development model. Units and leaders master individual and collective tasks required to execute the unit's designed capabilities and accomplish its mission. Army forces conduct training and education in the Army in three training domains: institutional, operational, and self-development. Army training and education methods evolve.

Readiness Through Training

Ref: ADP 3-0, Operations (Nov '16), p. 1-12.

Effective training is the cornerstone of operational success. As General Mark A. Milley, Chief of Staff of the Army, wrote in his initial message to the Army, "Readiness for ground combat is—and will remain—the U.S. Army's #1 priority. We will always be ready to fight today, and we will always prepare to fight tomorrow." Through training and leader development, Soldiers, leaders, and units achieve the tactical and technical competence that builds confidence and allows them to conduct successful operations across the continuum of conflict. Achieving this competence requires specific, dedicated training on offensive, defensive, and stability or defense support of civil authorities tasks. Training continues in deployed units to sustain skills and to adapt to changes in an operational environment.

Army training includes a system of techniques and standards that allow Soldiers and units to determine, acquire, and practice necessary skills. Candid assessments, after action reviews, and applied lessons learned and best practices produce quality Soldiers and versatile units, ready for all aspects of a situation. Through training and experiential practice and learning, the Army prepares Soldiers to win in land combat. Training builds teamwork and cohesion within units. It recognizes that Soldiers ultimately fight for one another and their units. Training instills discipline. It conditions Soldiers to operate within the law of war and rules of engagement.

Army training produces formations that fight and win with overwhelming combat power against any enemy. However, the complexity of integrating all unified action partners' demands that Army forces maintain a high degree of preparedness at all times, as it is difficult to achieve proficiency quickly. Leaders at all levels seek and require training opportunities between the Regular Army and Reserve Components, and their unified action partners at home station, at combat training centers, and when deployed.

The Army as a whole must be flexible enough to operate successfully across the range of military operations. Units must be agile enough to adapt quickly and be able to shift with little effort from a focus on one portion of the continuum of conflict to focus on another portion. Change and adaptation that once required years to implement must now be recognized, communicated, and enacted far more quickly. Technology, having played an increasingly important role in increasing the lethality of the industrial age battlefield, will assume more importance and require greater and more rapid innovation in tomorrow's conflicts. No longer can responses to hostile asymmetric approaches be measured in months; solutions must be anticipated and rapidly fielded across the force—and then be adapted frequently and innovatively as the enemy adapts to counter new-found advantages.

U.S. responsibilities are global; therefore, Army forces prepare to operate in any environment. Army training develops confident, competent, and agile leaders and units. Commanders focus their training time and other resources on tasks linked to their mission. Because Army forces face diverse threats and mission requirements, commanders adjust their training priorities based on a likely operational environment. As units prepare for deployment, commanders adapt training priorities to address tasks required by actual or anticipated operations.

Refer to TLS5: The Leader's SMARTbook, 5th Ed. for complete discussion of Military Leadership (ADP/ADRP 6-22); Leader Development (FM 6-22); Counsel, Coach, Mentor (ATP 6-22.1); Army Team Building (ATP 6-22.6); Military Training (ADP/ADRP 7-0); Train to Win in a Complex World (FM 7-0); Unit Training Plans, Meetings, Schedules, and Briefs; Conducting Training Events and Exercises; Training Assessments, After Action Reviews (AARs); and more!

(Tactical Mission Fundamentals) II. The Army's Operational Concept 1-10c*

V. Homeland Defense and Decisive Action
Ref: ADRP 3-0, Operations (Nov '16), p. 3-5.

Homeland defense is the protection of United States sovereignty, territory, domestic population, and critical infrastructure against external threats and aggression or other threats as directed by the President (JP 3-27). The Department of Defense has lead responsibility for homeland defense. The strategy for homeland defense (and defense support of civil authorities) calls for defending the U.S. territory against attack by state and nonstate actors through an active, layered defense—a global defense that aims to deter and defeat aggression abroad and simultaneously protect the homeland. The Army supports this strategy with capabilities in the forward regions of the world, in the geographic approaches to U.S. territory, and within the U.S. homeland.

Homeland defense operations conducted in the land domain could be the result of extraordinary circumstances and decisions by the President. In homeland defense, Department of Defense and Army forces work closely with federal, state, territorial, tribal, local, and private agencies. Land domain homeland defense could consist of offensive and defensive tasks as part of decisive action.

Homeland defense is a defense-in-depth that relies on collection, analysis, and sharing of information and intelligence; strategic and regional deterrence; military presence in forward regions; and the ability to rapidly generate and project warfighting capabilities to defend the United States, its allies, and its interests. These means may include support to civil law enforcement; antiterrorism and force protection; counterdrug; air and missile defense; chemical, biological, radiological, nuclear, and high-yield explosives; and defensive cyberspace operations; as well as security cooperation with other partners to build an integrated, mutually supportive concept of protection.

Refer to HDS1: The Homeland Defense & DSCA SMARTbook (Protecting the Homeland / Defense Support to Civil Authority) for complete discussion. Topics and references include homeland defense (JP 3-28), defense support of civil authorities (JP 3-28), Army support of civil authorities (ADRP 3-28), multi-service DSCA TTPs (ATP 3-28.1/MCWP 3-36.2), DSCA liaison officer toolkit (GTA 90-01-020), key legal and policy documents, and specific hazard and planning guidance.

Refer to CTS1: The Counterterrorism, WMD & Hybrid Threat SMARTbook for further discussion. CTS1 topics and chapters include: the terrorist threat (characteristics, goals & objectives, organization, state-sponsored, international, and domestic), hybrid and future threats, forms of terrorism (tactics, techniques, & procedures), counterterrorism, critical infrastructure, protection planning and preparation, countering WMD, and consequence management (all hazards response).

Refer to CYBER: The Cyberspace & Electronic Warfare SMARTbook (in development). U.S. armed forces operate in an increasingly network-based world. The proliferation of information technologies is changing the way humans interact with each other and their environment, including interactions during military operations. This broad and rapidly changing operational environment requires that today's armed forces must operate in cyberspace and leverage an electromagnetic spectrum that is increasingly competitive, congested, and contested.

Tactical Mission Fundamentals
III. Tactical Mission Tasks

Ref: ADRP 1-02, Operational Terms and Military Symbols (Feb '15), chap. 1 and 9.

A tactical mission task is a specific activity performed by a unit while executing a form of tactical operation or form of maneuver. A tactical mission task may be expressed as either an action by a **friendly force** or **effects on an enemy force**. The tactical mission tasks describe the results or effects the commander wants to achieve. See following pages (pp. 1-12* to 1-13*) for tactical mission tasks.

Not all tactical mission tasks have symbols. Some tactical mission task symbols will include unit symbols, and the tactical mission task "delay until a specified time" will use an amplifier. However, no modifiers are used with tactical mission task symbols. Tactical mission task symbols are used in course of action sketches, synchronization matrixes, and maneuver sketches. They do not replace any part of the operation order.

A. Mission Symbols

Counterattack (dashed axis)	CATK	A form of attack by part or all of a defending force against an enemy attacking force, with the general objective of denying the enemy his goal in attacking (FM 3-0).
Cover	C C	A form of security operation whose primary task is to protect the main body by fighting to gain time while also observing and reporting information and preventing enemy ground observation of and direct fire against the main body.
Delay	D	A form of retrograde in which a force under pressure trades space for time by slowing down the enemy's momentum and inflicting maximum damage on the enemy without, in principle, becoming decisively engaged (JP 1-02, see delaying operation).
Guard	G G	A form of security operations whose primary task is to protect the main body by fighting to gain time while also observing and reporting information and preventing enemy ground observation of and direct fire against the main body. Units conducting a guard mission cannot operate independently because they rely upon fires and combat support assets of the main body.
Penetrate		A form of maneuver in which an attacking force seeks to rupture enemy defenses on a narrow front to disrupt the defensive system (FM 3-0).
Relief in Place	RIP	A tactical enabling operation in which, by the direction of higher authority, all or part of a unit is replaced in an area by the incoming unit.
Retirement	R	A form of retrograde [JP 1-02 uses *operation*] in which a force out of contact with the enemy moves away from the enemy (JP 1-02).
Screen	S S	A form of security operations that primarily provides early warning to the protected force.
Withdraw	W	A planned operation in which a force in contact disengages from an enemy force (JP 1-02) [The Army considers it a form of retrograde.]

(Tactical Mission Fundamentals) III. Tactical Mission Tasks 1-11

B. Effects on Enemy Forces

Block	*Block* is a tactical mission task that denies the enemy access to an area or prevents his advance in a direction or along an avenue of approach.
	Block is also an engineer obstacle effect that integrates fire planning and obstacle effort to stop an attacker along a specific avenue of approach or prevent him from passing through an engagement area.
Canalize	*Canalize* is a tactical mission task in which the commander restricts enemy movement to a narrow zone by exploiting terrain coupled with the use of obstacles, fires, or friendly maneuver.
Contain	*Contain* is a tactical mission task that requires the commander to stop, hold, or surround enemy forces or to cause them to center their activity on a given front and prevent them from withdrawing any part of their forces for use elsewhere.
Defeat (No graphic)	*Defeat* occurs when an enemy has temporarily or permanently lost the physical means or the will to fight. The defeated force is unwilling or unable to pursue his COA, and can no longer interfere to a significant degree. Results from the use of force or the threat of its use.
Destroy	*Destroy* is a tactical mission task that physically renders an enemy force combat-ineffective until it is reconstituted. Alternatively, to destroy a combat system is to damage it so badly that it cannot perform any function or be restored to a usable condition without being entirely rebuilt.
Disrupt	*Disrupt* is a tactical mission task in which a commander integrates direct and indirect fires, terrain, and obstacles to upset an enemy's formation or tempo, interrupt his timetable, or cause his forces to commit prematurely or attack in a piecemeal fashion.
	Disrupt is also an engineer obstacle effect that focuses fire planning and obstacle effort to cause the enemy to break up his formation and tempo, interrupt his timetable, commit breaching assets prematurely, and attack in a piecemeal effort.
Fix	*Fix* is a tactical mission task where a commander prevents the enemy from moving any part of his force from a specific location for a specific period. Fixing an enemy force does not mean destroying it. The friendly force has to prevent the enemy from moving in any direction.
	Fix is also an engineer obstacle effect that focuses fire planning and obstacle effort to slow an attacker's movement within a specified area, normally an engagement area.
Interdict	*Interdict* is a tactical mission task where the commander prevents, disrupts, or delays the enemy's use of an area or route. Interdiction is a shaping operation conducted to complement and reinforce other ongoing offensive or defensive
Isolate	*Isolate* is a tactical mission task that requires a unit to seal off—both physically and psychologically—an enemy from his sources of support, deny him freedom of movement, and prevent him from having contact with other enemy forces.
Neutralize	*Neutralize* is a tactical mission task that results in rendering enemy personnel or materiel incapable of interfering with a particular operation.
Turn	*Turn* is a tactical mission task that involves forcing an enemy element from one avenue of approach or movement corridor to another.
	Turn is also a tactical obstacle effect that integrates fire planning and obstacle effort to divert an enemy formation from one avenue of approach to an adjacent avenue of approach or into an engagement area.

1-12 (Tactical Mission Fundamentals) III. Tactical Mission Tasks

C. Actions by Friendly Forces

Task	Graphic	Description
Attack by Fire		*Attack-by-fire* is a tactical mission task in which a commander uses direct fires, supported by indirect fires, to engage an enemy without closing with him to destroy, suppress, fix, or deceive him.
Breach		*Breach* is a tactical mission task in which the unit employs all available means to break through or secure a passage through an enemy defense, obstacle, minefield, or fortification.
Bypass		*Bypass* is a tactical mission task in which the commander directs his unit to maneuver around an obstacle, position, or enemy force to maintain the momentum of the operation while deliberately avoiding combat with an enemy force.
Clear		*Clear* is a tactical mission task that requires the commander to remove all enemy forces and eliminate organized resistance within an assigned area.
Control	*No graphic*	*Control* is a tactical mission task that requires the commander to maintain physical influence over a specified area to prevent its use by an enemy or to create conditions for successful friendly operations.
Counterrecon	*No graphic*	*Counterreconnaissance* is a tactical mission task that encompasses all measures taken by a commander to counter enemy reconnaissance and surveillance efforts.
Disengage	*No graphic*	*Disengage* is a tactical mission task where a commander has his unit break contact with the enemy to allow the conduct of another mission or to avoid decisive engagement.
Exfiltrate	*No graphic*	*Exfiltrate* is a tactical mission task where a commander removes soldiers or units from areas under enemy control by stealth, deception, surprise, or clandestine means.
Follow and Assume		*Follow and assume* is a tactical mission task in which a second committed force follows a force conducting an offensive operation and is prepared to continue the mission if the lead force is fixed, attritted, or unable to continue. The follow-and-assume force is not a reserve but is committed to accomplish specific tasks.
Follow and Support		*Follow and support* is a tactical mission task in which a committed force follows and supports a lead force conducting an offensive operation. The follow-and-support force is not a reserve but is a force committed to specific tasks.
Occupy		*Occupy* is a tactical mission task that involves moving a friendly force into an area so that it can control that area. Both the force's movement to and occupation of the area occur without enemy opposition.
Reduce	*No graphic*	*Reduce* is a tactical mission task that involves the destruction of an encircled or bypassed enemy force.
Retain		*Retain* is a tactical mission task in which the cdr ensures that a terrain feature controlled by a friendly force remains free of enemy occupation or use. The commander assigning this task must specify the area to retain and the duration of the retention, which is time- or event-driven.
Secure		*Secure* is a tactical mission task that involves preventing a unit, facility, or geographical location from being damaged or destroyed as a result of enemy action. This task normally involves conducting area security operations.
Seize		*Seize* is a tactical mission task that involves taking possession of a designated area by using overwhelming force. An enemy force can no longer place direct fire on an objective that has been seized.
Support by Fire		*Support-by-fire* is a tactical mission task in which a maneuver force moves to a position where it can engage the enemy by direct fire in support of another maneuvering force. The primary objective of the support force is normally to fix and suppress the enemy so he cannot effectively fire on the maneuvering force.

Tactical Doctrinal Taxonomy

Ref: ADRP 3-90, Offense and Defense (Aug '12), fig. 2-1, p. 2-3.

The following shows the Army's tactical doctrinal taxonomy for the four elements of decisive action (in accordance with ADRP 3-0) and their subordinate tasks. The commander conducts tactical enabling tasks to assist the planning, preparation, and execution of any of the four elements of decisive action. Tactical enabling tasks are never decisive operations in the context of the conduct of offensive and defensive tasks. (They are also never decisive during the conduct of stability tasks.) The commander uses tactical shaping tasks to assist in conducting combat operations with reduced risk.

Elements of Decisive Action (and subordinate tasks)

Offensive Tasks
- **Movement to Contact**
  - Search and attack
  - Cordon and search
- **Attack**
  - Ambush*
  - Counterattack*
  - Demonstration*
  - Spoiling attack*
  - Feint*
  - Raid*
  - *Also known as special purpose attacks*
- **Exploitation**
- **Pursuit**

Forms of Maneuver
- Envelopment
- Frontal attack
- Infiltration
- Penetration
- Turning Movement

Defensive Tasks
- **Area defense**
- **Mobile defense**
- **Retrograde operations**
  - Delay
  - Withdrawal
  - Retirement

Forms of the Defense
- Defense of linear obstacle
- Perimeter defense
- Reverse slope defense

Stability Tasks
- Civil security
- Civil control
- Restore essential services
- Support to governance
- Support to economic and infrastructure development

Defense Support to Civil Authorities
- Provide support for domestic disasters
- Provide support for domestic CBRN incidents
- Provide support for domestic law enforcement agencies
- Provide other designated support

Tactical Enabling Tasks

Reconnaissance Operations
- Zone
- Area
- Route
- Recon in force

Security Operations
- Screen
- Guard
- Cover
- Area (also route & convoy)
- Local

Troop Movement
- Administrative movement
- Approach march
- Road march

Encirclement Operations

Passage of Lines

Relief in Place

Mobility Operations
- Breaching operations
- Clearing operations (area and route)
- Gap-crossing operations
- Combat roads and trails
- Forward airfields and landing zones
- Traffic operations

Tactical Mission Tasks

Actions by Friendly Forces
- Attack-by-Fire
- Breach
- Bypass
- Clear
- Control
- Counterreconnaissance
- Disengage
- Exfiltrate
- Follow and Assume
- Follow and Support
- Occupy
- Reduce
- Retain
- Secure
- Seize
- Support-by-Fire

Effects on Enemy Force
- Block
- Canalize
- Contain
- Defeat
- Destroy
- Disrupt
- Fix
- Interdict
- Isolate
- Neutralize
- Suppress
- Turn

1-14 (Tactical Mission Fundamentals) III. Tactical Mission Tasks

IV. Understand, Visualize Describe, Direct, Lead, Assess

Ref: ADRP 5-0, The Operations Process (Mar '12), chap. I.

Commander's Activities

Commanders are the most important participants in the operations process. While staffs perform essential functions that amply the effectiveness of operations, **commanders drive the operations process** through understanding, visualizing, describing, directing, leading, and assessing operations.

See p. 1-24 for an overview and discussion of the operations process.

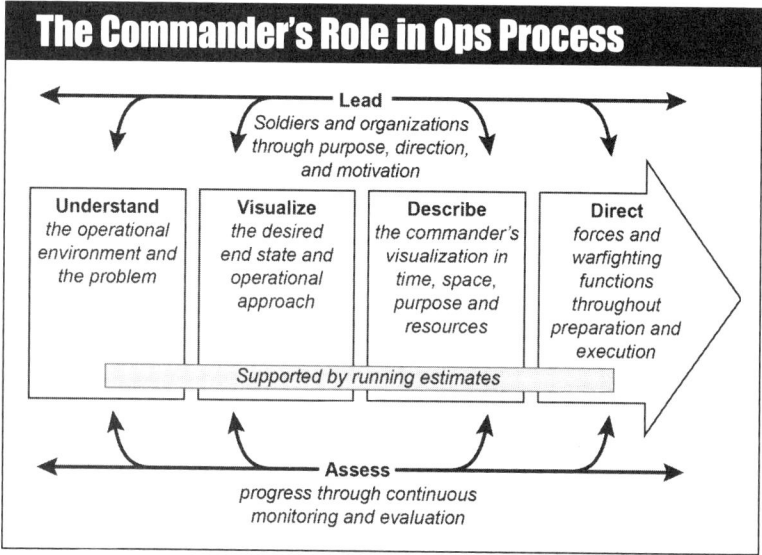

Ref: ADRP 5-0, The Operations Process, fig. 1-2, p. 1-3.

I. Understand

Understanding is fundamental to the commander's ability to establish a situation's context. It is essential to effective decision making during planning and execution. Analysis of the operational and mission variable provides the information used to develop understanding and frame the problem. In addition, conceptual and detailed planning assist commanders in developing their initial understanding of the operational environment and the problem. To develop a better understanding of an operational environment, commanders circulate within the area of operations as often as possible, collaborating with subordinate commanders and with Soldiers. Using personal observations and inputs from others (to include running estimates from the staff), commanders improve their understanding of their operational environment throughout the operations process.

Information collection (to include reconnaissance and surveillance) is indispensable to building and improving the commander's understanding. Formulating CCIRs, keeping them current, determining where to place key personnel, and arranging for liaison also contribute to improving the commander's understanding. Greater understanding enables commanders to make better decisions throughout the conduct of operations.

(Tactical Mission Fundamentals) IV. Commander's Activities 1-15

II. Visualize

As commanders begin to understand their operational environment and the problem, they start visualizing a desired end state and potential solutions to solve the problem. Collectively, this is known as commander's visualization—the mental process of developing situational understanding, determining a desired end state, and envisioning an operational approach by which the force will achieve that end state (ADP 5-0). Assignment of a mission provides the focus for developing the commander's visualization that, in turn, provides the basis for developing plans and orders. During preparation and execution, the commander's visualization helps commanders determine if, when, and what to decide, as they adapt to changing conditions.

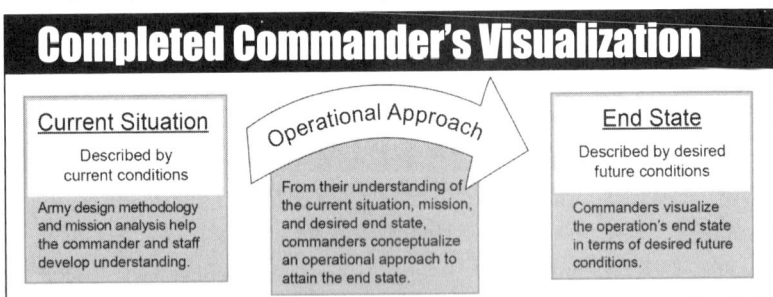

Ref: ADRP 5-0, The Operations Process, fig. 1-3, p. 1-4.

In building their visualization, commanders first seek to understand those conditions that represent the current situation. Next, commanders envision a set of desired future conditions that represents the operation's end state. Commanders complete their visualization by conceptualizing an operational approach—a description of the broad actions the force must take to transform current conditions into those desired at end state (JP 5-0).

Commanders apply the Army design methodology and use the elements of operational art when developing and describing their commander's visualization. They also actively collaborate with higher, subordinate and adjacent commanders, the staff, and unified action partners to assist them in building their visualization. Unified action partners are those military forces, governmental and nongovernmental organizations, and elements of the private sector that Army forces plan, coordinate, synchronize, and integrate with during the conduct of operations (ADRP 3-0). Because of the dynamic nature of military operations, commanders must continuously validate their visualization throughout the operations process.

See pp. *1-18 to *1-19 for further discussion.

III. Describe

After commanders visualize an operation, they describe it to their staffs and subordinates to facilitate shared understanding and purpose. During planning, commanders ensure subordinates understand their visualization well enough to begin course of action development. During execution, commanders describe modifications to their visualization in updated planning guidance and directives resulting in fragmentary orders that adjust the original order. Commanders describe their visualization in doctrinal terms, refining and clarifying it, as circumstances require. Commanders express their visualization in terms of:

- Commander's intent
- Planning guidance, including an operational approach
- Commander's critical information requirements
- Essential elements of friendly information

See pp. 1-20 to 1-21 for further discussion of the above elements.

IV. Direct

Commanders direct all aspects of operations by establishing their commander's intent, setting achievable objectives, and issuing clear tasks to subordinate units. Throughout the operations process, commanders direct forces by—
- Preparing and approving plans and orders
- Establishing command and support relationships
- Assigning and adjusting tasks, control measures, and task organization
- Positioning units to maximize combat power
- Positioning key leaders at critical places and times to ensure supervision
- Allocating resources to exploit opportunities and counter threats
- Committing the reserve as required

Refer to AODS5: The Army Operations & Doctrine SMARTbook for complete discussion of the fundamentals, principles and tenets of Army operations and organization (ADP/ADRP 3-0 Operations, 2016); chapters on each of the six warfighting functions: mission command (ADP/ADRP 6-0), movement and maneuver (ADPs 3-90, 3-07, 3-28, 3-05), intelligence (ADP/ADRP 2-0), fires (ADP/ADRP 3-09), sustainment (ADP/ADRP 4-0), and protection (ADP/ADRP 3-37); Doctrine 2015 guide and glossary of terms.

V. Lead

Through leadership, commanders provide purpose, direction, and motivation to subordinate commanders, their staff, and Soldiers. In many instances, a commander's physical presence is necessary to lead effectively. Where the commander locates within the area of operations is an important leadership consideration. Commanders balance their time between leading the staff through the operations process and providing purpose, direction, and motivation to subordinate commanders and Soldiers away from the command post.

Refer to TLS5: The Leader's SMARTbook, 5th Ed. for complete discussion of Military Leadership (ADP/ADRP 6-22); Leader Development (FM 6-22); Counsel, Coach, Mentor (ATP 6-22.1); Army Team Building (ATP 6-22.6); Military Training (ADP/ADRP 7-0); Train to Win in a Complex World (FM 7-0); Unit Training Plans, Meetings, Schedules, and Briefs; Conducting Training Events and Exercises; Training Assessments, After Action Reviews (AARs); and more!

VI. Assess

Commanders continuously assess the situation to better understand current conditions and determine how the operation is progressing. Continuous assessment helps commanders anticipate and adapt the force to changing circumstances. Commanders incorporate the assessments of the staff, subordinate commanders, and unified action partners into their personal assessment of the situation. Based on their assessment, commanders modify plans and orders to adapt the force to changing circumstances.

Refer to BSS5: The Battle Staff SMARTbook, 5th Ed. for further discussion. BSS5 covers the operations process (ADRP 5-0); commander's activities (Understand, Visualize, Describe, Direct, Lead, Assess); the military decisionmaking process and troop leading procedures (FM 6-0: MDMP/TLP); integrating processes and continuing activities (IPB, targeting, risk management); plans and orders (WARNOs/FRAGOs/OPORDs); mission command, command posts, liaison; rehearsals & after action reviews; and operational terms & symbols.

Tactical Msn Fundamentals

The Army Operational Framework
Ref: ADRP 3-0, Operations (Nov '16), pp. 4-4 to 4-8.

Army leaders are responsible for clearly articulating their concept of operations in time, space, purpose, and resources. They do this through an operational framework and associated vocabulary. An operational framework is a cognitive tool used to assist commanders and staffs in clearly visualizing and describing the application of combat power in time, space, purpose, and resources in the concept of operations (ADP 1-01). An operational framework establishes an area of geographic and operational responsibility for the commander and provides a way to visualize how the commander will employ forces against the enemy.

The operational framework has four components. First, commanders are assigned an area of operations for the conduct of operations. Second, a commander can designate a deep, close, and support areas to describe the physical arrangement of forces in time and space. Third, within this area, commanders conduct decisive, shaping, and sustaining operations to articulate the operation in terms of purpose. Finally, commanders designate the main and supporting efforts to designate the shifting prioritization of resources.

Area of Operations

An area of operations is an operational area defined by the joint force commander for land and maritime forces that should be large enough to accomplish their missions and protect their forces (JP 3-0). For land operations, an area of operations includes subordinate areas of operations assigned by Army commanders to their subordinate echelons as well. In operations, commanders use control measures to assign responsibilities, coordinate fires and maneuver, and control combat operations. A control measure is a means of regulating forces or warfighting functions (ADRP 6-0). One of the most important control measures is the area of operations. The Army commander or joint force land component commander is the supported commander within an area of operations designated by the joint force commander for land operations. Within their areas of operations, commanders integrate and synchronize combat power. To facilitate this integration and synchronization, commanders designate targeting priorities, effects, and timing within their areas of operations.

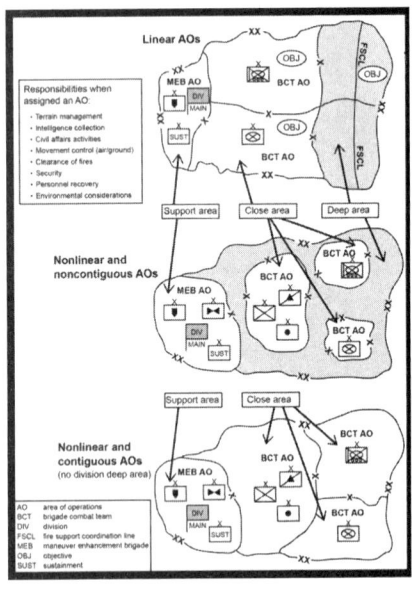

Area of Influence

Commanders consider a unit's area of influence when assigning it an area of operations. An area of influence is a geographical area wherein a commander is directly capable of influencing operations by maneuver or fire support systems normally under the commander's command or control (JP 3-0).

Understanding the area of influence helps the commander and staff plan branches to the current operation in which the force uses capabilities outside the area of operations.

Area of Interest

An area of interest is that area of concern to the commander, including the area of influence, areas adjacent thereto, and extending into enemy territory. This area also includes areas occupied by enemy forces who could jeopardize the accomplishment of the mission (JP 3-0). An area of interest for stability or DSCA tasks may be much larger than that area associated with the offense and defense.

*1-18 (Tactical Mission Fundamentals) IV. Commander's Activities

Deep, Close and Support Areas

- A **deep area** is the portion of the commander's area of operations that is not assigned to subordinate units. Operations in the deep area involve efforts to prevent uncommitted enemy forces from being committed in a coherent manner. A commander's deep area generally extends beyond subordinate unit boundaries out to the limits of the commander's designated area of operations. The purpose of operations in the deep area is frequently tied to other events distant in time, space, or both time and space.
- The **close area** is the portion of a commander's area of operations assigned to subordinate maneuver forces. Operations in the close area are operations that are within a subordinate commander's area of operations. Commanders plan to conduct decisive operations using maneuver and fires in the close area, and they position most of the maneuver force within it. Within the close area, depending on the echelon, one unit may conduct the decisive operation while others conduct shaping operations. A close operation requires speed and mobility to rapidly concentrate overwhelming combat power at the critical time and place and to exploit success.
- In operations, a commander may refer to a **support area**. The support area is the portion of the commander's area of operations that is designated to facilitate the positioning, employment, and protection of base sustainment assets required to sustain, enable, and control operations.

Decisive–Shaping–Sustaining Operations

Decisive, shaping, and sustaining operations lend themselves to a broad conceptual orientation.

- The **decisive operation** is the operation that directly accomplishes the mission. It determines the outcome of a major operation, battle, or engagement. The decisive operation is the focal point around which commanders design an entire operation. Multiple subordinate units may be engaged in the same decisive operation. Decisive operations lead directly to the accomplishment of a commander's intent. Commanders typically identify a single decisive operation, but more than one subordinate unit may play a role in a decisive operation.
- A **shaping operation** is an operation that establishes conditions for the decisive operation through effects on the enemy, other actors, and the terrain. Information operations, for example, may integrate Soldier and leader engagement tasks into the operation to reduce tensions between Army units and different ethnic groups through direct contact between Army leaders and local leaders. In combat, synchronizing the effects of aircraft, artillery fires, and obscurants to delay or disrupt repositioning forces illustrates shaping operations. Shaping operations may occur throughout the area of operations and involve any combination of forces and capabilities. Shaping operations set conditions for the success of the decisive operation. Commanders may designate more than one shaping operation.
- A **sustaining operation** is an operation at any echelon that enables the decisive operation or shaping operations by generating and maintaining combat power. Sustaining operations differ from decisive and shaping operations in that they focus internally (on friendly forces) rather than externally (on the enemy or environment).

Main and Supporting Efforts

Commanders designate main and supporting efforts to establish clear priorities of support and resources among subordinate units.

- The **main effort** is a designated subordinate unit whose mission at a given point in time is most critical to overall mission success. It is usually weighted with the preponderance of combat power. Typically, commanders shift the main effort one or more times during execution. Designating a main effort temporarily prioritizes resource allocation. When commanders designate a unit as the main effort, it receives priority of support and resources in order to maximize combat power. Commanders may designate a unit conducting a shaping operation as the main effort until the decisive operation commences. However, the unit with primary responsibility for the decisive operation then becomes the main effort upon the execution of the decisive operation.
- A **supporting effort** is a designated subordinate unit with a mission that supports the success of the main effort. Commanders resource supporting efforts with the minimum assets necessary to accomplish the mission. Forces often realize success of the main effort through success of supporting efforts.

Tactical Msn Fundamentals

Describe

Ref: ADRP 5-0, The Operations Process (Mar '12), pp. 1-4 to 1-6.

After commanders visualize an operation, they describe it to their staffs and subordinates to facilitate shared understanding and purpose. During planning, commanders ensure subordinates understand their visualization well enough to begin course of action development. During execution, commanders describe modifications to their visualization in updated planning guidance and directives resulting in fragmentary orders that adjust the original order. Commanders describe their visualization in doctrinal terms, refining and clarifying it, as circumstances require. Commanders express their visualization in terms of:

- Commander's intent
- Planning guidance, including an operational approach
- Commander's critical information requirements
- Essential elements of friendly information

A. Commander's Intent

The commander's intent is a clear and concise expression of the purpose of the operation and the desired military end state that supports mission command, provides focus to the staff, and helps subordinate and supporting commanders act to achieve the commander's desired results without further orders, even when the operation does not unfold as planned (JP 3-0). During planning, the initial commander's intent drives course of action development. In execution, the commander's intent guides disciplined initiative as subordinates make decisions when facing unforeseen opportunities or countering threats. Commanders develop their intent statement personally. It must be easy to remember and clearly understood by commanders and staffs two echelons lower in the chain of command. The more concise the commander's intent, the easier it is to recall and understand.

B. Planning Guidance

Commanders provide planning guidance to the staff based upon their visualization. Planning guidance must convey the essence of the commander's visualization, including a description of the operational approach. Effective planning guidance reflects how the commander sees the operation unfolding. It broadly describes when, where, and how the commander intends to employ combat power to accomplish the mission, within the higher commander's intent. Broad and general guidance gives the staff and subordinate leaders' maximum latitude; it lets proficient staffs develop flexible and effective options.

Commanders use their experience and judgment to add depth and clarity to their planning guidance. They ensure staffs understand the broad outline of their visualization while allowing them the latitude necessary to explore different options. This guidance provides the basis for the concept of operations without dictating the specifics of the final plan. As with their intent, commanders may modify planning guidance based on staff and subordinate input and changing conditions. (See ATTP 5-0.1 for a detailed discussion of developing and issue planning guidance).

C. Commander's Critical Information Requirements (CCIR)

A commander's critical information requirement is an information requirement identified by the commander as being critical to facilitating timely decision-making. The two key elements are friendly force information requirements and priority intelligence requirements (JP 3-0). A commander's critical information requirement (CCIR) directly influences decision making and facilitates the successful execution of military operations. Commanders decide to designate an information requirement as a CCIR based on likely decisions and their visualization of the course of the operation. A CCIR may support one or more decisions. During planning, staffs recommend information requirements for commanders to designate as CCIRs. During preparation and execution, they recommend changes to CCIRs based on assessment. A CCIR is:

- Specified by a commander for a specific operation
- Applicable only to the commander who specifies it
- Situation dependent—directly linked to a current or future mission
- Time-sensitive

Always promulgated by a plan or order, commanders limit the number of CCIRs to focus the efforts of limited collection assets. The fewer the CCIRs, the easier it is for staffs to remember, recognize, and act on each one. This helps staffs and subordinates identify information the commander needs immediately. While most staffs provide relevant information, a good staff expertly distills that information. It identifies answers to CCIRs and gets them to the commander immediately. It also identifies vital information that does not answer a CCIR but that the commander nonetheless needs to know. A good staff develops this ability through training and experience. Designating too many CCIRs limits the staff's ability to immediately recognize and react to them. Excessive critical items reduce the focus of collection efforts.

The list of CCIRs constantly changes. Commanders add and delete them throughout an operation based on the information needed for specific decisions. Commanders determine their own CCIRs, but they may select some from staff nominations. Once approved, a CCIR falls into one of two categories: priority intelligence requirements (PIRs) and friendly force information requirements (FFIRs).

Priority Intelligence Requirement (PIR)

A priority intelligence requirement is an intelligence requirement, stated as a priority for intelligence support, which the commander and staff need to understand the adversary or the operational environment (JP 2-0). PIRs identify the information about the enemy and other aspects of the operational environment that the commander considers most important. Lessons from recent operations show that intelligence about civil considerations may be as critical as intelligence about the enemy. Thus, all staff sections may recommend information about civil considerations as PIRs. The intelligence officer manages PIRs for the commander through planning requirements and assessing collection.

Refer to ATTP 2-0.1.

Friendly Force Information Requirement (FFIR)

A friendly force information requirement is information the commander and staff need to understand the status of friendly force and supporting capabilities (JP 3-0). FFIRs identify the information about the mission, troops and support available, and time available for friendly forces that the commander considers most important. In coordination with the staff, the operations officer manages FFIRs for the commander.

D. Essential Elements of Friendly Information (EEFI)

Commanders also describe information they want protected as essential elements of friendly information. An essential element of friendly information is a critical aspect of a friendly operation that, if known by the enemy, would subsequently compromise, lead to failure, or limit success of the operation and therefore should be protected from enemy detection. Although EEFIs are not CCIRs, they have the same priority. EEFIs establish elements of information to protect rather than one to collect. Their identification is the first step in the operations security process and central to the protection of information.

Elements of Combat Power (Direct)

Ref: ADRP 3-0, Operations (Nov '16), chap. 5 (and figure 5-1).

Operations executed through simultaneous offensive, defensive, stability, or defense support of civil authorities tasks require continuously generating and applying combat power, often for extended periods. Combat power is the total means of destructive, constructive, and information capabilities that a military unit or formation can apply at a given time. To an Army commander, Army forces generate combat power by converting potential into effective action. Combat power includes all capabilities provided by unified action partners that are integrated and synchronized with the commander's objectives to achieve unity of effort in sustained operations.

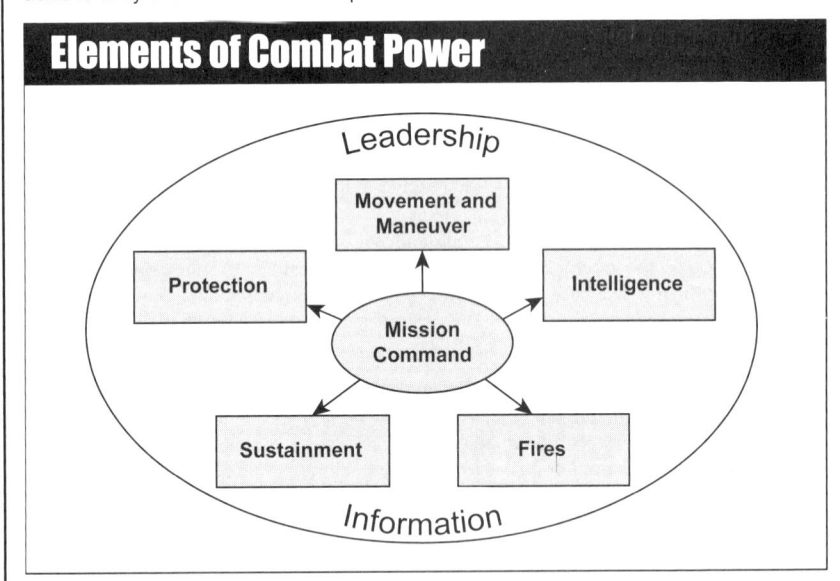

Elements of Combat Power

Leadership, Movement and Maneuver, Intelligence, Mission Command, Protection, Sustainment, Fires, Information

To execute combined arms operations, commanders conceptualize capabilities in terms of combat power. Combat power has eight elements: leadership, information, mission command, movement and maneuver, intelligence, fires, sustainment, and protection. These elements facilitate Army forces accessing joint and multinational fires and assets. The Army collectively describes the last six elements as the warfighting functions. Commanders apply combat power through the warfighting functions using leadership and information.

Generating and maintaining combat power throughout an operation is essential to success. Factors contributing to generating combat power include employing reserves, rotating committed forces, operating in cyberspace, and focusing joint support. Also, training forces on the conduct of operations, both when deployed and when not deployed, helps commanders to maintain and sustain combat power.

Refer to AODS5: The Army Operations & Doctrine SMARTbook for complete discussion of the fundamentals, principles and tenets of Army operations and organization (ADP/ADRP 3-0 Operations, 2016); chapters on each of the six warfighting functions: mission command (ADP/ADRP 6-0), movement and maneuver (ADPs 3-90, 3-07, 3-28, 3-05), intelligence (ADP/ADRP 2-0), fires (ADP/ADRP 3-09), sustainment (ADP/ADRP 4-0), and protection (ADP/ADRP 3-37); Doctrine 2015 guide and glossary of terms.

The Six Warfighting Functions

1. Mission Command
The mission command warfighting function is the related tasks and systems that develop and integrate those activities enabling a commander to balance the art of command and the science of control in order to integrate the other warfighting functions. Commanders, assisted by their staffs, integrate numerous processes and activities within the headquarters and across the force as they exercise mission command.
Refer to AODS5, chap. 2 for further discussion.

2. Movement and Maneuver
The movement and maneuver warfighting function is the related tasks and systems that move and employ forces to achieve a position of relative advantage over the enemy and other threats. Direct fire and close combat are inherent in maneuver. The movement and maneuver warfighting function includes tasks associated with force projection related to gaining a position of advantage over the enemy. Movement is necessary to disperse and displace the force as a whole or in part when maneuvering. Maneuver is the employment of forces in the operational area.
Refer to AODS5, chap. 3 for further discussion.

3. Intelligence
The intelligence warfighting function is the related tasks and systems that facilitate understanding the enemy, terrain, and civil considerations. This warfighting function includes understanding threats, adversaries, and weather. It synchronizes information collection with the primary tactical tasks of reconnaissance, surveillance, security, and intelligence operations. Intelligence is driven by commanders and is more than just collection. Developing intelligence is a continuous process that involves analyzing information from all sources and conducting operations to develop the situation.
Refer to AODS5, chap. 4 for further discussion.

4. Fires
The fires warfighting function is the related tasks and systems that provide collective and coordinated use of Army indirect fires, air and missile defense, and joint fires through the targeting process. Army fires systems deliver fires in support of offensive and defensive tasks to create specific lethal and nonlethal effects on a target.
Refer to AODS5, chap. 5 for further discussion.

5. Sustainment
The sustainment warfighting function is the related tasks and systems that provide support and services to ensure freedom of action, extend operational reach, and prolong endurance. The endurance of Army forces is primarily a function of their sustainment. Sustainment determines the depth and duration of Army operations. It is essential to retaining and exploiting the initiative. Sustainment provides the support necessary to maintain operations until mission accomplishment.
Refer to AODS5, chap. 6 for further discussion.

6. Protection
The protection warfighting function is the related tasks and systems that preserve the force so the commander can apply maximum combat power to accomplish the mission. Preserving the force includes protecting personnel (combatants and noncombatants) and physical assets of the United States and multinational military and civilian partners, to include the host nation. The protection warfighting function enables the commander to maintain the force's integrity and combat power. Protection determines the degree to which potential threats can disrupt operations and then counters or mitigates those threats.
Refer to AODS5, chap. 7 for further discussion.

Tactical Msn Fundamentals

Activities of the Operations Process

Ref: ADP 5-0, The Operations Process (Mar '12), pp. 2 to 6.

The Army's framework for exercising mission command is the operations process -- the major mission command activities performed during operations: planning, preparing, executing, and continuously assessing the operation.

The Operations Process (Underlying Logic)

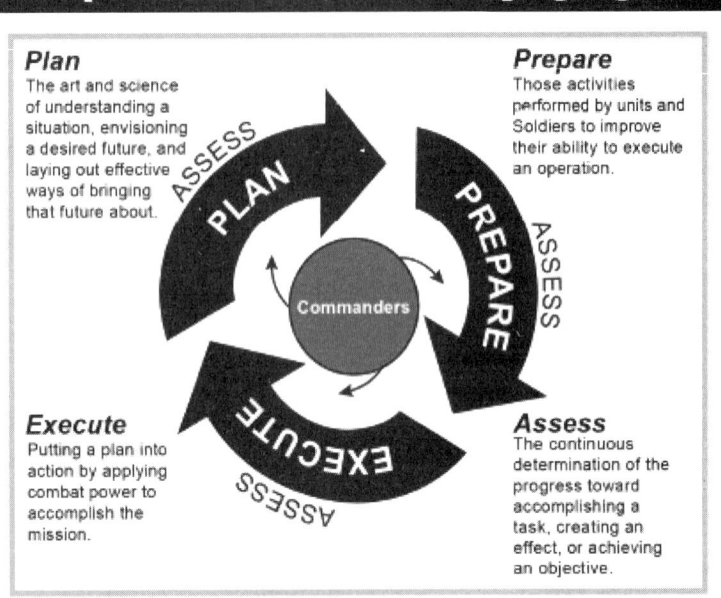

Plan
The art and science of understanding a situation, envisioning a desired future, and laying out effective ways of bringing that future about.

Prepare
Those activities performed by units and Soldiers to improve their ability to execute an operation.

Execute
Putting a plan into action by applying combat power to accomplish the mission.

Assess
The continuous determination of the progress toward accomplishing a task, creating an effect, or achieving an objective.

Central idea...

Commanders, supported by their staffs, use the **operations process** to drive the conceptual and detailed planning necessary to understand, visualize, and describe their operational environment; make and articulate decisions; and direct, lead, and assess military operations.

Principles

guided by...
- Commanders drive the operations process
- Apply critical and creative thinking
- Build and maintain situational understanding
- Encourage collaboration and dialogue

Refer to BSS5: The Battle Staff SMARTbook, 5th Ed. for further discussion. BSS5 covers the operations process (ADRP 5-0); commander's activities (Understand, Visualize, Describe, Direct, Lead, Assess); the military decisionmaking process and troop leading procedures (FM 6-0: MDMP/TLP); integrating processes and continuing activities (IPB, targeting, risk management); plans and orders (WARNOs/FRAGOs/OPORDs); mission command, command posts, liaison; rehearsals & after action reviews; and operational terms & symbols.

Chap 1
V. Troop Leading Procedures (TLP)

Tactical Msn Fundamentals

Ref: FM 6-0 (C2), Commander and Staff Organization and Operations (Apr '16), chap. 10.

Troop leading procedures extend the MDMP to the small-unit level. The MDMP and TLP are similar but not identical. They are both linked by the basic Army problem solving methodology explained. Commanders with a coordinating staff use the MDMP as their primary planning process. Company-level and smaller units lack formal staffs and use TLP to plan and prepare for operations. This places the responsibility for planning primarily on the commander or small-unit leader.

Leaders project their presence and guidance through troop leading procedures. TLP is the process a leader goes through to prepare the unit to accomplish a tactical mission. It begins when the mission is received. (Photo by Jeong, Hae-jung).

Troop leading procedures are a dynamic process used by small-unit leaders to analyze a mission, develop a plan, and prepare for an operation (ADP 5-0). These procedures enable leaders to maximize available planning time while developing effective plans and preparing their units for an operation. TLP consist of eight steps. The sequence of the steps of TLP is not rigid. Leaders modify the sequence to meet the mission, situation, and available time. Leaders perform some steps concurrently, while other steps may be performed continuously throughout the operation.

Leaders use TLP when working alone or with a small group to solve tactical problems. For example, a company commander may use the executive officer, first sergeant, fire support officer, supply sergeant, and communications sergeant to assist during TLP.

I. Performing Troop Leading Procedures

TLP provide small unit leaders a framework for planning and preparing for operations. This section discusses each step of TLP.

Army leaders begin TLP when they receive the initial WARNO or perceive a new mission. As each subsequent order arrives, leaders modify their assessments, update tentative plans, and continue to supervise and assess preparations. In some situations, the higher headquarters may not issue the full sequence of WARNOs; security considerations or tempo may make it impractical. In other cases, Army leaders may initiate TLP before receiving a WARNO based on existing plans and orders (contingency plans or be-prepared missions), and an understanding of the situation.

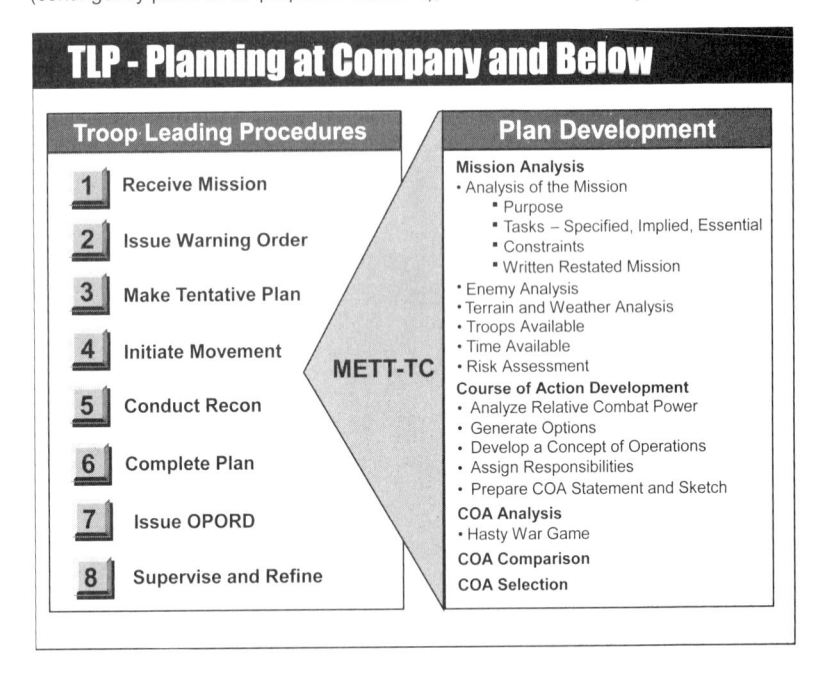

TLP - Planning at Company and Below

Troop Leading Procedures
1. Receive Mission
2. Issue Warning Order
3. Make Tentative Plan
4. Initiate Movement
5. Conduct Recon
6. Complete Plan
7. Issue OPORD
8. Supervise and Refine

METT-TC

Plan Development

Mission Analysis
- Analysis of the Mission
  - Purpose
  - Tasks – Specified, Implied, Essential
  - Constraints
  - Written Restated Mission
- Enemy Analysis
- Terrain and Weather Analysis
- Troops Available
- Time Available
- Risk Assessment

Course of Action Development
- Analyze Relative Combat Power
- Generate Options
- Develop a Concept of Operations
- Assign Responsibilities
- Prepare COA Statement and Sketch

COA Analysis
- Hasty War Game

COA Comparison

COA Selection

1. Receive The Mission

Receipt of a mission may occur in several ways. It may begin with the initial WARNO from higher or when a leader receives an OPORD. Frequently, leaders receive a mission in a FRAGO over the radio. Ideally, they receive a series of WARNOs, the OPORD, and a briefing from their commander. Normally after receiving an OPORD, leaders are required to give a confirmation brief to their higher commander to ensure they understand the higher commander's concept of operations and intent for his unit.

Upon receipt of mission, Army leaders perform an initial assessment of the situation (METT-TC analysis) and allocate the time available for planning and preparation. (Preparation includes rehearsals and movement.) This initial assessment and time allocation form the basis of their initial WARNO. Army leaders issue the initial WARNO quickly to give subordinates as much time as possible to plan and prepare.

Ideally, a battalion headquarters issues at least three WARNOs to subordinates when conducting the MDMP. WARNOs are issued upon receipt of mission, completion of mission analysis, and when the commander approves a COA. WARNOs serve a function in planning similar to that of fragmentary orders (FRAGOs) during execution.

Troop Leading Procedures and the MDMP
Ref: FM 6-0 (C2), Commander and Staff Organization and Operations (Apr '16), chap. 10.

Troop leading procedures extend the MDMP to the small-unit level. The MDMP and TLP are similar but not identical. The type, amount, and timeliness of information passed from higher to lower headquarters directly impact the lower unit leader's TLP. The solid arrows depict when a higher headquarters' planning event could start TLP of a subordinate unit. However, events do not always occur in the order shown. For example, TLP may start with receipt of a warning order (WARNO), or they may not start until the higher headquarters has completed the MDMP and issues an operation order (OPORD). WARNO's from higher headquarters may arrive at any time during TLP. Leaders remain flexible. They adapt TLP to fit the situation rather than try to alter the situation to fit a preconceived idea of how events should flow.

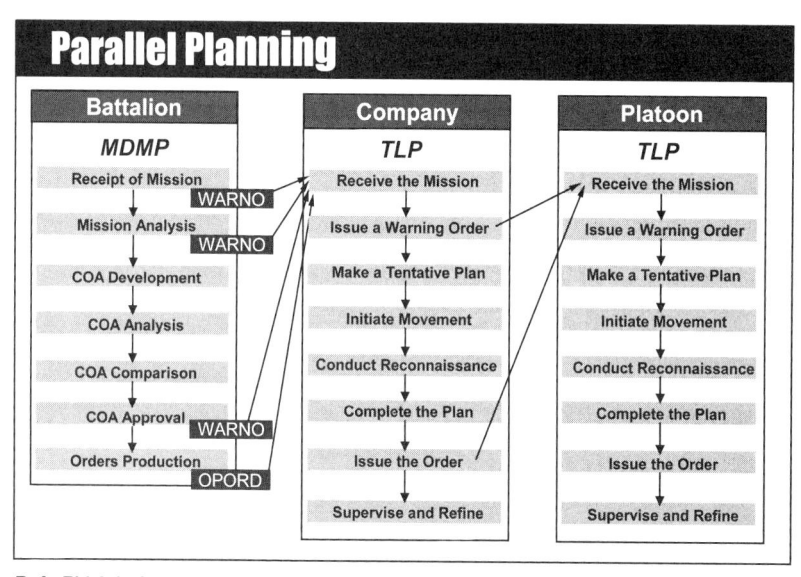

Parallel Planning

Battalion MDMP	Company TLP	Platoon TLP
Receipt of Mission → WARNO	Receive the Mission	Receive the Mission
Mission Analysis → WARNO	Issue a Warning Order	Issue a Warning Order
COA Development	Make a Tentative Plan	Make a Tentative Plan
COA Analysis	Initiate Movement	Initiate Movement
COA Comparison	Conduct Reconnaissance	Conduct Reconnaissance
COA Approval → WARNO	Complete the Plan	Complete the Plan
Orders Production → OPORD	Issue the Order	Issue the Order
	Supervise and Refine	Supervise and Refine

Ref: FM 6-0, Commander and Staff Organization and Operations, fig. 10-1, p. 10-2.

Normally, the first three steps (receive the mission, issue a WARNO, and make a tentative plan) of TLP occur in order. However, the sequence of subsequent steps is based on the situation. The tasks involved in some steps (for example, initiate movement and conduct reconnaissance) may occur several times. The last step, supervise and refine, occurs throughout.

A tension exists between executing current operations and planning for future operations. The small-unit leader must balance both. If engaged in a current operation, there is less time for TLP. If in a lull, transition, or an assembly area, there is more time to use TLP thoroughly. In some situations, time constraints or other factors may prevent leaders from performing each step of TLP as thoroughly as they would like. For example, during the step make a tentative plan, small-unit leaders often develop only one acceptable course of action (COA) vice multiple COA's. If time permits, leaders may develop, compare, and analyze several COA's before arriving at a decision on which one to execute.

A. Perform an Initial Assessment

The initial assessment addresses the factors of mission, enemy, terrain and weather, troops and support available, time available, and civil considerations (METT-TC). The order and detail in which Army leaders analyze the factors of METT-TC is flexible. It depends on the amount of information available and the relative importance of each factor. For example, they may concentrate on the mission, enemy, and terrain, leaving weather and civil considerations until they receive more detailed information.

Often, Army leaders will not receive their final unit mission until the WARNO is disseminated after COA approval or after the OPORD. Effective leaders do not wait until their higher headquarters completes planning to begin their planning. Using all information available, Army leaders develop their unit mission as completely as they can. They focus on the mission, commander's intent, and concept of operations of their higher and next higher headquarters. They pick out the major tasks their unit will probably be assigned and develop a mission statement based on information they have received. At this stage, the mission may be incomplete. For example, an initial mission statement could be, "First platoon conducts an ambush in the next 24 hours." While not complete, this information allows subordinates to start preparations. Leaders complete a formal mission statement during TLP step 3 (make a tentative plan) and step 6 (complete the plan).

B. Allocate the Available Time

Based on what they know, Army leaders estimate the time available to plan and prepare for the mission. They begin by identifying the times at which major planning and preparation events, including rehearsals, must be complete. Reverse planning helps them do this. Army leaders identify the critical times specified by higher headquarters and work back from them, estimating how much time each event will consume.

Leaders ensure that all subordinate echelons have sufficient time for their own planning and preparation needs. A general rule of thumb for leaders at all levels is to use no more than one-third of the available time for planning and issuance of the OPORD. Leaders allocate the remaining two-thirds of it to subordinates.

2. Issue a Warning Order (WARNO)

As soon as Army leaders finish their initial assessment of the situation and available time, they issue a WARNO. Leaders do not wait for more information. They issue the best WARNO possible with the information at hand and update it as needed with additional WARNOs.

The WARNO contains as much detail as possible. It informs subordinates of the unit mission and gives them the leader's time line. Army leaders may also pass on any other instructions or information they think will help subordinates prepare for the new mission. This includes information on the enemy, the nature of the higher headquarters plan, and any specific instructions for preparing their units. The most important thing is that leaders not delay in issuing the initial WARNO. As more information becomes available, leaders can -- and should -- issue additional WARNOs.

Warning Order (WARNO)

Normally an initial WARNO issued below battalion level includes:
- Mission or nature of the operation
- Time and place for issuing the OPORD
- Units or elements participating in the operation
- Specific tasks not addressed by unit SOP
- Time line for the operation

Allocate Available Time

Based on what they know, Army leaders estimate the time available to plan and prepare for the mission. They begin by identifying the times at which major planning and preparation events, including rehearsals, must be complete. Reverse planning helps them do this. Army leaders identify the critical times specified by higher headquarters and work back from them, estimating how much time each event will consume. Critical times might include aircraft loading times, the line of departure (LD) time, or the start point (SP) time for movement. By working backwards, Army leaders arrive at the time available to plan and prepare for the operation.

Leaders ensure that all subordinate echelons have sufficient time for their own planning and preparation needs. A general rule of thumb for leaders at all levels is to use no more than one-third of the available time for planning and issuance of the OPORD. Leaders allocate the remaining two-thirds of it to subordinates.

Below is a sample time schedule for an infantry company. This tentative schedule is adjusted as TLP progresses.

    0600 - Execute mission
    0530 - Finalize or adjust the plan based on leader's recon
    0400 - Establish the objective rallying point; begin leaders recon
    0200 - Begin movement
    2100 - Conduct platoon inspections
    1900 - Conduct rehearsals
    1800 - Eat meals (tray packs)
    1745 - Hold backbriefs (squad leaders to platoon leaders)
    1630 - Issue platoon OPORDs
    1500 - Hold backbriefs (platoon leaders to company commander)
    1330 - Issue company OPORD
    1045 - Conduct reconnaissance
    1030 - Update company WARNO
    1000 - Receive battalion OPORD
    0900 - Receive battalion WARNO; issue company WARNO

An example patrol timeline is below:

    11:00 Patrol secures OBJ [END]
    10:45 Patrol starts actions on OBJ
    10:30 Patrol passes through release point
    10:10 Patrol departs ORP
    10:00 Leader's recon returns to ORP
    09:30 Leader's recon departs ORP
    09:15 Patrol occupies ORP
    08:15 Patrol conducts passage of line through FLOT
    08:00 PL links up with forward unit commander/guide
    07:45 Patrol occupies AA
    07:30 Patrol begins movement to AA.
    07:15 Leaders conducts PCI
    07:00 Patrol conducts rehearsals
    06:30 PL issues OPORD
    06:00 PL issues WARNO [START]

3. Make a Tentative Plan

Once they have issued the initial WARNO, Army leaders develop a tentative plan. This step combines MDMP steps 2 through 6: mission analysis, COA development, COA analysis, COA comparison, and COA approval. At levels below battalion, these steps are less structured than for units with staffs. Often, leaders perform them mentally. They may include their principal subordinates-especially during COA development, analysis, and comparison. However, Army leaders, not their subordinates, select the COA on which to base the tentative plan.

A. Mission Analysis

To frame the tentative plan, Army leaders perform mission analysis. This mission analysis follows the METT-TC format, continuing the initial assessment performed in TLP step 1. FM 6 0 discusses the factors of METT-TC.

Note: See facing page (p. 1-31) for discussion and an outline of METT-TC.

METT-TC

- M - Mission
- E - Enemy
- T - Terrain and Weather
- T - Troops and Support Available
- T - Time Available
- C - Civil Considerations

The product of this part of the mission analysis is the restated mission. The restated mission is a simple, concise expression of the essential tasks the unit must accomplish and the purpose to be achieved. The mission statement states who (the unit), what (the task), when (either the critical time or on order), where (location), and why (the purpose of the operation).

B. Course of Action Development

Mission analysis provides information needed to develop COAs. The purpose of COA development is simple: to determine one or more ways to accomplish the mission. At lower echelons, the mission may be a single task. Most missions and tasks can be accomplished in more than one way. However, in a time-constrained environment, Army leaders may develop only one COA. Normally, they develop two or more. Army leaders do not wait for a complete order before beginning COA development. They develop COAs as soon as they have enough information to do so. Usable COAs are suitable, feasible, acceptable, distinguishable, and complete. To develop them, leaders focus on the actions the unit takes at the objective and conducts a reverse plan to the starting point.

Note: See The Battle Staff SMARTbook for further discussion of COA Development.

COA Development

1. Analyze relative combat power
2. Generate options
3. Array forces
4. Develop the concept of operations
5. Assign responsibilities
6. Prepare COA statement and sketch

II. METT-TC (Mission Variables)
Ref: Adapted from ADRP 5-0, The Operations Process (Mar '12), pp. 1-7 to 1-9.

Mission variables describe characteristics of the area of operations, focusing on how they might affect a mission. Incorporating the analysis of the operational variables into METT–TC ensures Army leaders consider the best available relevant information about conditions that pertain to the mission.

M - Mission
Army leaders analyze the higher headquarters WARNO or OPORD to determine how their unit contributes to the higher headquarters mission:
- Higher Headquarters Mission and Commander's Intent
- Higher Headquarters Concept of Operations
- Specified, Implied, and Essential Tasks
- Constraints

The product of this part of the mission analysis is the restated mission. The restated mission is a simple, concise expression of the essential tasks the unit must accomplish and the purpose to be achieved. The mission statement states who (the unit), what (the task), when (either the critical time or on order), where (location), and why (the purpose of the operation).

E - Enemy
With the restated mission as the focus, Army leaders continue the analysis with the enemy. For small unit ops, Army leaders need to know about the enemy's composition, disposition, strength, recent activities, ability to reinforce, and possible COAs.

T - Terrain and Weather
This aspect of mission analysis addresses the military aspects of terrain (OKOCA):
- Observation and Fields of Fire
- Key Terrain
- Obstacles
- Cover and Concealment
- Avenues of Approach

There are five military aspects of weather: visibility, winds, precipitation, cloud cover, and temperature/humidity (see FM 34-130). The analysis considers the effects on soldiers, equip., and supporting forces, such as air and artillery support.

Note: See pp. 1-32 to 1-33 for additional information on OCOKA.

T - Troops and Support Available
Perhaps the most important aspect of mission analysis is determining the combat potential of one's own force. Army leaders know the status of their soldiers' morale, their experience and training, and the strengths and weaknesses.

T - Time Available
Army leaders not only appreciate how much time is available; they understand the time-space aspects of preparing, moving, fighting, and sustaining. They view their own tasks and enemy actions in relation to time. They know how long it takes under such conditions to prepare for certain tasks (prepare orders, rehearsals, etc).

C - Civil Considerations
Civil considerations are how the man-made infrastructure, civilian institutions, and attitudes and activities of the civilian leaders, populations, and organizations within an area of operations influence the conduct of military operations (FM 6-0). Civil considerations are analyzed in terms of six factors (**ASCOPE**): areas, structures, capabilities, organizations, people and events.

III. OCOKA - Military Aspects of the Terrain

Ref: FM 34-130 Intelligence Preparation of the Battlefield, pp. 2-10 to 2-21.

Terrain analysis consists of an evaluation of the military aspects of the battlefield's terrain to determine its effects on military operations. The military aspects of terrain are often described using the acronym OCOKA.

O - Observation and Fields of Fire

Observation. Observation is the ability to see the threat either visually or through the use of surveillance devices. Factors that limit or deny observation include concealment and cover.

Fields of fire. A field of fire is the area that a weapon or group of weapons may effectively cover with fire from a given position. Terrain that offers cover limits fields of fire.

Terrain that offers both good observation and fields of fire generally favors defensive COAs.

The evaluation of observation and fields of fire allows you to:

- Identify potential engagement areas, or "fire sacks" and "kill zones"
- Identify defensible terrain and specific system or equipment positions
- Identify where maneuvering forces are most vulnerable to observation and fire

Evaluate observation from the perspective of electronic and optical line-of-sight (LOS) systems as well as unaided visual observation. Consider systems such as weapon sights, laser range finders, radars, radios, and jammers.

While ground based systems usually require horizontal LOS, airborne systems use oblique and vertical LOS. The same is true of air defense systems.

If time and resources permit, prepare terrain factor overlays to aid in evaluating observation and fields of fire. Consider the following:

- Vegetation or building height and density
- Canopy or roof closure
- Relief features, including micro-relief features such as defiles (elevation tinting techniques are helpful).
- Friendly and threat target acquisition and sensor capabilities
- Specific LOSs

C - Concealment and Cover

Concealment is protection from observation. Woods, underbrush, snowdrifts, tall grass, and cultivated vegetation provide concealment.

Cover is protection from the effects of direct and indirect fires. Ditches, caves, river banks, folds in the ground, shell craters, buildings, walls, and embankments provide cover.

The evaluation of concealment and cover aids in identifying defensible terrain, possible approach routes, assembly areas, and deployment and dispersal areas. Use the results of the evaluation to:

- Identify and evaluate AAs
- Identify defensible terrain and potential battle positions
- Identify potential assembly and dispersal areas

O - Obstacles

Obstacles are any natural or man-made terrain features that stop, impede, or divert military movement.

An evaluation of obstacles leads to the identification of mobility corridors. This in turn helps identify defensible terrain and AAs. To evaluate obstacles:

- Identify pertinent obstacles in the AI
- Determine the effect of each obstacle on the mobility of the evaluated force
- Combine the effects of individual obstacles into an integrated product

If DMA products are unavailable, and time and resources permit, prepare terrain factor overlays to aid in evaluating obstacles. Some of the factors to consider are:
- Vegetation (tree spacing/diameter)
- Surface drainage (stream width, depth, velocity, bank slope, & height)
- Surface materials (soil types and conditions that affect mobility)
- Surface configuration (slopes that affect mobility)
- Obstacles (natural and man-made; consider obstacles to flight as well as ground mobility)
- Transportation systems (bridge classifications and road characteristics such as curve radius, slopes, and width)
- Effects of actual or projected weather such as heavy precipitation or snow

K - Key Terrain

Key terrain is any locality or area the seizure, retention, or control of which affords a marked advantage to either combatant. Key terrain is often selected for use as battle positions or objectives. Evaluate key terrain by assessing the impact of its seizure, by either force, upon the results of battle.

A common technique is to depict key terrain on overlays and sketches with a large "K" within a circle or curve that encloses and follows the contours of the designated terrain. On transparent overlays use a color, such as purple, that stands out.

In the offense, key terrain features are usually forward of friendly dispositions and are often assigned as objectives. Terrain features in adjacent sectors may be key terrain if their control is necessary for the continuation of the attack or the accomplishment of the mission. If the mission is to destroy threat forces, key terrain may include areas whose seizure helps ensure the required destruction. Terrain that gives the threat effective observation along an axis of friendly advance may be key terrain if it is necessary to deny its possession or control by the threat.

In the defense, key terrain is usually within the AO and within or behind the selected defensive area.

Some examples of such key terrain are:
- Terrain that gives good observation over AAs to and into the defensive position
- Terrain that permits the defender to cover an obstacle by fire
- Important road junctions or communication centers that affect the use of reserves, sustainment, or LOCs

Additional Considerations:
- **Key terrain varies with the level of command.** For example, to an army or theater commander a large city may afford marked advantages as a communications center. To a division commander the high ground which dominates the city may be key terrain while the city itself may be an obstacle.
- **Terrain which permits or denies maneuver may be key terrain.**
- **Major obstacles are rarely key terrain features.** The high ground dominating a river rather than the river itself is usually the key terrain feature for the tactical commander. An exception is an obstacle such as a built-up area which is assigned as an objective.
- **Key terrain is decisive** terrain if it has an **extraordinary impact** on the mission.
- **Decisive terrain is rare and will not be present in every situation.**

A - Avenue of Approach (AA)

An **Avenue of Approach (AA)** is an air or ground route that leads an attacking force of a given size to its objective or to key terrain in its path.

During offensive operations, the evaluation of AAs leads to a recommendation on the best AAs to the command's objective and identification of avenues available to the threat for withdrawal or the movement of reserves.

During the defense, identify AAs that support the threat's offensive capabilities and avenues that support the movement and commitment of friendly reserves.

(Tactical Mission Fundamentals) V. Troop Leading Procedures 1-33

C. Analyze Courses of Action (Wargame)

For each COA, Army leaders think through the operation from start to finish. They compare each COA with the enemy's most probable COA. At small unit level, the enemy's most probable COA is what the enemy is most likely to do, given what friendly forces are doing at that instant. The leader visualizes a set of actions and reactions. The object is to determine what can go wrong and what decision the leader will likely have to make as a result.

D. Compare COAs and Make a Decision

Army leaders compare COAs by weighing the advantages, disadvantages, strengths, and weaknesses of each, as noted during the wargame. They decide which COA to execute based on this comparison and on their professional judgment. They take into account:

- Mission accomplishment
- Time to execute the operation
- Risk
- Results from unit reconnaissance
- Subordinate unit tasks and purposes
- Casualties incurred
- Posturing the force for future operations

4. Initiate Movement

Army leaders initiate any movement necessary to continue mission preparation or position the unit for execution, sometimes before making a tentative plan. They do this as soon as they have enough information to do so, or when the unit is required to move to position itself for a task. This is also essential when time is short. Movements may be to an assembly area, a battle position, a new AO, or an attack position. They may include movement of reconnaissance elements, guides, or quartering parties. Army leaders often initiate movement based on their tentative plan and issue the order to subordinates in the new location.

5. Conduct Reconnaissance

Whenever time and circumstances allow, Army leaders personally observe the AO for the mission. No amount of intelligence preparation of the battlefield (IPB) can substitute for firsthand assessment of METT-TC from within the AO. Unfortunately, many factors can keep leaders from performing a personal reconnaissance. The minimum action necessary is a thorough map reconnaissance, supplemented by imagery and intelligence products. In some cases, subordinates or other elements (such as scouts) may perform the reconnaissance for the leader while the leader completes other TLP steps.

Army leaders use the results of the wargame to identify information requirements. Reconnaissance operations seek to confirm or deny information that supports the tentative plan. They focus first on information gaps identified during mission analysis. Army leaders ensure their leader's reconnaissance complements the higher headquarters reconnaissance plan. The unit may conduct additional reconnaissance operations as the situation allows. This step may also precede making a tentative plan if there is not enough information available to begin planning. Reconnaissance may be the only way to develop the information required for planning.

6. Complete the Plan

During this step, Army leaders incorporate the result of reconnaissance into their selected COA to complete the plan or order. This includes preparing overlays, refining the indirect fire target list, coordinating combat service support and command and control requirements, and updating the tentative plan as a result of the reconnaissance. At lower levels, this step may entail only confirming or updating information contained in the tentative plan. If time allows, Army leaders make final coordination with adjacent units and higher headquarters before issuing the order.

7. Issue the Order

Small unit orders are normally issued verbally and supplemented by graphics and other control measures. The order follows the standard five-paragraph format OPORD format. Typically, Army leaders below company level do not issue a commander's intent. They reiterate the intent of their higher and next higher commander.

Note: See pp. 1-37 to 1-44 for a sample order formats.

The ideal location for issuing the order is a point in the AO with a view of the objective and other aspects of the terrain. The leader may perform a leader's reconnaissance, complete the order, and then summon subordinates to a specified location to receive it. Sometimes security or other constraints make it infeasible to issue the order on the terrain; then Army leaders use a sand table, detailed sketch, maps, and other products to depict the AO and situation.

8. Supervise and Refine

Throughout TLP, Army leaders monitor mission preparations, refine the plan, perform coordination with adjacent units, and supervise and assess preparations. Normally unit SOPs state individual responsibilities and the sequence of preparation activities. Army leaders supervise subordinates and inspect their personnel and equipment to ensure the unit is ready for the mission.

Army leaders refine their plan based on continuing analysis of their mission and updated intelligence. Most important, Army leaders know that they create plans to ensure all their subordinates focus on accomplishing the same mission within the commander's intent. If required, they can deviate from the plan and execute changes based on battlefield conditions and the enemy. Army leaders oversee preparations for operations. These include inspections, coordination, reorganization, fire support and engineer activities, maintenance, resupply, and movement. The requirement to supervise is continuous; it is as important as issuing orders. Supervision allows Army leaders to assess their subordinates' understanding of their orders and determine where additional guidance is needed. It is crucial to effective preparation.

Rehearsals

Note: See pp. 1-51 to 1-54 for a complete discussion of rehearsals.

A crucial component of preparation is the rehearsal. Rehearsals allow Army leaders to assess their subordinates' preparations. They may identify areas that require more supervision. Army leaders conduct rehearsals to:

- Practice essential tasks
- Identify weaknesses or problems in the plan
- Coordinate subordinate element actions
- Improve soldier understanding of the concept of operations
- Foster confidence among soldiers

Risk Management Process

Ref: ATP 5-19 (w/C1), Risk Management (Apr '14), chap. 1.

Risk Management is the process of identifying, assessing, and controlling risks arising from operational factors and making decisions that balance risk cost with mission benefits. (JP 3-0) *The Army no longer uses the term "composite risk management." Term replaced with joint term "risk management."*

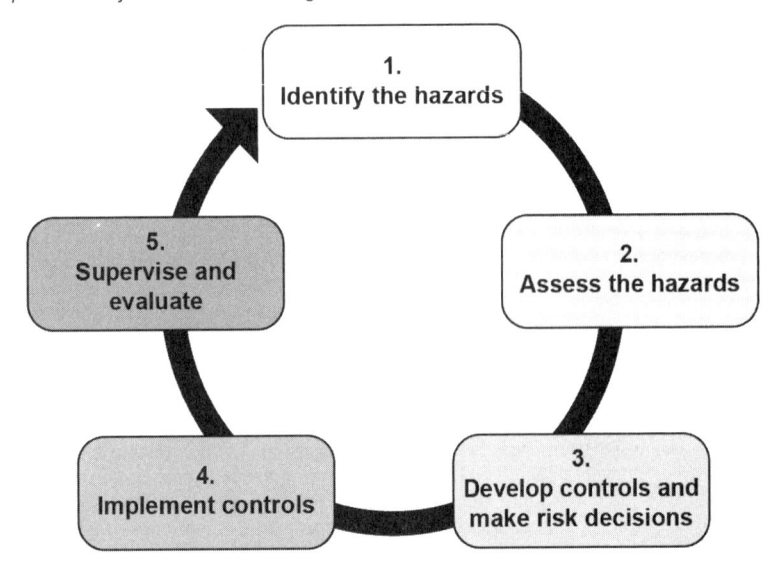

```
            1.
    Identify the hazards

5.                              2.
Supervise and              Assess the hazards
evaluate

            4.                  3.
    Implement controls    Develop controls and
                          make risk decisions
```

Ref: ATP 5-19, fig. 1-1. A cyclical, continuous process for managing risk.

1. Identify the hazards
A hazard is a condition with the potential to cause injury, illness, or death of personnel; damage to or loss of equipment or property; or mission degradation. Hazards exist in all environments—combat operations, stability operations, base support operations, training, garrison activities, and off-duty activities. The factors of mission, enemy, terrain and weather, troops and support available, time available, and civil considerations (METT-TC) serve as a standard format for identification of hazards, on-duty or off-duty.

2. Assess the hazards
This process is systematic in nature and uses charts, codes and numbers to present a methodology to assess probability and severity to obtain a standardized level of risk. Hazards are assessed and risk is assigned in terms of probability and severity of adverse impact of an event/occurrence.

3. Develop controls and make risk decisions
The process of developing and applying controls and reassessing risk continues until an acceptable level of risk is achieved or until all risks are reduced to a level where benefits outweigh the potential cost.

4. Implement controls
Leaders and staffs ensure that controls are integrated into SOPs, written and verbal orders, mission briefings, and staff estimates.

5. Supervise and evaluate

Chap 1: Tactical Mission Fundamentals

VI. Combat Orders

Ref: FM 3-21.8 The Infantry Rifle Platoon and Squad, pp. 5-4 to 5-5 and FM 6-0 (C2), Commander and Staff Organization and Operations (Apr '16), app. C.

Combat orders are the means by which the small unit leader receives and transmits information from the earliest notification that an operation will occur through the final steps of execution. WARNOs, OPORDs, and FRAGOs are absolutely critical to mission success. In a tactical situation, the small unit leaders work with combat orders on a daily basis, and they must have precise knowledge of the correct format for each type of order. At the same time, they must ensure that every Soldier in the unit understands how to receive and respond to the various types of orders.

Plans and orders are the means by which commanders express their visualization, commander's intent, and decisions. They focus on results the commander expects to achieve. Plans and orders form the basis commanders use to synchronize military operations. They encourage initiative by providing the "what" and "why" of a mission, and leave the how to accomplish the mission to subordinates. They give subordinates the operational and tactical freedom to accomplish the mission by providing the minimum restrictions and details necessary for synchronization and coordination.

Refer to BSS5: The Battle Staff SMARTbook, 5th Ed. for further discussion. BSS5 covers the operations process (ADRP 5-0); commander's activities (Understand, Visualize, Describe, Direct, Lead, Assess); the military decisionmaking process and troop leading procedures (FM 6-0: MDMP/TLP); integrating processes and continuing activities (IPB, targeting, risk management); plans and orders (WARNOs/FRAGOs/OPORDs); mission command, command posts, liaison; rehearsals & after action reviews; and operational terms & symbols.

Plans and orders:

- Permit subordinate commanders to prepare supporting plans and orders
- Implement instructions derived from a higher commander's plan or order
- Focus subordinates' activities
- Provide tasks and activities, constraints, and coordinating instructions necessary for mission accomplishment
- Encourage agility, speed, and initiative during execution
- Convey instructions in a standard, recognizable, clear, and simple format

The amount of detail provided in a plan or order depends on several factors, to include the experience and competence of subordinate commanders, cohesion and tactical experience of subordinate units, and complexity of the operation. Commanders balance these factors with their guidance and commander's intent, and determine the type of plan or order to issue. To maintain clarity and simplicity, plans and orders include annexes only when necessary and only when they pertain to the entire command. Annexes contain the details of support and synchronization necessary to accomplish the mission.

At the small unit level, orders are normally issued verbally and supplemented by graphics and other control measures. The order follows the standard five-paragraph format OPORD. Typically, Army leaders below company level do not issue a commander's intent. They reiterate the intent of their higher and next higher commander.

There are three types of orders of primary importance at the small unit level: the warning order (WARNO); the operation order (OPORD); and the fragmentary order (FRAGO).

The following information reflects a generic, standard form for each type of combat order. In many cases, we simply cut and paste the information into the format. However, some of this information will be generated from the vision of the commander.

Remember that when developing an order to consider two levels up and plan one level down. For example, a platoon leader will plan for the fires and maneuvers of his squads...while keeping in mind the overall mission of the company and battalion.

I. Warning Order (WARNO)

The WARNO serves as a notice of an upcoming mission and OPORD. It's also important because it allows troops to prepare mentally and physically. Experienced troops and leaders know from the mission statement what tasks will likely be required. They begin to ready any special equipment as well as their standard equipment. The troops also prepare themselves mentally, going over the tasks or lessons learned from previous experience and conducting battle drills or task rehearsals. Finally, the troops can pace themselves to some extent, getting sleep and food prior to the mission.

The warning order is a preliminary notice of an order or action which is to follow. It helps subordinate units and staffs prepare for new missions. WARNOs increase subordinates' planning time, provide details of the impending operations, and detail events that accompany preparation and execution.

The WARNO provides answers to the following questions:

- Who is involved in the mission?
- What is the task to be accomplished?
- Why are we performing this mission?
- When is the start time and location of the OPORD?

At a minimum, the WARNO states the situation, the mission, and coordinating instructions--such as the time and place of the OPORD. This allows subordinate leaders to prepare troops and equipment for the mission. (Photo by Jeong, Hae-jung).

A WARNO informs recipients of tasks they must do now or notifies them of possible future tasks. However, a WARNO does not authorize execution other than planning unless specifically stated. A WARNO follows the OPORD format. It may include some or all of the following:

- Series numbers, sheet numbers and names, editions, and scales of maps required (if changed from the current OPORD)
- The enemy situation and significant intelligence events
- The higher headquarters' mission
- Mission or tasks of the issuing headquarters
- The commander's intent statement
- Orders for preliminary actions, including intelligence, surveillance, and reconnaissance (ISR) operations
- Coordinating instructions (estimated timelines, orders group meetings, and the time to issue the OPORD)
- Service support instructions, any special equipment needed, regrouping of transport, or preliminary unit movement

II. Operations Order (OPORD)

An operation order is a directive issued by a commander to subordinate commanders for the purpose of effecting the coordinated execution of an operation (JP 1-02). It is the detailed plan of the mission, including the scheme of fire and maneuver, and the commander's intent. All Soldiers need to understand what is expected of them, what their specific role is in the mission, and how each fits into the "bigger picture." Rehearsals of actions on the objective allow each troop to see that big picture and where everyone will be physically located.

(Tactical Mission Fundamentals) VI. Combat Orders 1-39

Traditionally called the five paragraph field order, an OPORD contains, as a minimum, descriptions of the following:

- Task organization
- Situation
- Mission
- Execution
- Administrative and logistic support
- Command and signal for the specified operation
- OPORDs always specify an execution date and time

Much of paragraphs 1, 2, 4, and 5 of the OPORD are "cut and paste." However, that's not the case for Paragraph 3 – Execution. Paragraph 3a – Concept of the Operation is the very heart of the OPORD because it details the overall scheme of fire and maneuver. It answers the question, "How are we going to achieve this?"

The commander issues the OPORD and orients the OPORD to the terrain. Ideally, the order is given while standing on the terrain. This may be possible in the case of the defense, but is highly unlikely for any other type of combat operation. In those cases, a terrain model (or at least a topographical map) is used.

The commander's intent is the soul of the operation. The commander's intent answers the questions, "What are we going to do…and why are we doing this?" This is the most critical information of any mission—what, why, and how. Through the **commander's intent** statement, the commander explains in very concise terms what it is he or she expects to achieve. There is an art to this statement because it includes subtle nuances that must be understood by every troop on the mission.

See pp. 1-20 and 1-21 for further discussion of commander's intent.

III. Fragmentary Order (FRAGO)

The FRAGO is an adjustment to an existing OPORD. There are many reasons an order might need adjusting. Most commonly, a FRAGO is issued due to a significant change in the situation on the ground or for clarifying instructions.

It is issued after an OPORD to change or modify that order or to execute a branch or a sequel to that order.

FRAGOs differ from OPORDs only in the degree of detail provided. They address only those parts of the original OPORD that have changed. FRAGOs refer to previous orders and provide brief and specific instructions.

FRAGOs include all five OPORD paragraph headings. After each heading, state either new information or "no change." As such, this information depends on the specifics of the tactical situation. FRAGO may include:

- Updates to the enemy or friendly situation
- Changes to the scheme of maneuver
- Coordinating and clarifying instructions
- Expanding the mission tasks (branches and sequels)

Techniques for Issuing Orders
Ref: Adapted from FM 5-0 Army Planning and Orders Production, app. G, p. G-7.

There are several techniques for issuing orders: verbal, written, or electronically produced using matrices or overlays. The five-paragraph format is the standard for issuing combat orders. Orders may be generated and disseminated by electronic means to reduce the amount of time needed to gather and brief the orders group. When available preparation time or resources are constrained, commanders may use the matrix method of issuing orders.

At the small unit level, orders are normally issued verbally and supplemented by graphics and other control measures. The order follows the standard five-paragraph format OPORD format. Typically, Army leaders below company level do not issue a commander's intent. They reiterate the intent of their higher and next higher commander.

1. Verbal Orders
Verbal orders are used when operating in an extremely time-constrained environment. They offer the advantage of being passed quickly, but risk important information being overlooked or misunderstood. Verbal orders are usually followed up by written FRAGOs.

2. Graphics
Plans and orders generally include both text and graphics. Graphics convey information and instructions through military symbols (see FM 1-02). They complement the written portion of a plan or an order and promote clarity, accuracy, and brevity. The Army prefers depicting information and instructions graphically when possible. However, the mission statement and the commander's intent are always in writing.

3. Overlays
An overlay graphically portrays the location, size, and activity (past, current, or planned) of depicted units more consistently and accurately than text alone. An overlay enhances a viewer's ability to analyze the relationships of units and terrain. A trained viewer can attain a vision of a situation as well as insight into the identification of implied tasks, relationships, and coordination requirements that the written plan or order may not list or readily explain. Overlay graphics may be used on stand-alone overlays or overprinted maps. The issuing headquarters is responsible for the accuracy of control measures and for transposing graphics to and from the map scale used by subordinate headquarters.

4. Overlay Orders
An overlay order is a technique used to issue an order (normally a FRAGO) that has abbreviated instructions written on an overlay. Overlay orders combine a five-paragraph order with an operation overlay. Commanders may issue an overlay order when planning and preparation time is severely constrained and they must get the order to subordinate commanders as soon as possible. Commanders issue overlay orders by any suitable graphic method. An overlay order may consist of more than one overlay. A separate overlay or written annex can contain the service support coordination and organizations.

The Operations Order (OPORD) - A Small Unit Perspective

Editor's Note: This is an abbreviated OPORD sample based on small unit information needs. For a more complete and up-to-date OPORD (and WARNO/FRAGO) discussion and format, refer to BSS5: The Battle Staff SMARTbook.

Much of paragraphs 1, 2, 4, and 5 of the OPORD are "cut and paste." However, that's not the case for Paragraph 3 – Execution. Paragraph 3a – Concept of the Operation is the very heart of the OPORD because it details the overall scheme of fire and maneuver. It answers the question, "<u>How</u> are we going to achieve this?"

The commander issues the OPORD and orients the OPORD to the terrain. Ideally, the order is given while standing on the terrain. This may be possible in the case of the defense, but is highly unlikely for any other type of combat operation. In those cases, a terrain model (or at least a topographical map) is used.

Paragraph 1 - SITUATION
a. Enemy forces and battlefield conditions.
  (1) Weather and sunlight/moonlight data
  (2) Terrain using factors of OCOKA
  (3) Enemy forces
   Uniform identification
   Unit identification
   Recent activities
   Strengths/weaknesses
   Current location
   Most probable COA

b. Friendly forces
  (1) The larger unit's mission and commander's intent
  (2) Adjacent unit missions and locations
  (3) Identify fire support unit
  (4) Identify other supporting units

Paragraph 2 - MISSION
  Identify the task to be completed by the unit

Paragraph 3 - EXECUTION
  Commander's Intent
   Purpose
   Key tasks
   End state
  Note: See pp. 1-18 and 1-20 for complete discussion of Commander's Intent.

a. Concept of the operation
  (1) Scheme of maneuver
  (2) Fires
  (3) Engineer support

b. Tasks to maneuver units
- (1) Task for each element
- (2) Purpose for each element

c. Tasks to combat support units

d. Coordinating instructions
- (1) Movement instructions
    - SP Time and location
    - Order of march
    - Route of march
    - Rendezvous time and location (AA, ERP, ORP)
    - LOA and PL
- (2) Passage of lines
    - Linkup time and location
    - Passage point location
- (3) Priority Intelligence Requirements (PIR)
- (4) Troop safety
    - RFL and weapons control status

Paragraph 4 - SUSTAINMENT

a. Concept of support
- Location of combat field supply
- Location of aid station
- Scheme of support

b. Materiel and services
- (1) Supply
- (2) Transportation
- (3) Service
- (4) Maintenance

c. Medical evacuation procedures

d. Coordination for civilian personnel and EPW

Paragraph 5 - COMMAND & SIGNAL

a. Command
- (1) Location of leaders
- (2) Succession of command

b. Signal
- (1) SOI/CEOI in effect
- (2) Radio communications restrictions
    - Listening silence and time frame
    - Alternate frequencies and condition for frequency change
- (3) Visual and pyrotechnic signals
- (4) Brevity codes specific to the operation
- (5) Electronic protection, including COMSEC guidelines and procedures

On Point

Plans are the basis for any mission. To develop a plan (concept of the operation), the small unit leader summarizes how best to accomplish his mission within the scope of the commander's intent one and two levels up. The leader uses TLP to turn the concept into a fully developed plan and to prepare a concise, accurate operation order (OPORD). He assigns additional tasks (and outlines their purpose) for subordinate elements, allocates available resources, and establishes priorities to make the concept work.

Soldiers spell out the need and plan in terms of a "purpose" which is relayed through a "combat order". The WARNO alerts the unit of an upcoming mission. The OPORD provides greater detail, and most significantly includes the commander's intent and the scheme of the operation's fires and maneuver—the what, why, and how. The FRAGO allows the commander to adjust the mission once the OPORD has been issued.

As a practical matter, too much information can be a bad thing. Conversely, more information is always good. Both of these statements are correct. It's a paradox. While it is true that more information is good because an informed troop possesses the essential situational awareness to be successful on the battlefield.

While a 2-hour OPORD can be common at the company or battalion level, its rare at the platoon and squad level. Time plays a major factor in issuing the OPORD. Typically, a squad leader has about 15 minutes to issue the OPORD to his squad. If they have another 15 minutes, they can conduct a shoulder-to-shoulder rehearsal.

Tactical Mission Fundamentals
VII. Preparation & PCI

Ref: ADRP 5-0, The Operations Process (Mar '12), chap. 3 and FM 3-21.8 The Infantry Rifle Platoon and Squad, p. D-23.

I. Preparation

Preparation consists of those activities performed by units and Soldiers to improve their ability to execute an operation (ADP 5-0). Preparation creates conditions that improve friendly forces' opportunities for success. It requires commander, staff, unit, and Soldier actions to ensure the force is trained, equipped, and ready to execute operations. Preparation activities help commanders, staffs, and Soldiers understand a situation and their roles in upcoming operations.

Note: See following pages (pp. 1-46 to 1-47) for a listing of preparation activities, all of which involve actions at various levels by units, soldiers and staffs.

II. The Pre-Combat Inspection (PCI)

Pre-combat checks are detailed final checks that units conduct immediately before and during the execution of training and operations. These checks are usually included in unit SOPs. They are normally conducted as part of troop leading procedures and can be as simple or as complex as the training or operation dictates. Pre-combat checks start in garrison and many are completed in the assembly area or in the training location; for example, applying camouflage, setting radio frequencies and distributing ammunition.

Subordinate leaders MUST conduct an inspection of their troops prior to each combat mission. The PCI is an essential part of every mission. The objective of the PCI is to confirm the combat readiness of the unit. (Photo by Jeong, Hae-jung).

(Tactical Mission Fundamentals) VII. Preparation & PCI 1-45

Preparation Activities

Ref: ADRP 5-0, The Operations Process (Mar '12), pp. 3-1 to 3-5.

Preparation consists of those activities performed by units and Soldiers to improve their ability to execute an operation (ADP 5-0). Preparation creates conditions that improve friendly forces' opportunities for success. It requires commander, staff, unit, and Soldier actions to ensure the force is trained, equipped, and ready to execute operations. Preparation activities help commanders, staffs, and Soldiers understand a situation and their roles in upcoming operations.

Preparation Activities

Continue to coordinate and conduct liaison	Conduct rehearsals
Initiate information collection	Conduct plans-to-operations transitions
Initiate security operations	Refine the plan
Initiate troop movement	Integrate new Soldiers and units
Initiate sustainment preparations	Complete task organization
Initiate network preparations	Train
Manage terrain	Perform pre-operations checks and inspections
Prepare terrain	Continue to build partnerships and teams
Conduct confirmation briefs	

Ref: ADRP 5-0, The Operations Process, table 3-1, p. 3-1.

Preparation activities vary in accordance with the factors of METT-TC. For a listing and discussion of unit preparation activities from ADRP 5-0 (Aug '12), refer to The Battle Staff SMARTbook. The following list is adapted from FM 6-0 (Aug '03):

Reconnaissance Operations

During preparation, commanders take every opportunity to improve their situational understanding about the enemy and environment. Reconnaissance is often the most important part of this activity, providing data that contribute to answering the CCIR. As such, commanders conduct it with the same care as any other operation. They normally initiate reconnaissance operations before completing the plan.

Security Operations

Security operations during preparation prevent surprise and reduce uncertainty through security operations (see FM 3-90), local security, and operations security (OPSEC; see FM 3-13). These are all designed to prevent enemies from discovering the friendly force's plan and to protect the force from unforeseen enemy actions. Security elements direct their main effort toward preventing the enemy from gathering essential elements of friendly information (EEFI). As with reconnaissance, security is a dynamic effort that anticipates and thwarts enemy collection efforts. When successful, security operations provide the force time and maneuver space to react to enemy attacks.

Force Protection

Force protection consists of those actions taken to prevent or mitigate hostile actions against DoD personnel (to include family members), resources, facilities, and critical information. These actions conserve the force's fighting potential so it can be applied at the decisive time and place and incorporates the coordinated and synchronized offensive and defensive measures to enable the effective employment of the joint force while degrading opportunities for the enemy.

Revising and Refining the Plan
Plans are not static; commanders adjust them based on new information. During preparation, enemies are also acting and the friendly situation is evolving: Assumptions prove true or false. Reconnaissance confirms or denies enemy actions and dispositions. The status of friendly units changes. As these and other aspects of the situation change, commanders determine whether the new information invalidates the plan, requires adjustments to the plan, or validates the plan with no further changes.

Coordination and Liaison
Coordination is the action necessary to ensure adequately integrated relationships between separate organizations located in the same area. Coordination may include such matters as fire support, emergency defense measures, area intelligence, and other situations in which coordination is considered necessary (Army-Marine Corps). Coordination takes place continuously throughout operations and fall into two categories: external and internal. Available resources and the need for direct contact between sending and receiving headquarters determine when to establish liaison. The earlier liaison is established, the more effective the coordination.

Rehearsals
A rehearsal is a session in which a unit or staff practices expected actions to improve performance during execution. Rehearsals occur during preparation. *Is.*

Task Organizing
Task organizing is the process of allocating available assets to subordinate commanders and establishing their command and support relationships (FM 3-0). Receiving commands act to integrate units that are assigned, attached, under operational control (OPCON), or placed in direct support under a task organization.

Training
Training develops the teamwork, trust, and mutual understanding that commanders need to exercise mission command and forces need to achieve unity of effort. During repetitive, challenging training, commanders enhance their tactical skills and learn to develop, articulate, and disseminate their commander's intent.

Troop Movement
Troop movement is the movement of troops from one place to another by any available means (FM 3-90). Troop movements to position or reposition units for execution occur during preparation. Troop movements include assembly area reconnaissance by advance parties and route reconnaissance.

Pre-operation Checks and Inspections
Unit preparation includes completing pre-combat checks and inspections. These ensure that soldiers, units, and systems are as fully capable and ready to execute as time and resources permit. This preparation includes precombat training that readies soldiers and systems to execute the mission.

Logistic Preparation
Resupplying, maintaining, and issuing special supplies or equipment occurs during preparation. So does any repositioning of logistic assets. In addition, there are many other possible activities. These may include identifying and preparing forward bases, selecting and improving lines of communications, and identifying resources available in the area and making arrangements to acquire them. Commanders direct OPSEC measures to conceal preparations and friendly intentions.

Integrating New Soldiers and Units
Commanders and staffs ensure that new soldiers are assimilated into their units and new units into the force in a posture that allows them to contribute effectively. They also prepare new units and soldiers to perform their roles in the upcoming operation.

Tactical Msn Fundamentals

Conducting the PCI
Ref: FM 3-19 Military Police Leader's Handbook, appendix E.

The pre-combat inspection (PCI), or preoperation checks and inspections as FM 6-0 refers to them, is one of these critical preparation activities. Unit preparation includes completing precombat checks and inspections. These ensure that soldiers, units, and systems are as fully capable and ready to execute as time and resources permit. This preparation includes pre-combat training that readies soldiers and systems to execute the mission.

PCI is essential in that it checks:
1. Each troop's equipment necessary for mission accomplishment
2. Each troop's understanding of the mission purpose (commander's intent)
3. Each troop's understanding of how their task contributes to the mission

That means the PCI looks at equipment, asks questions regarding the commander's intent, and asks questions regarding the rehearsal. In this manner, the PCI protects the unit from missing any of critical steps of TLP.

The PCI is typically conducted in two stages. First, subordinate leaders conduct a PCI prior to movement to the AA to check equipment. Second, the PL conducts a PCI after the rehearsal to check equipment and mission knowledge.

The PCI is best achieved with a checklist. This keeps us from missing important key equipment and situational awareness.

Soldiers are inspected for what to bring and what *not* to bring. This will differ greatly depending on their role in the patrol. Every military unit must be able to shoot, move, and communicate. Those are the three basic Soldier skills. The PCI makes sure that each Soldier can do this.

A. Uniform and Gear
- Check that the troop is wearing the proper uniform and camouflage
- Check that boots are serviceable, comfortable, and appropriate
- Check that rain and cold weather gear is carried if needed
- Check water canteens and bladders are full and the troop is hydrated
- Check that first aid kits are present and complete
- Check that ID tags are worn, as well as special medical tags (allergies)
- Check that all specialty equipment is carried in either the LBE or rucksack
- Check all leaders for appropriate maps and compass/GPS
- Check all leaders for communication devices
- Check for secured gear by having the troop jump up and down

B. Communication Devices
- Check that extra batteries, antenna, mic, and basic radio kit are present
- Check that the radio is set to the proper channel and/or frequency
- Check the SOI/CEOI and ensure that each troop knows the call signs and code
- Check that all field phones are serviceable, clean, and in watertight containers
- Check for whistles, flares, color panels, and other communication devices

C. Weapon Systems

- Check that each weapon system is assigned to the appropriate troop
- Check that each weapon is serviceable, clean, and zeroed.
- Check that ammunition is serviceable and plentiful for each weapon
- Check that lubrication is present, as well as field cleaning kits
- Check optical devices (day and night) are serviceable
- Check that extra batteries are carried for optical devices

D. Specialty Equipment

- Check for first aid kits
- Check for protective gear—body armor, eye wear, kneepads, etc
- Check for screening smoke canisters
- Check for wire breaching/marking equipment if appropriate
- Check for mines/explosives if appropriate
- Check for anti-armor weapons if appropriate
- Check for rappelling/climbing/crossing gear if appropriate
- Check for pioneering tools if appropriate

E. Mission Knowledge

Small unit leaders should conduct a confirmation brief after issuing the oral OPORD to ensure subordinates know the mission, the commander's intent, the concept of the operation, and their assigned tasks. Confirmation briefs can be conducted face-to-face or by radio, depending on the situation. Face-to face-is the desired method, because all section and squad leaders are together to resolve questions, and it ensures that each leader knows what the adjacent squad is doing.

1. Commander's Intent

- Check that each troop understands the mission purpose
- Check that each troop understands the key tasks we must achieve
- Check that each troop understands the end state of success

2. Mission Tasks

- Check that each troop understands the mission statement
- Check that each troop understands their assigned task(s)
- Check that each troop knows how to identify the enemy
- Check that each troop knows the expected light, weather, and terrain conditions
- Check that each troop knows where other friendly troops are located
- Check that each troop knows his leader, and SOI/CEOI information

Commanders must allocate sufficient time for subordinate leaders to execute pre-combat checks and inspections to standard.

- OPORD briefed. Leaders and soldiers know what is expected of them
- Safety checks and briefings completed
- All required TADSS are on hand and operational; for example, MILES equipment zeroed
- Before-operations PMCS completed on vehicles, weapons, communications, and NBC equipment
- Leaders and equipment inspected; for example, compasses, maps, strip maps, and binoculars
- Soldiers and equipment inspected and camouflaged; for example, weapons, ID cards, driver's licenses
- Soldier packing lists checked and enforced
- Medical support present and prepared
- Communications checks completed
- Ammunition (Class V) drawn, accounted for, prepared, and issued
- Vehicle load plans checked and confirmed; cargo secured
- Rations (Class I) drawn and issued
- Quartering party briefed and dispatched
- OPFOR personnel deployed and ready to execute their OPORD

On Point

Mission failure is often due to equipment malfunction or a lack of mission knowledge. This is completely preventable. US military studies have shown that the PCI is conducted in only 40 percent of missions.

Leaders are responsible for ensuring that the combat vehicles and Soldiers in their unit are prepared to begin combat operations. A single item can doom the mission to failure. Leaders must conduct PCI. It takes only a few minutes. Subordinate leaders check each troop's equipment and specialty equipment prior to moving to the AA. After rehearsals, personally check every Soldier's equipment a second time and check each Soldier's knowledge of the mission.

Chap 1 — Tactical Mission Fundamentals

VIII. Rehearsals

Ref: FM 6-0 (C2), Commander and Staff Organization and Operations (Apr '16), chap. 12. For complete discussion of rehearsals, refer to BSS5: The Battle Staff SMARTbook.

Rehearsals allow leaders and their Soldiers to practice executing key aspects of the concept of operations. These actions help Soldiers orient themselves to their environment and other units before executing the operation. Rehearsals help Soldiers to build a lasting mental picture of the sequence of key actions within the operation. Rehearsals are the commander's tool to ensure staffs and subordinates understand the commander's intent and the concept of operations. They allow commanders and staffs to identify shortcomings (errors or omissions) in the plan not previously recognized. Rehearsals also contribute to external and internal coordination as the staff identifies additional coordinating requirements.

For units to be effective and efficient in combat, rehearsals need to become habitual in training. (Photo by Jeong, Hae-jung).

Effective and efficient units habitually rehearse during training. Commanders at every level routinely train and practice various rehearsal types and techniques. Local standard operating procedures (SOPs) identify appropriate rehearsal types, techniques, and standards for their execution.

Refer to BSS5: The Battle Staff SMARTbook, 5th Ed. for further discussion. BSS5 covers the operations process (ADRP 5-0); commander's activities (Understand, Visualize, Describe, Direct, Lead, Assess); the military decisionmaking process and troop leading procedures (MDMP & TLP); integrating processes and continuing activities (IPB, targeting, risk management); plans and orders (WARNOs/FRAGOs/OPORDs); mission command, command posts, liaison; rehearsals & after action reviews; and operational terms & symbols.

I. Methods of Rehearsals

Ref: FM 6-0 (C1), Commander and Staff Organization and Operations (May '15), pp. 12-2 to 12-6.

Techniques for conducting rehearsals are limited only by the commander's imagination and available resources. Generally, six techniques are used for executing rehearsals.

A. Full-dress Rehearsal

A full-dress rehearsal produces the most detailed understanding of the operation. It involves every participating soldier and system. If possible, organizations execute full-dress rehearsals under the same conditions-weather, time of day, terrain, and use of live ammunition-that the force expects to encounter during the actual operation

- **Time.** Full-dress rehearsals are the most time consuming of all rehearsal types. For companies and smaller units, the full-dress rehearsal is the most effective technique for ensuring all involved in the operation understand their parts. However, brigade and task force commanders consider the time their subordinates need to plan and prepare when deciding whether to conduct a full-dress rehearsal.
- **Echelons involved.** A subordinate unit can perform a full-dress rehearsal as part of a larger organization's reduced-force rehearsal.
- **OPSEC.** Moving a large part of the force may attract enemy attention. Commanders develop a plan to protect the rehearsal from enemy surveillance and reconnaissance. One method is to develop a plan, including graphics and radio frequencies, that rehearses selected actions but does not compromise the actual OPORD. Commanders take care to not confuse subordinates when doing this.
- **Terrain.** Terrain management for a full-dress rehearsal can be difficult if it is not considered during the initial array of forces. The rehearsal area must be identified, secured, cleared, and maintained throughout the rehearsal.

B. Key Leader Rehearsal

Circumstances may prohibit a rehearsal with all members of the unit. A key leader rehearsal involves only key leaders of the organization and its subordinate units. Often commanders use this technique to rehearse fire control measures for an engagement area during defensive operations. Commanders often use a reduced-force rehearsal to prepare key leaders for a full-dress rehearsal.

- **Time.** A reduced-force rehearsal normally requires less time than a full-dress rehearsal. Commanders consider the time their subordinates need to plan and prepare when deciding whether to conduct a reduced-force rehearsal.
- **Echelons involved.** A small unit can perform a full-dress rehearsal as part of a larger organization's reduced-force rehearsal.
- **OPSEC.** A reduced-force rehearsal is less likely to present an OPSEC vulnerability than a full-dress rehearsal because the number of participants is smaller. However, the number of radio transmissions required is the same as for a full-dress rehearsal and remains a consideration.
- **Terrain.** Terrain management for the reduced-force rehearsal can be just as difficult as for the full-dress rehearsal. The rehearsal area must be identified, secured, cleared, and maintained throughout the rehearsal.

C. Terrain-model Rehearsal (or "Digital" Terrain-model)

The terrain-model rehearsal is the most popular rehearsal technique. It takes less time and fewer resources than a full-dress or reduced-force rehearsal. When possible, commanders place the terrain model where it overlooks the actual terrain of the AO. (reverse slope for OPSEC, though). The model's orientation coincides with that of the terrain. The size of the terrain model can vary from small (using markers to represent units) to large (on which the participants can walk).

- **Time.** Often, the most time-consuming part of this technique is constructing the terrain model.
- **Echelons involved.** Because a terrain model is geared to the echelon conducting the rehearsal, multiechelon rehearsals using this technique are difficult.
- **OPSEC.** This rehearsal can present an OPSEC vulnerability if the area around the site is not secured. The collection of cdrs & vehicles can draw enemy attention.
- **Terrain.** Terrain management is less difficult than with the previous techniques. An optimal location overlooks the terrain where the operation will be executed. With today's digital capabilities, users can construct terrain models in virtual space.

D. Sketch-map Rehearsal

Commanders can use the sketch-map technique almost anywhere, day or night. The procedures are the same as for a terrain-model rehearsal, except the commander uses a sketch map in place of a terrain model. Effective sketches are large enough for all participants to see as each participant walks through execution of the operation. Participants move markers on the sketch to represent unit locations and maneuvers.

- **Time.** Sketch-map rehearsals take less time than terrain-model rehearsals and more time than map rehearsals.
- **Echelons involved.** Because a sketch map is geared to the echelon conducting the rehearsal, multiechelon rehearsals using this technique are difficult.
- **OPSEC.** This rehearsal can present an OPSEC vulnerability if the area around the site is not secured. The collection of cdrs & vehicles can draw enemy attention.
- **Terrain.** This technique requires less space than a terrain model rehearsal. A good site is easy for participants to find, yet concealed from the enemy. An optimal location overlooks the terrain where the unit will execute the operation.

E. Map Rehearsal

A map rehearsal is similar to a sketch-map rehearsal, except the commander uses a map and operation overlay of the same scale used to plan the operation.

- **Time.** The most time-consuming part is the rehearsal itself. A map rehearsal is normally the easiest technique to set up, since it requires only maps and current operational graphics.
- **Echelons involved.** Because a map is geared to the echelon conducting the rehearsal, multiechelon rehearsals using this technique are difficult.
- **OPSEC.** This rehearsal can present an OPSEC vulnerability if the area around the site is not secured. The collection of cdrs & vehicles can draw enemy attention.
- **Terrain.** This technique requires the least space. An optimal location overlooks the terrain where the ops will be executed, but is concealed from the enemy.

F. Network Rehearsal

Units conduct network rehearsals over wide-area networks or local area networks. Commanders and staffs practice these rehearsals by talking through critical portions of the operation over communications networks in a sequence the commander establishes. The organization rehearses only the critical parts of the operation. CPs can also rehearse battle tracking.

- **Time.** If the organization does not have a clear SOP and if all units are not up on the net, this technique can be very time consuming.
- **Echelons involved.** This technique lends itself to multiechelon rehearsals. Participation is limited only by cdr's desires and the availability of INFOSYSs.
- **OPSEC.** If a network rehearsal is executed from current unit locations, the volume of the communications transmissions and potential compromise of information through enemy monitoring can present an OPSEC vulnerability.
- **Terrain.** If a network rehearsal is executed from unit locations, terrain considerations are minimal.

II. Rehearsal Types

Ref: FM 6-0 (C2), Commander and Staff Organization and Operations (Apr '16), pp. 12-1 to 12-2.

Each rehearsal type achieves a different result and has a specific place in the preparation timeline.

A. Backbrief

A back brief is a briefing by subordinates to the commander to review how subordinates intend to accomplish their mission. Normally, subordinates perform back briefs throughout preparation. These briefs allow commanders to clarify the commander's intent early in subordinate planning. Commanders use the back brief to identify any problems in the concept of operations.

The back brief differs from the confirmation brief (a briefing subordinates give their higher commander immediately following receipt of an order) in that subordinate leaders are given time to complete their plan. Back briefs require the fewest resources and are often the only option under time-constrained conditions. Subordinate leaders explain their actions from start to finish of the mission. Back briefs are performed sequentially, with all leaders reviewing their tasks. When time is available, back briefs can be combined with other types of rehearsals. Doing this lets all subordinate leaders coordinate their plans before performing more elaborate drills.

B. Combined Arms Rehearsal

A combined arms rehearsal is a rehearsal in which subordinate units synchronize their plans with each other. A maneuver unit headquarters normally executes a combined arms rehearsal after subordinate units issue their operation order. This rehearsal type helps ensure that subordinate commanders' plans achieve the higher commander's intent.

C. Support Rehearsal

The support rehearsal helps synchronize each war fighting function with the overall operation. This rehearsal supports the operation so units can accomplish their missions. Throughout preparation, units conduct support rehearsals within the framework of a single or limited number of war fighting functions. These rehearsals typically involve coordination and procedure drills for aviation, fires, engineer support, or casualty evacuation. Support rehearsals and combined arms rehearsals complement preparations for the operation. Units may conduct rehearsals separately and then combine them into full-dress rehearsals. Although these rehearsals differ slightly by warfighting function, they achieve the same result.

D. Battle Drill or SOP Rehearsal

A battle drill is a collective action rapidly executed without applying a deliberate decision making process. A battle drill or SOP rehearsal ensures that all participants understand a technique or a specific set of procedures. Throughout preparation, units and staffs rehearse battle drills and SOPs. These rehearsals do not need a completed order from higher headquarters. Leaders place priority on those drills or actions they anticipate occurring during the operation. For example, a transportation platoon may rehearse a battle drill on reacting to an ambush while waiting to begin movement.

All echelons use these rehearsal types; however, they are most common for platoons, squads, and sections. They are conducted throughout preparation and are not limited to published battle drills. All echelons can rehearse such actions as a command post shift change, an obstacle breach lane-marking SOP, or a refuel-on-the-move site operation.

IX. The After Action Review (AAR)

Ref: FM 6-0 (C2), Commander and Staff Organization and Operations (Apr '16), chap. 16 and A Leader's Guide to After Action Reviews (Aug '12).

An after action review (AAR) is a guided analysis of an organization's performance, conducted at appropriate times during and at the conclusion of a training event or operation with the objective of improving future performance. It includes a facilitator, event participants, and other observers (ADRP 7-0, Training Units and Developing Leaders, Aug '12).

AARs are a key part of the training process, but they are not cure-alls for unit-training problems. Leaders must still make on-the-spot corrections and take responsibility for training soldiers and units. (Photo by Jeong, Hae-jung).

AARs are a professional discussion of an event that enables Soldiers/units to discover for themselves what happened and develop a strategy (e.g., retraining) for improving performance. They provide candid insights into strengths and weaknesses from various perspectives and feedback, and focus directly on the commander's intent, training objectives and standards. Leaders know and enforce standards for collective and individual tasks.

Refer to BSS5: The Battle Staff SMARTbook, 5th Ed. for further discussion. BSS5 covers the operations process (ADRP 5-0); commander's activities (Understand, Visualize, Describe, Direct, Lead, Assess); the military decisionmaking process and troop leading procedures (FM 6-0: MDMP/TLP); integrating processes and continuing activities (IPB, targeting, risk management); plans and orders (WARNOs/FRAGOs/OPORDs); mission command, command posts, liaison; rehearsals & after action reviews; and operational terms & symbols.

(Tactical Mission Fundamentals) IX. The AAR 1-55 *

Leaders must avoid creating the environment of a critique during AARs. Because Soldiers and leaders participating in an AAR actively self-discover what happened and why, they learn and remember more than they would from a critique alone. A critique only gives one viewpoint and frequently provides little opportunity for discussion of events by participants. Leaders make on-the-spot corrections and take responsibility for training Soldiers and units.

Types Of After Action Reviews

Two types of after action reviews exist: formal and informal. Commanders generally conduct formal action reviews after completing a mission. Normally, only informal after action reviews are possible during the conduct of operations.

Types of After-Action Reviews

Formal Reviews	Informal Reviews
■ Conducted by either internal or external leaders and external observer and controllers (OC)	■ Conducted by internal chain of command
■ Takes more time to prepare	■ Takes less time to prepare
■ Uses complex training aids	■ Uses simple training aids
■ Scheduled - events and / or tasks are identified beforehand	■ Conducted as needed. Primarily based on leaders assessment
■ Conducted where best supported	■ Held at the training site

Ref: A Leader's Guide to After Action Reviews, p. 5.

A. Formal

Leaders plan formal after action reviews when they complete an operation or otherwise realize they have the need, time, and resources available. Formal after action reviews require more planning and preparation than informal after action reviews. Formal after action reviews require site reconnaissance and selection; coordination for aids (such as terrain models and large-scale maps); and selection, setup, maintenance, and security of the after action review site. During formal after action reviews, the after action review facilitator (unit leader or other facilitator) provides an overview of the operation and focuses the discussion on topics the after action review plan identifies. At the conclusion, the facilitator reviews identified and discussed key points and issues, and summarizes strengths and weaknesses.

B. Informal

Leaders use informal after action reviews as on-the-spot coaching tools while reviewing Soldier and unit performance during or immediately after execution. Informal after action reviews involve all Soldiers. These after action reviews provide immediate feedback to Soldiers, leaders, and units after execution. Ideas and solutions leaders gathered during informal after action reviews can be applied immediately as the unit continues operations. Successful solutions can be identified and transferred as lessons learned.

The After Action Review (AAR)

Ref: FM 6-0 (C2), Commander and Staff Organization and Operations (Apr '16), pp. 16-3 to 16-4.

Formal and informal after action reviews generally follow the same format:

1. Review what was supposed to happen

The facilitator and participants review what was supposed to happen. This review is based on the commander's intent for the operation, unit operation or fragmentary orders (FRAGORDs), the mission, and the concept of operations.

2. Establish what happened

The facilitator and participants determine to the extent possible what actually happened during execution. Unit records and reports form the basis of this determination. An account describing actual events as closely as possible is vital to an effective discussion. The assistant chief of staff, intelligence (G-2 [S-2]) provides input about the operation from the enemy's perspective.

3. Determine what was right or wrong with what happened

Determine what was right or wrong with what happened. Participants establish the strong and weak points of their performance. The facilitator guides discussions so that the conclusions the participants reach are operationally sound, consistent with Army standards, and relevant to the operational environment.

4. Determine how the task should be done differently the next time

The facilitator helps the chain of command lead the group in determining how participants might perform the task more effectively. The intended result is organizational and individual learning that can be applied to future operations. If successful, this learning can be disseminated as lessons learned.

Leaders understand that not all tasks will be performed to standard. In their initial planning, they allocate time and other resources for retraining after execution or before the next operation. Retraining allows participants to apply the lessons learned from after action reviews and implement corrective actions. Retraining should be conducted at the earliest opportunity to translate observations and evaluations from after action reviews into performance in operations. Commanders ensure Soldiers understand that training is incomplete until the identified corrections in performance have been achieved.

After action reviews are often tiered as multi-echelon leader development tools. Following a session involving all participants, senior commanders may continue after action reviews with selected leaders as extended professional discussions. These discussions usually include a more specific review of leader contributions to the operation's results. Commanders use this opportunity to help subordinate leaders master current skills and prepare them for future responsibilities. After action reviews are opportunities for knowledge transfer through teaching, coaching, and mentoring.

Commanders conduct a final after action review during recovery after an operation. This after action review may include a facilitator. Unit leaders review and discuss the operation. Weaknesses or shortcomings identified during earlier after action reviews are identified again and discussed. If time permits, the unit conducts training to correct these weaknesses or shortcomings in preparation for future operations.

Lessons learned can be disseminated in at least three ways. First, participants may make notes to use in retraining themselves and their sections or units. Second, facilitators may gather their own and participants' notes for collation and analysis before dissemination and storage for others to use. Dissemination includes forwarding lessons to other units conducting similar operations as well as to the Center for Army Lessons Learned, doctrinal proponents, and generating force agencies. Third, units should publicize future successful applications of lessons as lessons learned.

III. AARs - A Small Unit Perspective

Ideally, the AAR is conducted by an outside source, such as an observer/controller. If there is no outside source, the AAR is moderated by the appropriate Army leader. A representation of the terrain covered in the mission is essential. This can be a terrain model, a map, or even video and photographs of the mission. The point is to get everyone thinking on the same sheet of music in terms of where and when.

An after-action review (AAR) is an assessment conducted after an event or major activity that allows participants to learn what and why something happened, and most importantly, how the unit can improve through change. This professional discussion enables units and their leaders to understand why things happened during the progression of an operation, and to learn from that experience. This learning is what enables units and their leaders to adapt to their operational environment. The AAR does not have to be performed at the end of the activity. Rather, it can be performed after each identifiable event (or whenever feasible) as a live learning process.

The AAR is a professional discussion that includes the participants and focuses directly on the tasks and goals. While it is not a critique, the AAR has several advantages over a critique:

- It does not judge success or failure
- It attempts to discover why things happened
- It focuses directly on the tasks and goals that were to be accomplished
- It encourages participants to raise important lessons in the discussion
- More Soldiers participate so more of the project or activity can be recalled and more lessons can be learned and shared

Be sure to include everyone's perspective, not just the leader's.

1. The Plan - Review What Was Supposed to Happen

This phase details "what was supposed to happen." In fact, it is nothing more than a restatement of the operation order (OPORD) in a concise form. The patrol leader (PL) stands in front of the team and quickly restates the commander's intent and the scheme of maneuver.

Remember this is not "what I think might have happened" or "what I really wished had happened." There should be no conjecture or grandstanding at this point. It's simply a restatement of the plan in as much detail as necessary, and no more.

For the first phase, the plan, it is a good idea to begin an AAR by asking each of the element leaders what mission they were tasked to do and to let them answer in their own words. The leader will listen to the clarity of focus from each of these subordinate leaders for their grasp of the mission as a whole and the understanding of their element's role in the mission.

2. The Performance - Establish What Happened

The second phase of the AAR details "what really happened." Those team members who were on the ground and taking part in the action state this part of the AAR. The PL is generally discouraged from speaking during this phase unless they have a unique perspective or information that is of use.

The point isn't to put ego on the line and taking up time arguing or finger pointing, yet the patrol has to get a realistic measure of their performance. This can be a painful, even embarrassing process. But why didn't the mission go off just as planned? Or did it? That is what patrol is looking for in this phase of the AAR—the gap between the plan and the actual performance.

For the second phase, the performance, the initial movement is a good place to start within the execution of the mission. Did the patrol hit their start time? How about the time hacks for phase lines or reaching their objective rally point (ORP)? If not, this may explain any "accordion" effect in the time frames or the physical movement of the formation.

Actions in the ORP must be addressed. Was a leader's recon forward of the ORP required? Were the plans finalized in the ORP and, if so, did the leaders feel confident that the members of their element understood these plans? Often, when plans are changed in the ORP, the lines of communication tend to have gaps. Only with clearly delineated responsibilities and element integrity can these communication barriers be overcome.

Actions on the objective (OBJ) include all considerations from the release point—that magical place on the map where the commander relinquishes control to his subordinate leaders—up to the point of reconsolidation. To list every consideration would be too much information for this single chapter, but most issues fall within each element's ability to "shoot, move, and communicate." More exactly, this refers to the ability of each element to engage the enemy within their sectors of fire, to move toward and across the OBJ, and to coordinate these efforts with friendly elements to their flanks and rear. Most of the AAR is spent focusing on the actions on the OBJ and with good reason. It is on the objective that efforts are either realized in success or lost in defeat.

3. The Issues - Determine What Was Right or Wrong

Once the patrol members have established the gap between the plan and the performance, they set out to identify the reasons for the discrepancy. In truth, Soldiers usually stumble across and identify the issues during the performance phase. However, it will be necessary to restate these issues so that everyone can see the big picture. This responsibility typically falls to the observer-controller during training, but may be executed by the PL in lieu of an observer.

Combat leaders identify the issues in the third phase. It is a good idea to allow some venting—whether negative or positive. Even though a moderator may lose some control of the flow of information, it is a good idea to invite everyone to speak his mind during this phase.

A technique, known as "nut shelling," requires everyone to make a concise statement on the exercise, if time permits. To begin, simply single out the most vocal member of the group and ask the Soldier what they think needs to be the focus of future training (or more simply, what was the most significant thing to go wrong with the mission). Then ask the next troop what was one thing they saw that was well executed or coordinated. Move around the group, alternating these two simple questions back and forth. Listen closely for consensus on these areas. A neat variation of this technique is to make those troops most vocal about negative issues give a positive observation of the mission and to ask the most positive troops to give a negative observation of the mission.

4. The Fix - Determine What Should Be Done Differently

The last step of the AAR is to identify who—by name—is responsible for correcting the deficiencies that contributed to the gap between the plan and the performance. Do not make the mistake of assuming that all of the responsibility falls to the PL. Subordinate leaders and even experienced troops can lead training or take responsibility for equipment operation/procurement. The PL assigns these tasks accordingly.

Finally, before the AAR draws to a close, identify those individuals who will be responsible for **the fix** of each issue. Often this means an individual is assigned a leadership role in training the troops on the identified task or deficiency. Just as common is to fix or acquire necessary equipment. Sometimes, a fix simply means to research the problem and report back to the team and/or leaders with the findings.

On Point

Leaders are responsible for training their units and making their units adapt. The AAR is one of the primary tools used to accomplish this. It does this by providing feedback, which should be direct and on the spot. Each time an incorrect performance is observed, it should be immediately corrected so it does not interfere with future tasks. During major events or activities, it is not always easy to notice incorrect performances. An AAR should be planned at the end of each activity or event. In doing so, feedback can be provided, lessons can be learned, and ideas and suggestions can be generated to ensure the next project or activity will be an improved one.

To maximize the effectiveness of AARs, leaders should plan and rehearse before training begins. After-action review planning is a routine part of unit near-term planning (six to eight weeks out). During planning, leaders assign OC responsibilities and identify tentative times and locations for AARs. This ensures the allocation of time and resources to conduct AARs and reinforces the important role AARs play in realizing the full benefit of training.

Because soldiers and leaders participating in an AAR actively discover what happened and why, they learn and remember more than they would from a critique alone. A critique only gives one viewpoint and frequently provides little opportunity for discussion of events by participants. Soldier observations and comments may not be encouraged. The climate of the critique, focusing only on what is wrong, prevents candid discussion of training events and stifles learning and team building.

The art of an AAR is in obtaining mutual trust so people will speak freely. Problem solving should be practical and Soldiers should not be preoccupied with status, territory, or second guessing "what the leader will think." There is a fine line between keeping the meeting from falling into chaos where little is accomplished, to people treating each other in a formal and polite manner that masks issues (especially with the leader).

The AAR facilitator should—

- Remain unbiased throughout the review
- Ask open-ended questions to draw out comments from all
- Do not allow personal attacks
- Focus on learning and continuous improvement
- Strive to allow others to offer solutions rather than offering them yourself
- Find solutions and recommendations to make the unit better

To avoid turning an AAR into a critique or lecture—

- Ask why certain actions were taken
- Ask how Soldiers reacted to certain situations
- Ask when actions were initiated
- Ask leading and thought-provoking questions
- Exchange "war stories" (lessons learned)
- Ask Soldiers to provide their own point of view on what happened
- Relate events to subsequent results
- Explore alternative courses of actions that might have been more effective
- Handle complaints positively
- When the discussion turns to errors made, emphasize the positive and point out the difficulties of making tough decisions
- Summarize
- Allow junior leaders to discuss the events with their Soldiers in private
- Follow up on needed actions

Chap 2 — The Offense

Ref: ADRP 3-90, Offense and Defense (Aug '12), chap. 3.

Offensive actions are combat operations conducted to defeat and destroy enemy forces and seize terrain, resources, and population centers. They impose the commander's will on the enemy. A commander may also conduct offensive actions to deprive the enemy of resources, seize decisive terrain, deceive or divert the enemy, develop intelligence, or hold an enemy in position. This chapter discusses the basics of the offense. The basics discussed in this chapter apply to all offensive tasks.

The commander seizes, retains, and exploits the initiative when conducting offensive actions. Specific operations may orient on a specific enemy force or terrain feature as a means of affecting the enemy. Even when conducting primarily defensive actions, wresting the initiative from the enemy requires offensive actions.

Offensive operations are combat operations conducted to defeat and destroy enemy forces and seize terrain, resources, and population centers. They impose the commander's will on the enemy. (Photo by Jeong, Hae-jung).

Effective offensive operations capitalize on accurate intelligence regarding the enemy, terrain and weather, and civil considerations. Commanders maneuver their forces to advantageous positions before making contact. However, commanders may shape conditions by deliberately making contact to develop the situation and mislead the enemy. In the offense, the decisive operation is a sudden, shattering action against enemy weakness that capitalizes on speed, surprise, and shock. If that operation does not destroy the enemy, operations continue until enemy forces disintegrate or retreat to where they are no longer a threat.

I. Primary Offensive Tasks

An offensive task is a task conducted to defeat and destroy enemy forces and seize terrain, resources, and population centers (ADRP 3-0). The four primary offensive tasks are movement to contact, attack, exploitation, and pursuit.

A. Movement to Contact

Movement to contact is an offensive task designed to develop the situation and to establish or regain contact. The goal is to make initial contact with a small element while retaining enough combat power to develop the situation and mitigate the associated risk. A movement to contact also creates favorable conditions for subsequent tactical actions. The commander conducts a movement to contact when the enemy situation is vague or not specific enough to conduct an attack. Forces executing this task seek to make contact with the smallest friendly force feasible. A movement to contact may result in a meeting engagement. Once contact is made with an enemy force, the commander has five options: attack, defend, bypass, delay, or withdraw. The Army includes search and attack and cordon and search operations as part of movement to contact operations.

B. Attack

An attack is an offensive task that destroys or defeats enemy forces, seizes and secures terrain, or both. Attacks incorporate coordinated movement supported by fires. They may be either decisive or shaping operations. Attacks may be hasty or deliberate, depending on the time available for assessing the situation, planning, and preparing. However, based on mission variable analysis, the commander may decide to conduct an attack using only fires. An attack differs from a movement to contact because, in an attack, the commander knows part of the enemy's disposition. This knowledge enables the commander to better synchronize and employ combat power more effectively in an attack than in a movement to contact.

Subordinate forms of the attack have special purposes and include the ambush, counterattack, demonstration, feint, raid, and spoiling attack. The commander's intent and the mission variables of mission, enemy, terrain and weather, troops and support available, time available, and civil considerations (METT-TC) determine which of these forms of attack are employed. The commander can conduct each of these forms of attack, except for a raid, as either a hasty or a deliberate operation.

See pp. 2-13 to 2-18.

C. Exploitation

Exploitation is an offensive task that usually follows the conduct of a successful attack and is designed to disorganize the enemy in depth. Exploitations seek to disintegrate enemy forces to the point where they have no alternative but to surrender or take flight. Exploitations take advantage of tactical opportunities. Division and higher headquarters normally plan exploitations as branches or sequels.

See pp. 2-19 to 2-22.

D. Pursuit

A pursuit is an offensive task designed to catch or cut off a hostile force attempting to escape, with the aim of destroying it. A pursuit normally follows a successful exploitation. However, any offensive task can transition into a pursuit, if enemy resistance has broken down and the enemy is fleeing the battlefield. Pursuits entail rapid movement and decentralized control. Bold action, calculated initiative, and accounting for the associated risk are required in the conduct of a pursuit.

See pp. 2-23 to 2-28.

II. Purposes of Offensive Operations

Ref: Adapted from FM 3-0 Operations (2008) and ADRP 3-90 (Aug '12).

The main purpose of the offense is to defeat, destroy, or neutralize the enemy force. Additionally, commanders conduct offensive tasks to secure decisive terrain, to deprive the enemy of resources, to gain information, to deceive and divert the enemy, to hold the enemy in position, to disrupt the enemy's attack, and to set up the conditions for future successful operations.

1. Defeat, Destroy, or Neutralize the Enemy force

Well-executed offensive operations dislocate, isolate, disrupt, and destroy enemy forces. If destruction is not feasible, offensive operations compel enemy forces to retreat. Offensive maneuver seeks to place the enemy at a positional disadvantage. This allows friendly forces to mass overwhelming effects while defeating parts of the enemy force in detail before the enemy can escape or be reinforced. When required, friendly forces close with and destroy the enemy in close combat. Ultimately, the enemy surrenders, retreats in disorder, or is eliminated altogether.

2. Seize Decisive Terrain

Offensive maneuver may seize terrain that provides the attacker with a decisive advantage. The enemy either retreats or risks defeat or destruction. If enemy forces retreat or attempt to retake the key terrain, they are exposed to fires and further friendly maneuver.

3. Deprive the Enemy of Resources

At the operational level, offensive operations may seize control of major population centers, seats of government, production facilities, and transportation infrastructure. Losing these resources greatly reduces the enemy's ability to resist. In some cases, Army forces secure population centers or infrastructure and prevent irregular forces from using them as a base or benefitting from the resources that they generate.

4. To Gain Information

Enemy deception, concealment, and security may prevent friendly forces from gaining necessary intelligence. Some offensive operations are conducted to develop the situation and discover the enemy's intent, disposition, and capabilities.

5. Disrupt the Enemy's Attack

Offensive operations distract enemy ISR. They may cause the enemy to shift reserves away from the friendly decisive operation.

6. Set up the Conditions for Future Successful Operations

III. Forms of Maneuver
Ref: FM 3-90 Tactics, pp. 3-11 to 3-32.

Forms of maneuver are distinct tactical combinations of fire and movement with a unique set of doctrinal characteristics that differ primarily in the relationship between the maneuvering force and the enemy. The commander generally chooses one form on which he builds a course of action (COA). The higher commander rarely specifies the specific form of offensive maneuver. However, his guidance and intent, along with the mission that includes implied tasks, may impose constraints such as time, security, and direction of attack.

A. Envelopment
The envelopment is a form of maneuver in which an attacking force seeks to avoid the principal enemy defenses by seizing objectives to the enemy rear to destroy the enemy in his current positions. At the tactical level, envelopments focus on seizing terrain, destroying specific enemy forces, and interdicting enemy withdrawal routes. Envelopments avoid the enemy front, where he is protected and can easily concentrate fires. Single envelopments maneuver against one enemy flank; double envelopments maneuver against both. Either variant can develop into an encirclement.

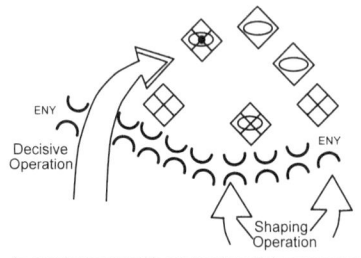

An envelopment avoids enemy strength by maneuver around or over enemy defenses. The decisive operation is directed against the enemy flanks or rear.

To envelop the enemy, commanders find or create an assailable flank. Sometimes the enemy exposes a flank by advancing, unaware of friendly locations. In other conditions, such as a fluid battle involving forces in noncontiguous AOs, a combination of air and indirect fires may create an assailable flank by isolating the enemy on unfavorable terrain.

An envelopment may result in an encirclement. Encirclements are operations where one force loses its freedom of maneuver because an opposing force is able to isolate it by controlling all ground lines of communications. An offensive encirclement is typically an extension of either a pursuit or envelopment. A direct pressure force maintains contact with the enemy, preventing his disengagement and reconstitution. Meanwhile, an encircling force maneuvers to envelop the enemy, cutting his escape routes and setting inner and outer rings. The outer ring defeats enemy attempts to break through to his encircled force. The inner ring contains the encircled force. All available means, including obstacles, should be used to contain the enemy. Then friendly forces use all available fires to destroy him. Encirclements often occur in nonlinear offensive operations.

B. Turning Movement
A turning movement is a form of maneuver in which the attacking force seeks to avoid the enemy's principal defensive positions by seizing objectives to the rear and causing the enemy to move out of current positions or divert major forces to meet the threat. A major threat to his rear forces the enemy to attack or withdraw rearward, thus "turning" him out of his defensive positions. Turning movements typically require greater depth than other maneuver forms. Deep fires take on added importance. They protect the enveloping force and attack the enemy.

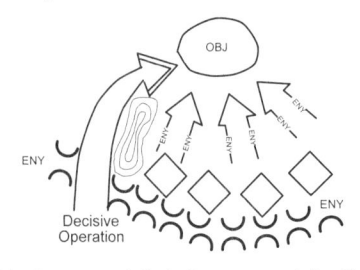

A turning movement attacks the enemy rear to "turn" him out of position and force him to fight to the rear of his flanks.

2-4 The Offense

C. Infiltration

An infiltration is a form of maneuver in which an attacking force conducts undetected movement through or into an area occupied by enemy forces to occupy a position of advantage in the enemy rear while exposing only small elements to enemy defensive fires. The need to avoid being detected and engaged may limit the size and strength of infiltrating forces. Infiltration rarely defeats a defense by itself. Cdrs direct infiltrations to attack lightly defended positions or stronger positions from the flank and rear, to secure key terrain to support the decisive operation, or to disrupt enemy sustaining operations.

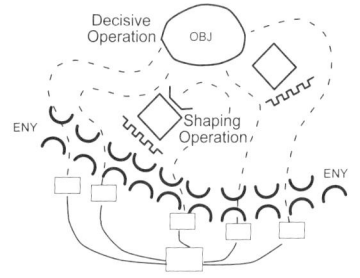

An infiltration uses covert movement of forces through enemy lines to attack position in the enemy rear.

D. Penetration

A penetration is a form of maneuver in which an attacking force seeks to rupture enemy defenses on a narrow front to disrupt the defensive system. It is used when enemy flanks are not assailable or time does not permit another form of maneuver. Successful penetrations create assailable flanks and provide access to enemy rear areas. Penetrations frequently are directed into the front of the enemy defense, and risk more friendly casualties than envelopments, turning movements and infiltrations.

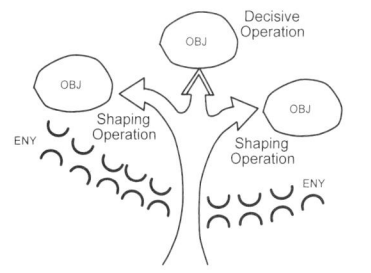

A penetration has three stages: initial rupture, rolling up the flanks, and continuing the attack to secure a deep objective.

E. Frontal Attack

A frontal attack is a form of maneuver in which an attacking force seeks to destroy a weaker enemy force or fix a larger enemy force in place over a broad front. At the tactical level, an attacking force can use a frontal attack to rapidly overrun a weaker enemy force. A frontal attack strikes the enemy across a wide front and over the most direct approaches. Commanders normally use it when they possess overwhelming combat power and the enemy is at a clear disadvantage. Commanders mass the effects of direct and indirect fires, shifting indirect and aerial fires just before the assault. Success depends on achieving an advantage in combat power throughout the attack.

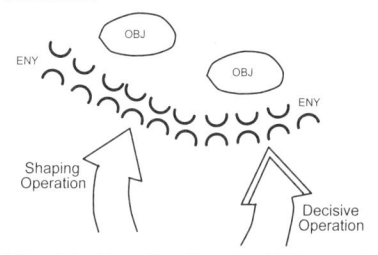

A frontal attack is conducted across a wide front over the most direct approach.

The frontal attack is frequently the most costly form of maneuver, since it exposes the majority of the attackers to the concentrated fires of the defenders. As the most direct form of maneuver, however, the frontal attack is useful for overwhelming light defenses, covering forces, or disorganized enemy resistance. It is often the best form of maneuver for hasty attacks and meeting engagements, where speed and simplicity are essential to maintain tempo and the initiative. Commanders may direct a frontal attack as a shaping operation and another form of maneuver as the decisive operation. Commanders may also use the frontal attack during an exploitation or pursuit. Commanders of large formations conducting envelopments or penetrations may direct subordinate elements to conduct frontal attacks as either shaping operations or the decisive operation.

IV. Common Offensive Control Measures

Ref: ADRP 3-90, Offense and Defense (Aug '12), pp. 3-4 to 3-7.

This section defines common control measures that a commander uses to synchronize the effects of combat power. The commander uses the minimum control measures required to successfully complete the mission while providing subordinates the flexibility needed to respond to changes in the situation.

Assault Position. An assault position is a covered and concealed position short of the objective from which final preparations are made to assault the objective. Ideally, it offers both cover and concealment. These final preparations can involve tactical considerations, such as a short halt to coordinate the final assault, reorganize to adjust to combat losses, or make necessary adjustments in the attacking force's dispositions. They can also involve technical items, such as engineers conducting their final prepare-to-fire checks on obstacle clearing systems and the crews of plow- and roller-equipped tanks removing their locking pins. An assault position may be located near a final coordination line (FCL) or a probable line of deployment (PLD).

Assault Time. The assault time establishes the moment to attack the initial objectives throughout the geographical scope of the operation. It is imposed by the higher headquarters in operations to achieve simultaneous results from several different units. It synchronizes the moment the enemy feels the effects of friendly combat power. It is similar to the time-on-target control method for fire mission processing used by the field artillery. A commander uses it instead of a time of attack because of the different distances that different elements of the force must traverse, known obstacles, and differences in unit tactical mobility.

Attack by Fire Position. An attack by fire position designates the general position from which a unit conducts the tactical task of attack by fire. The purpose of these positions is to mass the effects of direct fire systems from one or multiple locations on the enemy. An attack by fire position does not indicate the specific site. Attack by fire positions are rarely applicable to units larger than company size.

Attack Position. The attack position is the last position an attacking force occupies or passes through before crossing the line of departure. An attack position facilitates the deployment and last-minute coordination of the attacking force before it crosses the line of departure (LD). It is located on the friendly side of the LD and offers cover and concealment. It is used primarily at battalion level and below. Whenever possible, units move through their attack position without stopping. A unit occupies an attack position for a variety of reasons, such as waiting for specific results from preparation fires or when it is necessary to conduct additional coordination, such as a forward passage of lines.

Axis of Advance. An axis of advance designates the general area through which the bulk of a unit's combat power must move. When developing the axis of advance, the commander also establishes bypass criteria. Bypass criteria are measures during the conduct of an offensive operation established by higher headquarters that specify the conditions and size under which enemy units and contact may be avoided.

Battle Handover Line. The battle handover line is a designated phase line on the ground where responsibility transitions from the stationary force to the moving force and vice versa. The common higher commander of the two forces establishes the battle handover line (BHL) after consulting both commanders. The stationary commander determines the location of the line. The BHL is forward of the forward edge of the battle area (FEBA) in the defense or the forward line of own troops (FLOT) in the offense.

Direction of Attack. The direction of attack is a specific direction or assigned route a force uses and does not deviate from when attacking. It is a restrictive control measure. The commander's use of a direction of attack maximizes control over the subordinate unit movement, and is often used during night attacks, infiltrations, and when attacking through smoke.

Final Coordination Line. The final coordination line is a phase line close to the enemy position used to coordinate the lifting or shifting of supporting fires with the final deployment of maneuver elements. Final adjustments to supporting fires necessary to reflect the actual versus the planned tactical situation take place prior to crossing this line. It should be easily recognizable on the ground. The FCL is not a fire support coordination measure.

Limit of Advance. The limit of advance is a phase line used to control forward progress of the attack. The attacking unit does not advance any of its elements or assets beyond the limit of advance, but the attacking unit can push its security forces to that limit.

Line of Departure. The line of departure is a phase line crossed at a prescribed time by troops initiating an offensive task. The purpose of the LD is to coordinate the advance of the attacking force, so that its elements strike the enemy in the order and at the time desired. The LD also marks where the unit transitions from movement to maneuver. The commander can also use it to facilitate the coordination of fires. Generally, it should be perpendicular to the direction the attacking force will take on its way to the objective.

Objective. An objective is a location on the ground used to orient operations, phase operations, facilitate changes of direction, and provide for unity of effort. An objective can be terrain- or force-oriented. Terrain objectives should be easily identifiable on the ground to facilitate their recognition. The commander determines force-oriented objectives based on known enemy positions. The commander normally assigns subordinate commanders only their final objectives, but can assign intermediate objectives as necessary.

Point of Departure. The point of departure is the point where the unit crosses the line of departure and begins moving along a direction of attack. Units conducting patrols and other operations in a low-visibility environment commonly use a point of departure as a control measure. Like a LD, it marks the point where the unit transitions from movement to maneuver under conditions of limited visibility.

Probable Line of Deployment. A probable line of deployment is a phase line that designates the location where the commander intends to deploy the unit into assault formation before beginning the assault. The PLD is used primarily at battalion level and below when the unit does not cross the LD in its assault formation. It is usually a linear terrain feature perpendicular to the direction of attack and recognizable under conditions of limited visibility. The PLD should be located outside the range where the enemy can place the attacking force under effective direct fire. It has no use except as it relates to the enemy.

Rally Point. A rally point is an easily identifiable point on the ground at which units can reassemble and reorganize if they become dispersed. Alternatively, it is an easily identifiable point on the ground at which aircrews and passengers can assemble and reorganize following an incident requiring a forced landing (ADRP 1-02). The objective rally point is a rally point established on an easily identifiable point on the ground where all elements of the infiltrating unit assemble and prepare to attack the objective. It is typically near the infiltrating unit's objective; however, there is no standard distance from the objective to the objective rally point. It should be far enough away from the objective so that the enemy will not detect the infiltrating unit's attack preparations.

Support by Fire Position. A support by fire position designates the general position from which a unit conducts the tactical mission task of support by fire. The purpose of these positions is to increase the supported force's freedom of maneuver by placing direct fires on an objective that is going to be assaulted by a friendly force. Support by fire positions are located within the maximum friendly direct-fire range of enemy positions. The commander selects them so that the moving assault force does not mask its supporting fires. For this reason, support by fire positions are normally located on the flank of the assault force, elevated above the objective if possible. Support by fire positions are rarely applicable to units larger than company size.

Time of Attack. The time of attack is the moment the leading elements of the main body cross the line of departure, or in a night attack, the point of departure. A commander uses it when conducting simultaneous operations where a shaping operation must accomplish its mission to set the conditions for the success of the decisive operation.

The Offense 2-5b

V. Transition

Ref: ADRP 3-90, Offense and Defense (Aug '12), pp. 3-21 to 3-23.

A transition occurs when the commander makes the assessment that the unit must change its focus from one element of decisive action to another.

A commander halts the offense only when it results in complete victory and the end of hostilities, reaches a culminating point, or the commander receives a change in mission from a higher commander. This **change in mission** may be a result of the interrelationship of the other instruments of national power, such as a political decision.

All offensive actions that do not achieve complete victory reach a **culminating point** when the balance of strength shifts from the attacking force to its opponent. Usually, offensive actions lose momentum when friendly forces encounter heavily defended areas that cannot be bypassed. They also reach a culminating point when the resupply of fuel, ammunition, and other supplies fails to keep up with expenditures, Soldiers become physically exhausted, casualties and equipment losses mount, and repairs and replacements do not keep pace with losses. Because of enemy surprise movements, offensive actions also stall when reserves are not available to continue the advance, the defender receives reinforcements, or the defender counterattacks with fresh troops. Several of these causes may combine to halt an offense. When this occurs, the attacking unit can regain its momentum, but normally this only happens after difficult fighting or after an operational pause.

The commander plans a **pause to replenish combat power** and **phases the operation** accordingly, if the commander cannot anticipate securing decisive objectives before subordinate forces reach their culminating points. Simultaneously, the commander attempts to prevent the enemy from knowing when friendly forces become overextended.

Transition to Defense

Once offensive actions begin, the attacking commander tries to sense when subordinates reach, or are about to reach, their respective culminating points. Before they reach this point, the commander must transition to a focus on the defense. The commander has more freedom to choose where and when to halt the attack, if the commander can sense that subordinate forces are approaching culmination. The commander can plan future activities to aid the defense, minimize vulnerability to attack, and facilitate renewal of the offense as the force transitions to branches or sequels of the ongoing operation. For example, some of the commander's subordinate units may move into battle positions before the entire unit terminates its offensive actions to start preparing for ensuing defensive tasks. The commander can echelon sustainment assets forward to establish a new echelon support area. This may also serve to prevent overburdening the extended lines of communications that result from advances beyond eight hours of travel from the echelon support area.

A lull in combat operations often accompanies a transition. The commander cannot forget about stability tasks because the civilian populations of the unit's AO tend to come out of their hiding positions and request assistance from friendly forces during these lulls. The commander must consider how to minimize civilian interference with the force's combat operations while protecting civilians from future hostile actions in accordance with the law of armed conflict. The commander must also consider the threat civilians pose to the force and its operations, if enemy agents or saboteurs are part of the civilian population.

A commander anticipating the termination of unit offensive actions prepares orders that include the time or circumstances under which the current offense transitions to the defense, the missions and locations of subordinate units, and control measures.

As the unit transitions from an offensive focus to a defensive focus, the commander–
- Maintains contact and surveillance of the enemy, using a combination of reconnaissance units and surveillance assets to develop the information required to plan future actions
- Establishes a security area and local security measures
- Redeploys artillery assets to ensure the support of security forces
- Redeploys forces for probable future employment
- Maintains or regains contact with adjacent units in a contiguous AO and ensures that units remain capable of mutual support in a noncontiguous AO
- Shifts the engineer emphasis from mobility to countermobility and survivability
- Consolidates and reorganizes
- Explains the rationale for transitioning from the offense to the unit's Soldiers

The commander conducts any required reorganization and resupply concurrently with other transition activities. This requires a transition in the sustainment effort. It shifts in emphasis from ensuring the force's ability to move forward (POL and forward repair of maintenance and combat losses) to ensuring the force's ability to defend on its chosen location (forward stockage of construction, barrier, and obstacle material, and ammunition). A transition is often a time when units can perform equipment maintenance. Additional assets may also be available for casualty evacuation and medical treatment because of a reduction in the tempo.

Transition to Stability

At some point in time the unit will probably transition from one phase of the major operations or campaign plan to another and begin executing a sequel to its previous operations order. The end of the offense action may not be the decisive act. The conduct of stability tasks may be the decisive operation in the major operation or campaign. The transition to a focus on stability tasks cannot be an afterthought. Setting the conditions for stability operations may have significant impact on the planning and execution of offensive tasks.

It is likely that a significant reorganization of the unit will occur to introduce those capabilities required by the changes in the mission variables of METT-TC. Depending on the specific operational environment the unit finds itself in, the appropriate official departmental publications dealing with other missions should be referenced to refresh previous training and education in those subjects. The mission command and protection functions remain important because it is likely that some Soldiers may want to relax discipline and safety standards as the stress of active offensive actions disappears.

During major combat operations, the commander transitions to a stability focus, if the unit's offensive actions are successful in destroying or defeating the enemy and the situation makes a focus on defensive actions inappropriate. The commander's concept of operations and intent drive the design of and planning for the conduct of stability tasks. Generally, a tactical commander will focus on meeting the immediate essential service and civil security needs of the civilian inhabitants of the area of operations in coordination with any existing host nation government and nongovernmental organizations before addressing the other three primary stability tasks. Also, the commander will probably change the rules of engagement, and these rules must be transmitted down to the squad and individual Soldier level.

When involved in other missions, such as peace operations, irregular warfare, and military engagement, unit offensive actions normally are closely related to the movement to contact tasks of search and attack or cordon and search. Offensive actions in these other types of tasks will normally employ restrictive rules of engagement throughout the mission, regardless of the dominate type of action at any specific moment. When executing tasks within these other missions, the emphasis on the stability element is much more dominate than the defensive element of decisive action.

VI. Characteristics of Offensive Operations

Characteristics of the offense include audacity, concentration, surprise, and rapid tempo. Effective offensive actions capitalize on accurate and timely intelligence and other relevant information regarding enemy forces, weather, and terrain. The commander maneuvers forces to advantageous positions before contact. Protection tasks, such as security operations, operations security, and information protection keep or inhibit the enemy from acquiring accurate information about friendly forces. Contact with enemy forces before the decisive operation is deliberate, designed to shape the optimum situation for the decisive operation. The decisive operation that conclusively determines the outcome of the major operation, battle, and engagement capitalizes on subordinate initiative and a common operational picture to expand throughout the area of operations (AO). Without hesitation, the commander violently executes both maneuver and fires—within the higher commander's intent—to break the enemy's will or destroy the enemy.

A. Audacity

Audacity means boldly executing a simple plan of action. Commanders display audacity by developing bold, inventive plans that produce decisive results. Commanders demonstrate audacity by violently applying combat power. They understand when and where to take risks, and they do not hesitate as they execute their plan. Commanders dispel uncertainty through action; they compensate for lack of information by seizing the initiative and pressing the battle.

B. Concentration

Concentration is the massing of overwhelming effects of combat power to achieve a single purpose. Commanders balance the necessity for concentrating forces to mass effects with the need to disperse them to avoid creating lucrative targets. Advances in ground and air mobility, target acquisition, and long-range precision munitions enable attackers to rapidly concentrate effects.

C. Surprise

In the offense, commanders achieve surprise by attacking the enemy at a time or place the enemy does not expect or in a manner that the enemy is unprepared for. Estimating the enemy commander's intent and denying that commander the ability to gain thorough and timely situational understanding is necessary to achieve surprise. Unpredictability and boldness help gain surprise. Surprise delays enemy reactions, overloads and confuses the enemy commander's command and control systems, induces psychological shock in enemy soldiers and leaders, and reduces the coherence of the defense. By diminishing enemy combat power, surprise enables attackers to exploit enemy paralysis and hesitancy. Operational and tactical surprise complement each other. Operational surprise creates the conditions for successful tactical operations. Tactical surprise can cause the enemy to hesitate or misjudge a situation. But tactical surprise is fleeting.

D. Tempo

Controlling or altering tempo is necessary to retain the initiative. At the operational level, a faster tempo allows attackers to disrupt enemy defensive plans by achieving results quicker than the enemy can respond. At the tactical level, a faster tempo allows attackers to quickly penetrate barriers and defenses and destroy enemy forces in depth before they can react.

Commanders adjust tempo as tactical situations, sustainment necessity, or operational opportunities allow to ensure synchronization and proper coordination, but not at the expense of losing opportunities to defeat the enemy. Rapid tempo demands quick decisions. It denies the enemy the chance to rest, and it continually creates opportunities.

Chap 2 The Offense

I. Movement to Contact

Ref: FM 3-90 Tactics, chap 4; FM 3-21.10 The Infantry Rifle Company, chap 4 and FM 3-21.8 The Infantry Rifle Platoon and Squad, pp. 7-18 to 7-24.

Movement to contact is a type of offensive operation designed to develop the situation and establish or regain contact. A commander conducts this type of offensive operation when the tactical situation is not clear or when the enemy has broken contact. A properly executed movement to contact develops the combat situation and maintains the commander's freedom of action after contact is gained. This flexibility is essential in maintaining the initiative.

Purposeful and aggressive movement, decentralized control, and the hasty deployment of combined arms formations from the march to attack or defend characterize the movement to contact. The fundamentals of a movement to contact are—

- Focus all efforts on finding the enemy
- Make initial contact with the smallest force possible, consistent with protecting the force
- Make initial contact with small, mobile, self-contained forces to avoid decisive engagement of the main body on ground chosen by the enemy. This allows the commander maximum flexibility to develop the situation
- Task-organize the force and use movement formations to deploy and attack rapidly in any direction
- Keep forces within supporting distances to facilitate a flexible response
- Maintain contact regardless of the course of action (COA) adopted once contact is gained

Meeting Engagement

The movement to contact results in a meeting engagement. A meeting engagement is the combat action that occurs when a moving element engages a stationary or moving enemy at an unexpected time and place. Meeting engagements are characterized by—

- Limited knowledge of the enemy
- Minimum time available for the leader to conduct actions on contact
- Rapidly changing situation
- Rapid execution of battle and crew drills

A meeting engagement is a combat action that occurs when a moving force engages an enemy at an unexpected time and place. Conducting an MTC results in a meeting engagement. The enemy force may be either stationary or moving. Such encounters often occur in small-unit operations when reconnaissance has been ineffective. The force that reacts first to the unexpected contact generally gains an advantage over its opponent. However, a meeting engagement may also occur when the opponents are aware of each other and both decide to attack immediately to obtain a tactical advantage or seize key or decisive terrain. A meeting engagement may also occur when one force attempts to deploy into a hasty defense while the other force attacks before its opponent can organize an effective defense. Acquisition systems may discover the enemy before the security force can gain contact. No matter how the force makes contact, seizing the initiative is the overriding imperative.

I. MTC - Organization

A movement to contact is organized with an offensive covering force or an advance guard as a forward security element and a main body as a minimum. Based on the factors of METT-TC, the commander may increase his security forces by having an offensive covering force and an advance guard for each column, as well as flank and rear security (normally a screen or guard).

The MTC may use multiple teams to find the enemy. When a team makes contact, they report the information. The commander decides when to commit the body of the main force.

A movement to contact is conducted using one of two techniques: approach march, or search and attack. The approach march technique is used when the enemy is expected to deploy using relatively fixed offensive or defensive formations, and the situation remains vague. The search and attack technique is used when the enemy is dispersed, when he is expected to avoid contact or quickly disengage and withdraw, or when the higher unit needs to deny him movement in an area of operation.

1. Search and Attack
Search and attack is a technique for conducting a MTC; this technique shares many of the same characteristics of an area security mission. Conducted primarily by Infantry forces and often supported by heavy forces, a commander employs this form of a MTC when the enemy is operating as small, dispersed element, or when the task is to deny the enemy the ability to move within a given area. The battalion is the echelon that normally conducts a search and attack. *(Note: See also p.2-12)*

2. Approach-March Technique
A unit normally uses this technique when it conducts a MTC as part of a battalion. Depending on its location in the formation and its assigned mission, the company can act as the advance guard, move as part of the battalion main body, or provide flank or rear guards for the battalion. *(Note: See also p.2-12)*

2-8 (The Offense) I. Movement to Contact

Prompt execution of battle drills at platoon level and below, and standard actions on contact for larger units can give that initiative to the friendly force.

II. Planning & Preparation

Small units (e.g., an infantry company) normally conducts MTC as part of a battalion or larger element; however, based on the METT-TC factors, it can conduct the operation independently. As an example, the company may conduct MTC prior to occupation of a screen line. Because the enemy situation is not clear, the company moves in a way that provides security and supports a rapid buildup of combat power against enemy units once they are identified. Two techniques for conducting a MTC are the search-and-attack technique and the approach-march technique. If no contact occurs, the company might be directed to conduct consolidation on the objective. The Infantry company commander analyzes the situation and selects the proper tactics to conduct the mission. He reports all information rapidly and accurately and strives to gain and maintain contact with the enemy. He retains freedom of maneuver by moving the company in a manner that--

- Ensures adequate force protection measures are always in effect
- Makes enemy contact (ideally visual contact) with the smallest element possible (ideally, a reconnaissance and surveillance [R&S] element). The commander plans for any forms of contact to identify enemy locations
- Rapidly develops combat power upon enemy contact
- Provides all-round security for the unit
- Supports the battalion concept

The higher commander will task-organize the subordinate units into reconnaissance (finding, fixing, and finishing) elements. He will assign specific tasks and purposes to his search and attack elements. It is important to note that within the concept of find, fix, and finish, all platoons could be the reconnaissance element. Depending on the size of the enemy they find, they could end up executing a reconnaissance mission, become the fixing element, or find that they are able to finish the enemy. Planning considerations for organizing include—

- The factors of METT-TC
- The requirement for decentralized execution.
- The requirement for mutual support. (The platoon leader must be able to respond to contact with his rifle squads or to mutually support another platoon within the company.)
- The Soldier's load. (The leader should ask, "Does the Soldier carry his rucksack, cache it, or leave it at a central point? How will the rucksacks be linked up with the Soldier?")
- Resupply and CASEVAC
- The employment of key weapons
- The requirement for patrol bases

(The Offense) I. Movement to Contact 2-9

III. Conducting the MTC - A Small Unit Perspective

1. From the assembly area (AA) the commander conducts a final communication check with the teams and dispatches them forward of the main body. The commander starts movement of the main body, keeping a distance great enough to avoid becoming absorbed when a team engages the enemy—and yet close enough to respond quickly with the reserve force.

2. Each team seeks visual contact with the enemy, being careful not to become engaged in a firefight. If the team enters a firefight with the enemy, the patrol leader (PL) loses the mobility necessary to develop the situation and collect intelligence for the commander of the MTC. If the team becomes decisively engaged, the PL reports this situation back to the commander.

3. The commander assesses the information coming from any force in contact with the enemy to determine if the main body should be committed to attacking the enemy, or if the team can effectively attack the enemy force, or if the enemy body should simply be bypassed.

Information is relayed back to the commander who remains with the main body of the MTC. The forces continue to develop the situation and send timely reports. (Photo by Jeong, Hae-jung).

4. The team in contact with the enemy acts as a shaping operation—meaning that the team doesn't conduct the main attack. Instead, the team fixes the enemy using an attack by fire to suppress and isolate the enemy force until the reserve force maneuvers to destroy the enemy. Alternatively, the team can be instructed to attack and destroy any smaller enemy force that would present a threat to the main body's flank or rear if bypassed. Even in this case, the team functions as a shaping operation within the force protection framework.

5. If the commander determines the detected enemy needs to be eliminated by the main body, he directs or leads the reserve force to crush the fixed enemy. This becomes the decisive operation—meaning it becomes the main attack. It consumes the attention and resources of the MTC until the engagement is concluded. If necessary, the commander pulls the other teams into the attack, attempting to isolate each subsequent enemy force so the enemy cannot coordinate their actions.

Once the commander has determined that a force is in contact with a key enemy element, the main body of the MTC is committed to the battle. The force fixes the enemy position, while the main body attacks aggressively. (Photo by Jeong, Hae-jung).

6. Once the engagement concludes, the commander again dispatches his teams forward and continues the movement—either to clear an axis of movement, or to defeat an identified enemy force. In any case, it is critical that the team seek and maintain visual contact with the enemy force and continues to develop the situation for the MTC commander.

7. The commander of the MTC looks for signs of culmination. That may come by means of hitting the established boundaries of his area of operation (AO) and limit of advance (LOA). Or the enemy may no longer be putting up resistance—such as fleeing and the number of engagements has drastically fallen. Or, the troops of the MTC may have become exhausted. In any case, at the sign of culmination, the commander must reconsolidate the friendly forces and assume a 360-degree security posture.

(The Offense) I. Movement to Contact 2-11

On Point

An Infantry company normally conducts MTC as part of a battalion or larger element; however, based on the METT-TC factors, it can conduct the operation independently. As an example, the company may conduct MTC prior to occupation of a screen line. Because the enemy situation is not clear, the company moves in a way that provides security and supports a rapid buildup of combat power against enemy units once they are identified. Two techniques for conducting a MTC are the search-and-attack technique and the approach-march technique. If no contact occurs, the company might be directed to conduct consolidation on the objective. The Infantry company commander analyzes the situation and selects the proper tactics to conduct the mission. He reports all information rapidly and accurately and strives to gain and maintain contact with the enemy.

Search and Attack

Search and attack is a technique for conducting a MTC; this technique shares many of the same characteristics of an area security mission (FM 3-0). Conducted primarily by Infantry forces and often supported by heavy forces, a commander employs this form of a MTC when the enemy is operating as small, dispersed element, or when the task is to deny the enemy the ability to move within a given area. The battalion is the echelon that normally conducts a search and attack. A brigade will assist its subordinate battalions by ensuring the availability of indirect fires and other support.

A commander conducts a search and attack for one or more of the following purposes:

- Protect the force--prevent the enemy from massing to disrupt or destroy friendly military or civilian operations, equipment, property, and key facilities.
- Collect information--gain information about the enemy and the terrain to confirm the enemy COA predicted by the IPB process. Help generate SA for the company and higher headquarters.
- Destroy the enemy and render enemy units in the AO combat ineffective.
- Deny the area--prevent the enemy from operating unhindered in a given area such as in any area he is using for a base camp or for logistics support.

Approach-March Technique

The Infantry company normally uses this technique when it conducts a MTC as part of the battalion. Depending on its location in the formation and its assigned mission, the company can act as the advance guard, move as part of the battalion main body, or provide flank or rear guards for the battalion. When planning for an approach-march MTC, the company commander needs certain information from the battalion commander. With this information, the company commander develops his scheme of maneuver and fire support plan. He provides this same information to the platoon leaders. As a minimum, he needs to know:

- The company's mission
- The friendly and enemy situations
- The route (axis of advance) and the desired rate of movement
- The control measures to be used
- The company's actions on contact
- The fire support plan
- The company's actions upon reaching the march objective, if one is used

Chap 2
The Offense
II. Attack

Ref: FM 3-90 Tactics, chap 5 and FM 3-21.10 (FM 7-10) The Infantry Rifle Company, chap 4. Note: See also chap. 6 for discussion of special purpose attacks: the ambush, raid and swarming attacks.

An attack is an offensive operation that destroys or defeats enemy forces, seizes and secures terrain, or both. When the commander decides to attack or the opportunity to attack occurs during combat operations, the execution of that attack must mass the effects of overwhelming combat power against selected portions of the enemy force with a tempo and intensity that cannot be matched by the enemy. The resulting combat should not be a contest between near equals.

The support team masses together combat power in weaponry. Machineguns, grenade launchers, and rockets are employed to suppress enemy defenses while the assault team moves forward to destroy key positions. (Dept. of Army photo by Arthur McQueen).

Platoons and squads normally conduct an attack as part of the Infantry company. An attack requires detailed planning, synchronization, and rehearsals to be successful. The company commander designates platoon objectives with a specific mission for his assault, support, and breach elements. To ensure synchronization, all leaders must clearly understand the mission, with emphasis on the purpose, of peer and subordinate elements. Leaders must also know the location of their subordinates and adjacent units during the attack.

Hasty vs. Deliberate Attacks

In addition to having different forms based on their purposes, attacks are characterized as hasty, or deliberate. The primary difference between the hasty and deliberate attack is the planning and coordination time available to allow the full integration and synchronization of all available combined arms assets.

I. Attack - Organization

The attack force breaks into two teams, the support team and the assault team. If the enemy is specifically known to have prior warning to our attack and is known to have the available resources to conduct spoiling attacks, the combat leader may also form a security team for protecting the flanks from an enemy spoiling attack. Otherwise, the attack uses just the support and assault teams.

The assault team includes a breach team. The breach team has the job of moving forward to neutralize the enemy's obstacles. They use defilades and smoke to obscure their activity. (Photo by Jeong, Hae-jung.)

A. Assault Team

The assault team includes an internal breach team that is tasked with the responsibility of cutting a path through the enemy's forward obstacles. This team typically consists of a 2-man security team and a 2-man engineer team. The breach team carries specialized equipment for cutting through wire, neutralizing land mines, and marking the path once they have created the breach in the enemy's obstacles. In most cases, the breach team will require ample amounts of smoke canisters to mask their activity.

B. Support Team

The support team employs mass casualty producing weapons and may also have an internal sub-team to conduct a feign attack. The feign attack is coordinated to advance at the same time as the breach team advances. The feign attack serves to pull attention away from the breach team to confuse the enemy as to the true location of the main assault force.

Like the breach team, the feign attack team employs smoke canisters and high rates of fire to make their attack look convincing—though the feign attack team does not attempt to pass through enemy obstacles. Instead, the feign attack remains just outside the enemy's obstacles and conducts an attack by fire to augment the support team fires.

II. Planning & Preparation

In an attack, friendly forces seek to place the enemy in a position where he can easily be defeated or destroyed. The commander seeks to keep the enemy off-balance while continually reducing the enemy's options. In an attack the commander focuses the maneuver effects, supported by the other warfighting functions (formerly labeled "battlefield operating systems"), on those enemy forces that prevent him from accomplishing his mission and seizing his objective.

Fire superiority is that degree of dominance in the fires of one force over another that permits that force to conduct maneuver at a given time and place without prohibitive interference by the enemy. The force must gain and maintain fire superiority at critical points during the attack. Having fire superiority allows the commander to maneuver his forces without prohibitive losses. The commander gains fire superiority by using a variety of tactics, techniques, and procedures. Achieving fire superiority requires the commander to take advantage of—

- The range and lethality of available weapon systems
- Offensive information operations to prevent the enemy commander from synchronizing the effects of his available combat power.
- Movement to place the enemy in a position of disadvantage where his weapons can be destroyed, one or more at a time, with little risk to friendly systems

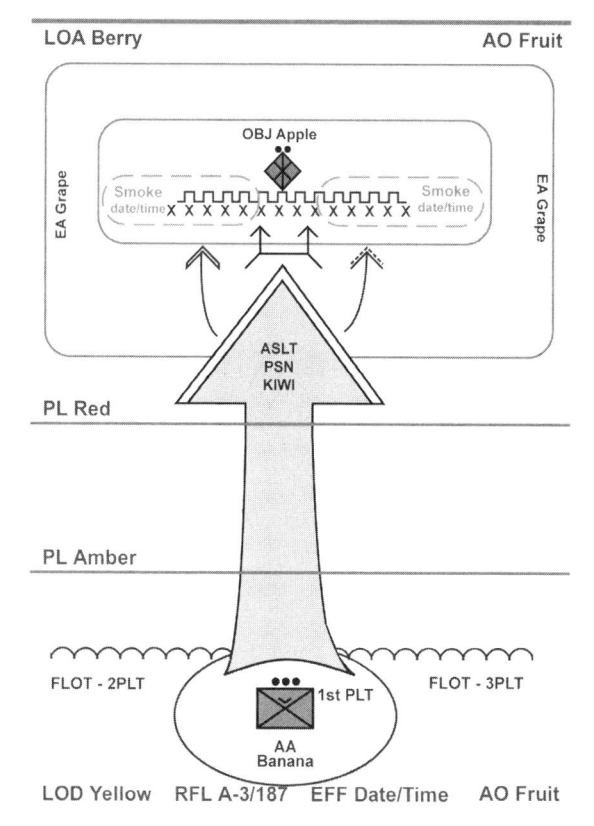

The progression of an attack takes the patrol from the assembly area through an attack lane up to the assault position. The support team deploys first, then the assault team and feign attack.

(The Offense) II. Attack 2-15

III. Conducting the Attack - A Small Unit Perspective

In the event a patrol is tasked to conduct an attack, the patrol may go through a 9-step process for the planning and execution of actions on the objective.

In the attack, the a small unit maneuvers along lines of least resistance using the terrain for cover and concealment. This indirect approach affords the best chance to achieve surprise on the enemy force. In the attack, the unit maneuvers along lines of least resistance using the terrain for cover and concealment. This indirect approach affords the best chance to achieve surprise on the enemy force.

1. Move to the Assault Position
An assault position is the last covered and concealed position before the OBJ. It differs from an ORP in that our patrol will *not* return to the assault position. We are not required to stop in the assault position; however, it is common to wait here so that friendly units and supporting fires can be synchronized against the enemy target. A leader's recon of the objective is optional, but with good intelligence, this is not required.

2. Form the Attack
At the designated time, the PL forms the patrol into its attack formation and order of march. This is the start of the attack. The attack will bypass any obstacles that have not been specifically assigned—unless they present a threat. The attack force passes through danger areas maintaining attack formation. Only as large a force as necessary is exposed to achieve each task. This conceals the size and intent of attack force.

3. Identify the Objective
The pointman determines the direction and distance of the OPFOR position (the OBJ). The PL leads the support team to its assigned position. This position allows the support team to observe the OPFOR defensive line and employ suppressive fires if the enemy detects the attack force.

4. Engage the Objective
The support team *does not* indiscriminately fire upon the OPFOR position if the attack still has the element of surprise! The assault's breach team moves up to the objective to neutralize the enemy's obstacles. If at any time the attack force is detected, the support team employs suppressive fire against enemy positions and may dispatch a feign attack.

5. Breach Obstacles
Using a defilade or similar cover, the assault's breach team moves forward. If the support team is not firing, the breach team uses stealth in approaching its objective. If suppressive fires are already underway, the breach team employs smoke to mask their movement. Simultaneously, the support team's feign attack employs a smoke screen at another point along the enemy's defensive line and attacks by fire. This will confuse the enemy as to the location of the breach team. The breach team uses visual tape during the day, or lights at night, to mark the breach of the enemy's obstacles.

6. Exploit the Breach
The breach team signals the assault leader once they have opened a path through the obstacles. Using the breach team as near side security, and moving under the cover fire of the support team, the entire assault team moves quickly through the breach.

Once the breach team opens a path through the enemy obstacles, the rest of the assault team moves forward to destroy key enemy positions. The support team coordinates suppressive fires during this time. (Photo by Jeong, Hae-jung.)

7. Clear the Objective
The assault team uses a high volume of fire and a small maneuver force to clear the first couple of fighting positions. Fireteams are assigned specific tasks, such as taking out bunkers or providing for left and right security. The support team continues to shift fires as the assault team clears more enemy fighting positions. The support team must keep the OPFOR suppressed and unable to form a counterattack. When the assault team clears enough of the enemy positions, the assault leader signals for the support team to move through the breach and join the assault team. Now the attack force assaults assigned targets, such as the OPFOR CP and communications nodes, or the attack may continue against the enemy's exposed flanks.

8. Reconsolidate and Reorganize
To regain the patrol's mass and strength, the PL forms the patrol into a 180° security position on the far side of the OBJ and prepares for a counterattack. Element leaders use liquid, ammo, casualties, equipment (LACE) considerations to account for personnel, ensure key weapon systems are manned, and redistribute ammunition, water, medical supplies, and batteries. The PL reports to higher command the progress of the attack and coordinates for supporting resources. Friendly casualties are evacuated, as are EPWs, according to the OPORD. In truth, reconsolidation and reorganization may occur even before all the objectives have been accomplished. It is an ongoing task and may be conducted multiple times.

9. Continue the Mission
After seizing the OBJ, the attack force prepares to transform into a new mission. The patrol might be required to defend, withdraw, or begin an exploitation or pursuit.

(The Offense) II. Attack 2-17

On Point

The attack is a primary form of offensive operation that seeks to impose our will through either a decisive defeat of an enemy force or by seizing essential terrain or facilities. This requires mass, mobility, and the element of surprise.

The massing of combat power (troops, weapons, and fires) is a luxury that the defense does not have. The defense must distribute its combat power over the larger area it intends to defend. True, these resources are coordinated to support one another, but there isn't a great deal of flexibility in the defense's use of combat power unless the defense is willing to surrender valuable territory or facilities.

Hasty or Deliberate

No clear distinction exists between hasty and deliberate attacks, because they are similar. However, the main difference between the two is the extent of planning and preparation conducted by the attacking force. Attacks range along a continuum defined at one end by FRAGOs, which direct the rapid execution of battle drills by forces immediately available. At the other end of the continuum, the company moves into a deliberate attack from a reserve position or assembly area with detailed knowledge of the enemy (a task organization designed specifically for the attack) and a fully rehearsed plan. Most attacks fall somewhere between these two ends of the continuum.

Characteristics of the Offense

The characteristics of the offense are surprise, tempo, concentration, and audacity. Due to the nature of modern offensive operations, flexibility is included in the following discussion of the offense. For each mission, the commander decides how to apply these characteristics to focus the effects of his combat power against enemy weakness. Detailed planning is critical to achieve a synchronized and effective operation. Instead of 'fighting the plan,' commanders should exploit enemy weaknesses.

The following considerations will help avoid common mistakes during the attack:

- Use stealth until contact has been made
- Do not break the attack formation for danger areas
- Bypass all unassigned obstacles
- Use only the force necessary to complete a task
- React with violence once contact has been made

The patrol leader (PL) uses the commander's intent to develop the attack plan. The principals of Mission, Enemy, Terrain & weather, Time available, Troops available, and Civilians on the battlefield (METT-TC) are considered and influence the attack plan's scheme of maneuver and fire support. Control measures are implemented to ensure that friendly fire incidents do not occur.

Maneuver Control Measures

Using a map overlay, the patrol leader identifies the maneuver control measures including the assembly area (AA), the line of departure (LD), the axis of advance, phase lines, the assault position, the objective (OBJ), and the limit of advance (LOA). Furthermore, the larger area of operations (AO) is indicated, giving special attention to any friendly unit's location and activity.

See pp. 2-5a to 2-5b for discussion and listing of common offensive control measures.

Fire Control Measures

Using a map overlay, the patrol leader identifies the fire control measures including the engagement area (EA), target reference points (TRP), the direction of fire, and the restrictive fire line (RFL). In addition to those terrain-based fire control measures, the patrol leader delineates the threat-based fire control measures. At a minimum, these include the engagement priorities, rules of engagement (ROE), and weapon safety postures.

The Offense
III. Exploitation

Ref: FM 3-90 Tactics, chap 6 and FM 3-21.10 (FM 7-10) The Infantry Rifle Company, p. 4-3.

Exploitation is a type of offensive operation that usually follows a successful attack and is designed to disorganize the enemy in depth. Commanders at all echelons exploit successful offensive actions. Attacks that succeed in annihilating a defending enemy are rare. Failure to aggressively exploit success at every turn may give the enemy time to reconstitute an effective defense by shifting his forces or by regaining the initiative through a counterattack. The commander designs his exploitation to maintain pressure on the enemy, compound and take advantage of his disorganization, shatter his will to resist, and seize decisive or key terrain.

Exploitations are conducted at the battalion level and higher. Exploitations seek to disintegrate enemy forces to where they have no alternative but surrender or flight. Companies and platoons may conduct movements to contact or attack as part of a higher unit's exploitation.

Exploitation operations seek to gain key terrain, and enemy facilities or resources. The seizure of weapon and ammunition caches is a legitimate goal of the exploitation. (Photo by Jeong, Hae-jung.)

Exploitation is the primary means of translating tactical success into operational advantage. It reinforces enemy force disorganization and confusion in the enemy's command and control (C2) system caused by tactical defeat. It is an integral part of the concept of the offense. The psychological effect of tactical defeat creates confusion and apprehension throughout the enemy C2 structure and reduces the enemy's ability to react. Exploitation takes advantage of this reduction in enemy capabilities to make permanent what would be only a temporary tactical effect if exploitation were not conducted. Exploitation may be decisive.

Conducting the Exploitation - A Small Unit Perspective

Commanders at all echelons exploit successful offensive actions. Attacks that succeed in annihilating a defending enemy are rare. Failure to aggressively exploit success at every turn may give the enemy time to reconstitute an effective defense by shifting his forces or by regaining the initiative through a counterattack. Therefore, every offensive operation not restricted by higher authority or lack of resources should be followed without delay by bold exploitation. The commander designs his exploitation to maintain pressure on the enemy, compound and take advantage of his disorganization, shatter his will to resist, and seize decisive or key terrain.

The expressed intention of the exploitation is to seize land or structures. Exploitation forces do this by capitalizing on the success of the prior mission—thereby making the exploitation mission an opportunistic venture. By following up tactical success with further tactical success, operational gains are achieved.

Exploitations are conducted at the battalion level and higher. Companies and platoons may conduct movements to contact or attack as part of a higher unit's exploitation.

I. Organization

The forces conducting an attack are also the forces that initially exploit that attack's success. Typically, the commander does not assign a subordinate unit the mission of exploitation before starting a movement to contact (MTC) or an attack. The commander reorganizes his unit internally to reflect the existing factors of METT-TC when the opportunity to exploit success occurs. He uses fragmentary orders (FRAGOs) to conduct actions on contact. If a commander needs additional resources to support the exploitation, he requests them from the appropriate headquarters. The additional resources may include intelligence, surveillance, and reconnaissance (ISR) assets to help identify targets for attack, as well as attack helicopters and controlled munitions, such as the Army tactical missile system, to attack identified targets. Each exploitation force should be large enough to protect itself from those enemy forces it expects to encounter. It should also be a reasonably self-sufficient combined arms force capable of operations beyond the supporting range of the main body.

The units that create an opportunity to exploit should not be expected to perform the exploitation to an extended depth. If the commander plans to exploit with a specific subordinate unit, he must specify the degree of damage or risk to that force he is willing to accept in the course of the current operation. If the initially attacking units incur significant losses of combat power, the commander should replace them as soon as possible. When the exploiting force's combat power weakens because of fatigue, disorganization, or attrition, or when it must hold ground or resupply, the commander should continue the exploitation with a fresh force. In both cases, the replacement force should have a high degree of tactical mobility so it can conduct the exploitation.

II. Planning & Preparation

The commander's ability to deny the enemy options by proactive use of his battlefield operating systems is critical to a successful exploitation. He does this by arranging his battlefield operating systems within his opponent's time and space relationship in accordance with the factors of METT-TC.

The commander must plan for the decentralized execution of an exploitation. His commander's intent is especially important because subordinates must be able to exercise initiative in a rapidly changing, fluid situation. The commander must state the purpose of the exploitation, which may be to force the retrograde of enemy forces from an area, encircle enemy forces so they cannot withdraw, or destroy enemy artillery and other fire support systems. The intent must describe the desired end state. That intent will also determine his decisive and shaping operations and guide the designation of his main effort at any given point.

A clear commander's intent provides subordinates with guidance on how to integrate their operations into the overall operations of the higher headquarters. Only subordinates who can act quickly can seize all opportunities to damage the enemy or accelerate the tempo of operations. A commander should place minimal restrictions on his subordinates. These may include clear instructions regarding the seizure of key terrain and the size of enemy forces that may be bypassed. Reliable, secure communications between the exploiting force, the follow and support force, and the commander facilitate coordination that can maximize the impact of the exploitation. However, all subordinates should have a clear picture of the desired end state to conduct operations that support it, even if communications are lost.

Planning for an exploitation begins during the preparation phase of all offensive operations. To avoid losing critical time during the transition from an MTC or an attack to an exploitation, the commander tentatively identifies forces, objectives, and AOs for subordinate units before the offensive operation begins. When the opportunity to exploit occurs, brigade and higher echelon commanders should initiate the exploitation, either as a branch of or a sequel to the existing operation. The commander's plan should attempt to avoid driving the enemy back in the direction of his own sustaining base.

III. Execution

Exploitation requires physical and mental aggressiveness to combat the friction of limited visibility, fatigue, bad weather, dangers of fratricide, and extended operations. It requires bold and aggressive reconnaissance, prompt use of firepower, and rapid employment of previously uncommitted units. Exploiting forces maneuver swiftly toward their objectives, sever escape routes, and strike at enemy command posts, communications nodes, reserves, artillery, and CS units to prevent the enemy from reorganizing an effective defense.

The exploitation is planned as a branch or a sequel to another combat operation. As such, it is difficult to detail the exact execution because much depends on the primary operation. However, exploitation has the intended goal of seizing enemy terrain, facilities, or resources. The exact target of the exploitation may not become clear until the first operation has achieved its objectives.

1. The commander and subordinate leaders must know what lies ahead. As the exploitation is conducted, small pockets of enemy resistance are commonly bypassed—because the destruction of the enemy force is *not* the objective. Instead, the exploitation focuses on the seizure of property. The force bypasses anything that does not present a threat to the rear or flanks.

2. If an obstacle cannot be bypassed, the commander may choose to isolate the enemy with disruptive fires and continue to press the exploitation. The commander may also choose to destroy the enemy precisely as done in a MTC. In any case, the commander does not want to leave an enemy force behind that would be capable of striking either the exploitation force or our sustainment force.

3. At some point a unit conducting an exploitation reaches a culminating point or transitions to a pursuit. Culmination can occur for the variety of reasons, such as friendly losses or the enemy's commitment of his reserve. The commander, when he makes an assessment that his force is approaching culmination, should transition to another type of operation. On the other hand, a pursuit enables the commander to complete his destruction of the enemy.

On Point

The exploitation is a follow-on form of attack. It seeks to gain territory, key facilities or enemy resources, such as supply caches. However, exploitation is not a primary form of attack. That means that the exploitation will always follow a successful attack, movement to contact (MTC), or defense. The force conducting such operations will be required to transform into the exploitation. It may be planned as a branch or sequel to any mission.

Defined Objective

The exploitation requires a defined objective, such as terrain or facility. Additionally, like the MTC, exploitation forces rely on the experience of the combat leader to know when to bypass the enemy, when to engage the enemy, and how to recognize signs of culmination—the point at which the mission will bear no more success.

The conduct of exploitation is very similar in organization and conduct as the MTC. However, whereas the MTC is dedicated to a given axis of advance or targets the enemy force, the exploitation seeks a specific objective at a given geographical location. This means that bypassing the enemy is not only acceptable, but in most cases it is preferred.

Branch or Sequel

Exploitation is planned only as a branch or sequel to another combat operation. It is not a primary form of offensive operations. Primary operations (attack, MTC, or defense) must be prepared to transition quickly into the exploitation. This requires well-informed subordinate leaders.

Exploitation is the primary means of translating tactical success into operational advantage. It reinforces enemy force disorganization and confusion in the enemy's command and control (C2) system caused by tactical defeat. It is an integral part of the concept of the offense. The psychological effect of tactical defeat creates confusion and apprehension throughout the enemy C2 structure and reduces the enemy's ability to react. Exploitation takes advantage of this reduction in enemy capabilities to make permanent what would be only a temporary tactical effect if exploitation were not conducted. Exploitation may be decisive.

Transition

Local exploitation by the committed force follows a successful attack. A unit conducts a local exploitation when it capitalizes on whatever tactical opportunities it creates in the course of accomplishing its assigned offensive mission. Whenever possible, the lead attacking unit transitions directly to the exploitation after accomplishing its mission in a local exploitation. If this is not feasible, the commander can pass fresh forces (follow and assume) into the lead. The commander acts quickly to capitalize on local successes. Although such local exploitations may appear insignificant, their cumulative effects can be decisive. Subordinate commanders, working within a higher commander's intent, can use their initiative to launch an exploitation. When a commander initiates a local exploitation, he informs his higher headquarters to keep that commander informed of his intentions. This prevents the inadvertent disruption of the higher echelon's battle or campaign and allows the higher headquarters to assess the possibility.

The Offense
IV. Pursuit

Ref: FM 3-90 Tactics, chap 7 and FM 3-21.10 (FM 7-10) The Infantry Rifle Company, p. 4-3.

The pursuit is a follow-on form of attack, or counterattack when conducted from a defense. Pursuits are normally conducted at the brigade or higher level. A pursuit typically follows a successful exploitation. Ideally, it prevents a fleeing enemy from escaping and then destroys him. Companies and platoons will participate in a larger unit's exploitation and may conduct attacks as part of the higher unit's operation. Therefore, it must be planned as a branch or a sequel of other operations. The primary operation must be prepared to transition into a pursuit at an appropriate time.

The pursuit seeks to find, fix, and finish a fleeing enemy force. To do this, an enveloping force moves quickly forward along a parallel route of the escaping enemy and suppresses the enemy in their escape routes. (Photo by Jeong, Hae-jung).

Unlike an exploitation, which may focus on seizing key or decisive terrain instead of the enemy force, the pursuit always focuses on destroying the fleeing enemy force. This is seldom accomplished by directly pushing back the hostile forces on their lines of communication (LOCs). The commander in a pursuit tries to combine direct pressure against the retreating forces with an enveloping or encircling maneuver to place friendly troops across the enemy's lines of retreat. This fixes the enemy in positions where he can be defeated in detail. If it becomes apparent that enemy resistance has broken down entirely and the enemy is fleeing the battlefield, any type of offensive operation can transition to a pursuit.

Pursuit operations begin when an enemy force attempts to conduct retrograde operations. At that point, it becomes most vulnerable to the loss of internal cohesion and complete destruction. A pursuit aggressively executed leaves the enemy trapped, unprepared, and unable to defend, faced with the options of surrendering or complete destruction. The rapid shifting of units, continuous day and night movements, hasty attacks, containment of bypassed enemy forces, large numbers of prisoners, and a willingness to forego some synchronization to maintain contact with and pressure on a fleeing enemy characterize this type of offensive operation. Pursuit requires swift maneuver and attacks by forces to strike the enemy's most vulnerable areas. A successful pursuit requires flexible forces, initiative by commanders at all levels, and the maintenance of a high operational tempo during execution.

The enemy may conduct a retrograde when successful friendly offensive operations have shattered his defense. In addition, the enemy may deliberately conduct a retrograde when—

- He is reacting to a threat of envelopment
- He is adjusting his battlefield dispositions to meet changing situations
- He is attempting to draw the friendly force into fire sacks, kill zones, or engagement areas
- He is planning to employ weapons of mass destruction

I. Organization

Normally, the commander does not organize specifically for a pursuit ahead of time, although he may plan for a pursuit as a branch or sequel to his offensive operation. Therefore, he must be flexible to react when the situation presents itself. The commander's maneuver and sustainment forces continue their ongoing activities while he readjusts their priorities to better support the pursuit. He acquires additional support from his higher headquarters in accordance with the factors of METT-TC.

For most pursuits, the commander organizes his forces into security, direct-pressure, encircling, follow and support, and reserve forces.

Given sufficient resources, there can be more than one encircling force. The follow and support force polices the battlefield to prevent the dissipation of the direct-pressure force's combat power. Appendix B addresses the duties of a follow and support force. The reserve allows the commander to take advantage of unforeseen opportunities or respond to enemy counterattacks.

There are two basic organizational options in conducting a pursuit; each involves a direct-pressure force. The first is a frontal pursuit that employs only a direct-pressure force. The second is a combination that uses a direct-pressure force and an encircling force. The combination pursuit is generally more effective. Either the direct-pressure force or the encircling force can conduct the decisive operation in a combination pursuit.

A. Frontal Pursuit

In a frontal pursuit, the commander employs only a direct-pressure force to conduct operations along the same retrograde routes used by the enemy. The commander chooses this option in two situations. The first is when he cannot create an encircling force with enough mobility to get behind the enemy force. The second is when he cannot create an encircling force capable of sustaining itself until it links up with the direct-pressure force. Either situation can occur because of restrictive terrain or because an enemy withdraws in a disciplined, cohesive formation and still has significant available combat power.

B. Combination Pursuit

In the pursuit, the most decisive effects result from combining the frontal pursuit with encirclement. In the combination pursuit, the direct-pressure force initiates a frontal pursuit immediately on discovering the enemy's initiation of a retrograde operation. This slows the tempo of the enemy's withdrawal (or fixes him in his current position if possible), and may destroy his rear security force. The direct-pressure force's actions help to set the conditions necessary for the success of the encircling force's operation by maintaining constant pressure. The encircling force conducts an envelopment or a turning movement to position itself where it can block the enemy's escape and trap him between the two forces, which leads to complete annihilation.

II. Planning & Preparation

The commander anticipates an enemy retrograde operation as either a branch or a sequel to the plan. The plan should identify possible direct-pressure, encircling, follow and support, and reserve forces and issue on-order or be-prepared missions to these forces. The commander should employ the maximum number of available combat troops in the pursuit. He bases the details of his plan on the enemy's anticipated actions, the combat formation of the attacking troops, and the amount of planning time available. The commander also considers—

- Possible routes the enemy might use to conduct his retrograde operations
- Availability of his intelligence, surveillance, and reconnaissance assets to detect enemy forces and acquire targets in depth
- Scheme of maneuver
- Availability and condition of pursuit routes
- Availability of forces to keep the pressure on the enemy until his destruction is complete
- Critical terrain features
- Use of reconnaissance and security forces
- Allocation of precision-guided munitions and aviation support
- Availability of CS and CSS resources

Pursuit planning must address the possibility of defending temporarily during operational pauses while making preparations to continue the pursuit or to consolidate gains. However, the use of an operational pause generally results in the abandonment of the pursuit because the enemy is able to use that time to organize a coherent defense.

The commander must specifically address how to detect the enemy retrograde operations; otherwise, the enemy may succeed in breaking contact. The commander relies on active reconnaissance, an understanding of enemy tactics, and knowledge of the current tactical situation. He must watch for signs that indicate the enemy is preparing to conduct a retrograde, such as when the enemy—

- Lacks the capability to maintain his position or cohesion
- Conducts limited local counterattacks
- Intensifies his reconnaissance and intelligence efforts
- Increases the amount of rearward movements and changes the type of elements conducting them, especially by fire support and reserves
- Prepares his facilities, installations, equipment, and supply stock-piles for demolition and destruction
- Decreases fire in intensity and effectiveness through the AO
- Increases his fires in one or more individual sectors of the front, which does not appear to be in accordance with the developing situation, and at a time when the amount of defensive fires seems to be decreasing

III. Conducting the Pursuit - A Small Unit Perspective

Pursuits are normally conducted at the brigade or higher level. A pursuit typically follows a successful exploitation. Ideally, it prevents a fleeing enemy from escaping and then destroys him. Companies and platoons will participate in a larger unit's exploitation and may conduct attacks as part of the higher unit's operation.

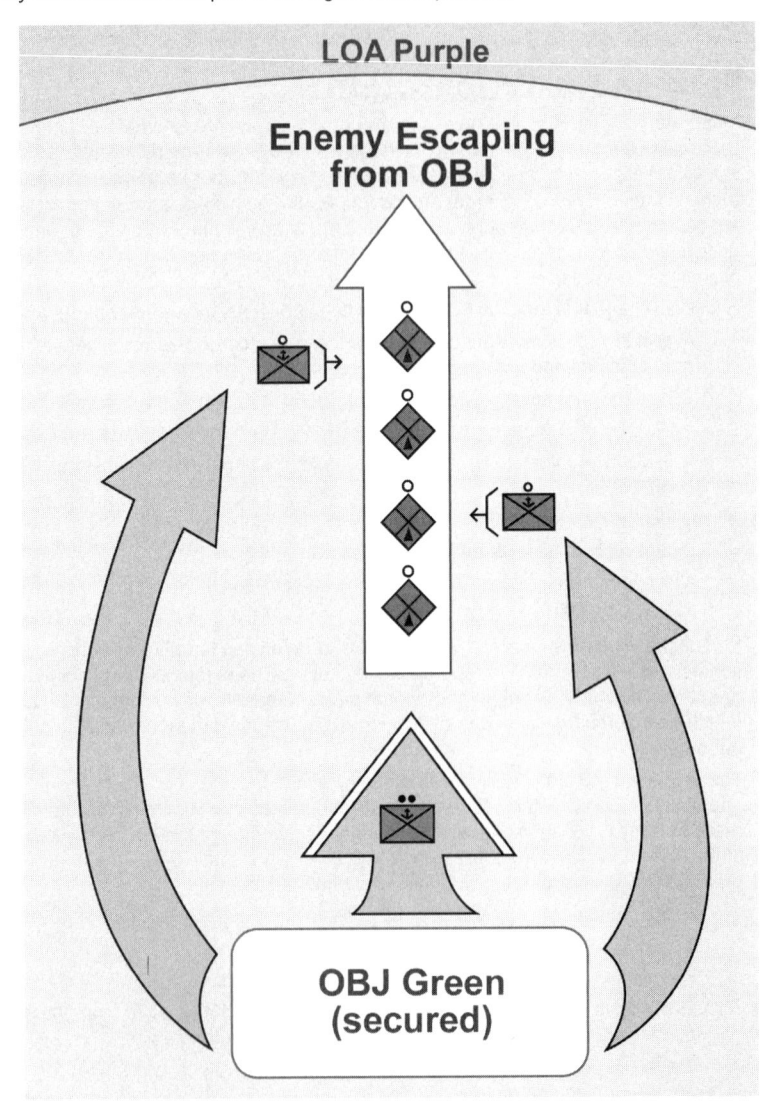

The pursuit deploys teams forward to envelop and fix the enemy in their escape route. The suppressed enemy cannot coordinate a defense, and the main force defeats them in detail. (Ref: FM 3-90, chap 7, fig. 7-2).

The pursuit is planned as a branch or a sequel to another combat operation. It is virtually impossible to detail the execution because much depends on the status of the enemy forces after the primary operation concludes. However, pursuit has the intended goal of destroying or capturing the enemy force.

Once the commander initiates a pursuit, he continues pursuing the enemy until a higher commander terminates the pursuit. Conditions under which a higher commander may terminate a pursuit include the following—

- The pursuing force annihilates or captures the enemy and resistance ceases
- The pursuing force fixes the enemy for follow-on forces
- The high commander makes an assessment that the pursuing force is about to reach a culminating point

The main body maneuvers aggressively while the enveloping force still has the enemy fixed. The main body attacks and defeats in detail each pocket of enemy. This process continues until culmination. (Photo by Jeong, Hae-jung).

1. The commander and subordinate leaders must know approximately where the enemy escape routes are located. A fixing force (typically the reserve of any combat operation) must move deep into enemy territory to identify the enemy's route of escape and relay this information back to the main force. The fixing force will suppress the enemy in individual pockets within the escape routes to prevent the enemy from coordinating an effective response with adjacent enemy units. The main force will aggressively maneuver up to and destroy the trapped, isolated enemy. Then the process repeats in order to trap and destroy as many enemy troops as possible.

2. The pursuit is focused on the destruction of the enemy force. As such, the pursuit force bypasses any terrain or facilities that do not present a threat to their rear or flanks. Do not leave an enemy force behind that would be capable of delaying the fixing force or the main force of the pursuit.

3. The commander must look for signs of culmination. As the enemy is pursued, they often are falling back on their own resources. For example, they could be falling back to designated rally points—which theoretically offer them defensive positions.

On Point

The pursuit is the classic "hammer and anvil" maneuver. The pursuit force must have some idea of where the enveloping force is, as well as where the enemy force is, before they can begin swinging the hammer of the main force.

When the commander initiates a pursuit, he often creates the encircling force from uncommitted or reserve elements. Normally, these forces do not have fire support assets allocated to them. The commander must plan how to redistribute his fire support assets to properly support the encircling force. Attack helicopters and close air support are well suited to support the encircling force.

A pursuit is often conducted as a series of encirclements in which successive portions of the fleeing enemy are intercepted, cut off from outside support, and captured or destroyed. The direct-pressure force conducts a series of hasty attacks to destroy the enemy's rear security force, maintain constant pressure on the enemy's main body, and slow the enemy's withdrawal. At every opportunity, the direct-pressure force fixes, slows down, and destroys enemy elements, provided such actions do not interfere with its primary mission of maintaining constant pressure on the enemy's main body. The direct pressure force can bypass large enemy forces if it can hand them off to follow and support units, or if they do not pose a risk to the direct-pressure force.

Mobility and Countermobility

Engineer mobility and countermobility assets are instrumental in sustaining the rate of advance and hindering the enemy's withdrawal. Engineers prepare the route of advance and support the lateral dispersion of units transitioning to the pursuit and the movement of the reserve. During the pursuit, the commander must plan for his engineers to provide assault bridging and emergency road repairs to sustain the tempo of the pursuit. The commander also plans to use his engineer assets to block any bypassed enemy's withdrawal routes by using antitank and command-operated mines, demolitions, and obstacles.

Risk

Conducting a pursuit is a calculated risk. Once the pursuit begins, the commander maintains contact with the enemy and pursues retreating enemy forces without further orders. The commander maintains the pursuit as long as the enemy appears disorganized and friendly forces continue to advance.

Transition

A pursuit often transitions into other types of offensive and defensive operations. If the enemy attempts to reorganize, forces conducting a pursuit execute hasty attacks. They conduct an exploitation to capitalize on the success of these attacks and then move back into pursuit. Forces conducting a pursuit may also transition into a defensive operation if the pursuing force reaches a culminating point. This usually occurs when the enemy introduces strong reinforcements to prepare for a counteroffensive.

Chap 2

V. Small Unit Offensive Tactical Tasks

Ref: FM 3-21.8 (FM 7-8) The Infantry Rifle Platoon and Squad, pp. 7-31 to 7-36.

Tactical tasks are specific activities performed by units as they conduct tactical operations or maneuver. At the platoon level, these tasks are the warfighting actions the platoon may be called on to perform in battle. This section provides discussion and examples of some common actions and tasks the platoon may perform during a movement to contact, a hasty attack, or a deliberate attack. It is extremely important to fully understand the purpose behind a task (what) because the purpose (why) defines what the platoon must achieve as a result of executing its mission. A task can be fully accomplished, but if battlefield conditions change and the platoon is unable to achieve the purpose, the mission is a failure.

Note: The situations used in this section to describe the platoon leader's role in the conduct of tactical tasks are examples only. They are not applicable in every tactical operation, nor are they intended to prescribe any specific method or technique the platoon must use in achieving the purpose of the operation. Ultimately, it is up to the commander or leader on the ground to apply both the principles discussed here, and his knowledge of the situation. An understanding of his unit's capabilities, the enemy he is fighting, and the ground on which the battle is taking place are critical when developing a successful tactical solution.

Task	Graphic	Description
Seize		*Seize* is a tactical mission task that involves taking possession of a designated area by using overwhelming force. An enemy force can no longer place direct fire on an objective that has been seized.
Suppress	No graphic provided in FM 3-90.	*Suppress* is a tactical mission task that results in the temporary degradation of the performance of a force or weapon system below the level needed to accomplish its mission.
Support by Fire		*Support-by-fire* is a tactical mission task in which a maneuver force moves to a position where it can engage the enemy by direct fire in support of another maneuvering force. The primary objective of the support force is normally to fix and suppress the enemy so he cannot effectively fire on the maneuvering force.
Clear		*Clear* is a tactical mission task that requires the commander to remove all enemy forces and eliminate organized resistance within an assigned area.
Attack by Fire		*Attack-by-fire* is a tactical mission task in which a commander uses direct fires, supported by indirect fires, to engage an enemy without closing with him to destroy, suppress, fix, or deceive him.

Tactical mission tasks describe the results or effects the commander wants to achieve - the what and why of a mission statement. For a more complete listing of tactical mission tasks, see pp. 1-11 to 1-14.

I. Seize

Seizing involves gaining possession of a designated objective by overwhelming force. Seizing an objective is complex. It involves closure with the enemy, under fire of the enemy's weapons to the point that the friendly assaulting element gains positional advantage over, destroys, or forces the withdrawal of the enemy.

A platoon may seize prepared or unprepared enemy positions from either an offensive or defensive posture. Examples include the following:

- A platoon seizes the far side of an obstacle as part of a company breach or seizes a building to establish a foothold in an urban environment
- A platoon seizes a portion of an enemy defense as part of a company deliberate attack
- A platoon seizes key terrain to prevent its use by the enemy

There are many inherent dangers in seizing an objective. They include the requirement to execute an assault, prepared enemy fires, a rapidly changing tactical environment, and the possibility of fratricide when friendly elements converge. These factors require the platoon leader and subordinate leaders to understand the following planning considerations.

Developing a clear and current picture of the enemy situation is very important. The platoon may seize an objective in a variety of situations, and the platoon leader will often face unique challenges in collecting and disseminating information on the situation. For example, if the platoon is the seizing element during a company deliberate attack, the platoon leader should be able to develop an accurate picture of the enemy situation during the planning and preparation for the operation. He must be prepared to issue modifications to the platoon as new intelligence comes in or as problems are identified in rehearsals.

In another scenario, the platoon leader may have to develop his picture of the enemy situation during execution. He must rely more heavily on reports from units in contact with the enemy and on his own development of the situation. In this type of situation, such as when the platoon is seizing an enemy combat security outpost during a movement to contact, the platoon leader must plan on relaying information as it develops. He uses clear, concise FRAGOs to explain the enemy situation, and gives clear directives to subordinates.

II. Suppress

The platoon maneuvers to a position on the battlefield where it can observe the enemy and engage him with direct and indirect fires. The purpose of suppressing is to prevent the enemy from effectively engaging friendly elements with direct or indirect fires. To accomplish this, the platoon must maintain orientation both on the enemy force and on the friendly maneuver element it is supporting. During planning and preparation, the platoon leader should consider—

- Conducting a line-of-sight analysis during his terrain analysis to identify the most advantageous positions from which to suppress the enemy
- Planning and integrating direct and indirect fires
- Determining control measures (triggers) for lifting, shifting, or ceasing direct fires
- Determining control measures for shifting or ceasing indirect fires
- Planning and rehearsing actions on contact
- Planning for large Class V expenditures. (The company commander and the platoon leader must consider a number of factors in assessing Class V require-

ments including the desired effects of the platoon direct fires; the composition, disposition, and strength of the enemy force; and the time required to suppress the enemy.)

- Determining when and how the platoon will reload ammunition during the fight while still maintaining suppression for the assaulting element

III. Support by Fire

The platoon maneuvers to a position on the battlefield from where it can observe the enemy and engage him with direct and indirect fires. The purpose of support by fire is to prevent the enemy from engaging friendly elements.

To accomplish this task, the platoon must maintain orientation both on the enemy force and on the friendly maneuver element it is supporting. The platoon leader should plan and prepare by—

- Conducting line-of-sight analysis to identify the most advantageous support-by-fire positions
- Conducting planning and integration for direct and indirect fires
- Determining triggers for lifting, shifting, or ceasing direct and indirect fires
- Planning and rehearsing actions on contact
- Planning for large Class V expenditures, especially for the weapons squad and support elements, because they must calculate rounds per minute. (The platoon leader and weapons squad leader must consider a number of factors in assessing Class V requirements, including the desired effects of platoon fires; the time required for suppressing the enemy; and the composition, disposition, and strength of the enemy force.)

A comprehensive understanding of the battlefield and enemy and friendly disposition is a crucial factor in all support-by-fire operations. The platoon leader uses all available intelligence and information resources to stay abreast of events on the battlefield. Additional considerations may apply. The platoon may have to execute an attack to secure the terrain from where it will conduct the support by fire. The initial support-by-fire position may not afford adequate security or may not allow the platoon to achieve its intended purpose. This could force the platoon to reposition to maintain the desired weapons effects on the enemy. The platoon leader must ensure the platoon adheres to these guidelines:

- Maintain communication with the moving element
- Be prepared to support the moving element with both direct and indirect fires
- Be ready to lift, shift, or cease fires when masked by the moving element
- Scan the area of operations and prepare to acquire and destroy any enemy element that threatens the moving element
- Maintain 360-degree security
- Use Javelins to destroy any exposed enemy vehicles
- Employ squads to lay a base of sustained fire to keep the enemy fixed or suppressed in his fighting positions
- Prevent the enemy from employing accurate direct fires against the protected force

IV. Clear

Note: See also pp. 7-3 to 7-7 for TTPs of entering and clearing buildings.

Clearing requires the platoon to remove all enemy forces and eliminate organized resistance within an assigned area. The platoon may be tasked with clearing an objective area during an attack to facilitate the movement of the remainder of the company, or may be assigned clearance of a specific part of a larger objective area. Infantry platoons are normally best suited to conduct clearance operations, which in many cases will involve working in restrictive terrain, to include clearing a —

- Defile, including choke points in the defile and high ground surrounding it
- Heavily wooded area
- Built-up or strip area. Refer to FM 3-06, Urban Operations, and FM 3-06.11, Combined Operations in Urban Terrain, for a detailed discussion of urban combat
- Road, trail, or other narrow corridor, which may include obstacles or other obstructions on the actual roadway and in surrounding wooded and built-up areas

General Terrain Considerations

The platoon leader must consider several important terrain factors when planning and executing the clearance task. Observation and fields of fire may favor the enemy. To be successful, the friendly attacking element must neutralize this advantage by identifying dead spaces where the enemy cannot see or engage friendly elements. It should also identify multiple friendly support-by-fire positions that are necessary to support a complex scheme of maneuver which cover the platoon's approach, the actual clearance task, and friendly maneuver beyond the restrictive terrain.

When clearing in support of tactical vehicles, cover and concealment are normally abundant for Infantry elements, but scarce for trail-bound vehicles. Lack of cover leaves vehicles vulnerable to enemy antiarmor fires. While clearing in support of mechanized vehicles, obstacles influence the maneuver of vehicles entering the objective area. The narrow corridors, trails, or roads associated with restrictive terrain can be easily obstructed with wire, mines, and log cribs.

Key terrain may include areas dominating the objective area, approaches, or exits, and any terrain dominating the area inside the defile, wooded area, or built-up area. Avenues of approach will be limited. The platoon must consider the impact of canalization.

Restrictive Terrain Considerations

Conducting clearance in restrictive terrain is both time consuming and resource intensive. During the planning process, the platoon leader evaluates the tactical requirements, resources, and other considerations for each operation.

During the approach, the platoon leader focuses on moving combat power into the restrictive terrain and posturing it to start clearing the terrain. The approach ends when the rifle squads complete their preparations to conduct an attack. The platoon leader—

- Establishes support-by-fire positions
- Destroys or suppresses any known enemy positions to allow elements to approach the restrictive terrain
- Provides more security by incorporating suppressive indirect fires and obscuring or screening smoke

The platoon leader provides support by fire for the rifle squads. He prepares to support the rifle squads where they enter the restrictive terrain by using—

- High ground on either side of a defile
- Wooded areas on either side of a trail or road
- Buildings on either side of a road in a built-up area
- Movement of rifle squads along axes to provide cover and concealment

Clearance begins as the rifle squads begin their attack in and around the restrictive terrain. Examples of where this maneuver may take place include—
- Both sides of a defile, either along the ridgelines or high along the walls of the defile
- Along the wood lines parallel to a road or trail
- Around and between buildings on either side of the roadway in a built-up area

The following apply during clearance:
- The squads provide a base of fire to allow the weapons squad or support-by-fire element to bound to a new support-by-fire position. This cycle continues until the entire area is cleared.
- Direct-fire plans should cover responsibility for horizontal and vertical observation, and direct fire
- Squads should clear a defile from the top down and should be oriented on objectives on the far side of the defile
- Engineers with manual breaching capability should move with the rifle squads. Engineers may also be needed in the overwatching element to reduce obstacles.

The platoon must secure the far side of the defile, built-up area, or wooded area until the company moves forward to pick up the fight beyond the restrictive terrain. If the restrictive area is large, the platoon may be directed to assist the passage of another element forward to continue the clearance operation. The platoon must be prepared to—
- Destroy enemy forces
- Secure the far side of the restrictive terrain
- Maneuver squads to establish support-by-fire positions on the far side of the restrictive terrain
- Support by fire to protect the deployment of the follow-on force assuming the fight
- Suppress any enemy elements that threaten the company while it exits the restrictive terrain
- Disrupt enemy counterattacks
- Protect the obstacle reduction effort
- Maintain observation beyond the restrictive terrain
- Integrate indirect fires as necessary

Enemy Analysis

Careful analysis of the enemy situation is necessary to ensure the success of clearing. The enemy evaluation should include the following:

- Enemy vehicle location, key weapons, and Infantry elements in the area of operations
- Type and locations of enemy reserve forces
- Type and locations of enemy OPs
- The impact of the enemy's CBRN and or artillery capabilities

Belowground Operations

Belowground operations involve clearing enemy trenches, tunnels, caves, basements, and bunker complexes. The platoon's base-of-fire element and maneuvering squads must maintain close coordination. The weapons squad or support-by-fire element focuses on protecting the squads as they clear the trench line, or maneuver to destroy individual or vehicle positions. The base-of-fire element normally concentrates on destroying key surface structures (especially command posts and crew-served weapons bunkers) and the suppression and destruction of enemy vehicles.

(The Offense) V. Small Unit Tactical Tasks 2-33

V. Attack by Fire

The platoon maneuvers to a position on the battlefield from where it can observe the enemy and engage him with direct and indirect fires at a distance to destroy or weaken his maneuvers. The platoon destroys the enemy or prevents him from repositioning. The platoon employs long-range fires from dominating terrain. It also uses flanking fires or takes advantage of the standoff range of the unit's weapons systems. The company commander may designate an attack-by-fire position from where the platoon will fix the enemy. An attack-by-fire position is most commonly employed when the mission or tactical situation focuses on destruction or prevention of enemy movement. In the offense, it is usually executed by supporting elements. During defensive operations, it is often a counterattack option for the reserve element.

Considerations

When the platoon is assigned an attack-by-fire position, the platoon leader obtains the most current intelligence update on the enemy and applies his analysis to the information. During planning and preparation, the platoon leader should consider—

- Conducting a line-of-sight analysis during terrain analysis to identify the most favorable locations to destroy or fix the enemy
- Conducting direct and indirect fire planning and integration
- Determining control measures (triggers) for lifting, shifting, or ceasing direct fires
- Determining control measures for shifting or ceasing indirect fires
- Planning and rehearsing actions on contact

Several other considerations may affect the successful execution of an attack by fire. The platoon may be required to conduct an attack against enemy security forces to seize the ground from where it will establish the attack-by-fire position. The initial attack-by-fire position may afford inadequate security or may not allow the platoon to achieve its task or purpose. This could force the platoon to reposition to maintain the desired weapons effects on the enemy force. Because an attack by fire may be conducted well beyond the direct fire range of other platoons, it may not allow the platoon to destroy the targeted enemy force from its initial positions. The platoon may begin to fix the enemy at extended ranges. Additional maneuver would then be required to close with the enemy force and complete its destruction. Throughout an attack by fire, the platoon should reposition or maneuver to maintain flexibility, increase survivability, and maintain desired weapons effects on the enemy. Rifle squad support functions may include:

- Seizing the attack-by-fire position before occupation by mounted sections
- Providing local security for the attack-by-fire position
- Executing timely, decisive actions on contact
- Using maneuver to move to and occupy attack-by-fire positions
- Destroying enemy security elements protecting the targeted force
- Employing effective direct and indirect fires to disrupt, fix, or destroy the enemy force

Chap 3: The Defense

Ref: ADRP 3-90, Offense and Defense (Aug '12), chap. 4.

While the offensive element of combat operations is more decisive, the defense is the stronger element. However, the conduct of defensive tasks alone normally cannot achieve a decision. Their purpose is to create conditions for a counteroffensive that allows Army forces to regain the initiative. Other reasons for conducting defensive actions include—
- Retaining decisive terrain or denying a vital area to the enemy
- Attriting or fixing the enemy as a prelude to offensive actions
- Surprise action by the enemy
- Increasing the enemy's vulnerability by forcing the enemy commander to concentrate subordinate forces

While the offense is the most decisive type of combat operation, the defense is the stronger type. The inherent strengths of the defense include the defender's ability to occupy his positions before the attack and use the available time to prepare his defenses. (Photo by Jeong, Hae-jung).

A defensive task is a task conducted to defeat an enemy attack, gain time, economize forces, and develop conditions favorable for offensive or stability tasks (ADRP 3-0). While the offensive element of combat operations is more decisive, the defense is the stronger element. The inherent strengths of the defense include the defender's ability to occupy positions before the attack and use the available time to prepare the defenses. The defending force ends its defensive preparations only when it retrogrades or begins to engage the enemy. Even during combat, the defending force takes the opportunities afforded by lulls in the action to improve its positions and repair combat damage. The defender maneuvers to place the enemy in a position of disadvantage and attacks the enemy at every opportunity, using fires, electronic warfare, and joint assets, such as close air support.

The static and mobile elements of the defense combine to deprive the enemy of the initiative. The defender contains the enemy while seeking every opportunity to transition to the offense.

I. Purposes of Defense Operations

Commanders choose to defend to create conditions for a counteroffensive that allows Army forces to regain the initiative. Other reasons for conducting a defense include to retain decisive terrain or deny a vital area to the enemy, to attrit or fix the enemy as a prelude to the offense, in response to surprise action by the enemy, or to increase the enemy's vulnerability by forcing the enemy to concentrate forces.

II. Defensive Tasks

There are three basic defensive tasks—area defense, mobile defense, and retrograde. These apply to both the tactical and operational levels of war, although the mobile defense is more often associated with the operational level. These three tasks have significantly different concepts and pose significantly different problems. Therefore, each defensive task must be dealt with differently when planning and executing the defense. Although the names of these defensive tasks convey the overall aim of a selected defense, each typically contains elements of the other and combines static and mobile elements.

Although on the defense, the commander remains alert for opportunities to attack the enemy whenever resources permit. Within a defensive posture, the defending commander may conduct a spoiling attack or a counterattack, if permitted to do so by the mission variables of mission, enemy, terrain and weather, troops and support available, time available, and civil considerations (METT-TC).

A. Area Defense

The area defense is a defensive task that concentrates on denying enemy forces access to designated terrain for a specific time rather than destroying the enemy outright. The focus of the area defense is on retaining terrain where the bulk of the defending force positions itself in mutually supporting, prepared positions. Units maintain their positions and control the terrain between these positions. The decisive operation focuses on fires into engagement areas, possibly supplemented by a counterattack.

Note: See pp. 3-11 to 3-18.

B. Mobile Defense

The mobile defense is a defensive task that concentrates on the destruction or defeat of the enemy through a decisive attack by a striking force. The mobile defense focuses on defeating or destroying the enemy by allowing enemy forces to advance to a point where they are exposed to a decisive counterattack by the striking force.

Note: See pp. 3-5 to 3-10.

C. Retrograde

The retrograde is a defensive task that involves organized movement away from the enemy. The enemy may force these operations, or a commander may execute them voluntarily. The higher commander of the force executing the retrograde must approve the retrograde operation before its initiation in either case. The retrograde is a transitional operation; it is not conducted in isolation. It is part of a larger scheme of maneuver designed to regain the initiative and defeat the enemy.

Note: See pp. 3-19 to 3-22.

III. Characteristics of the Defense

A feature of the defense is a striving to regain the initiative from the attacking enemy. The defending commander uses the characteristics of the defense to help accomplish that task. (Photo by Jeong, Hae-jung).

A. Disruption
Defenders disrupt the attackers' tempo and synchronization with actions designed to prevent them from massing combat power. Commanders employ disruptive actions to unhinge the enemy's preparations and attacks. Disruption methods include misdirecting or destroying enemy reconnaissance forces, breaking up formations, isolating units, and attacking or disrupting systems.

B. Flexibility
The conduct of the defense requires flexible plans. Commanders focus planning on preparations in depth, use of reserves, and the ability to shift the main effort. Commanders add flexibility by designating supplementary positions, designing counterattack plans, and preparing to counterattack.

C. Maneuver
Maneuver allows the defender to take full advantage of the area of operations and to mass and concentrate when desirable. Maneuver, through movement in combination with fire, allows the defender to achieve a position of advantage over the enemy to accomplish the mission.

D. Mass and Concentration
Defenders seek to mass the effects of overwhelming combat power where they choose and shift it to support the decisive operation. Commanders retain and, when necessary, reconstitute a reserve and maneuver to gain local superiority at the point of decision.

E. Operations in Depth
Simultaneous application of combat power throughout the area of operations improves the chances for success while minimizing friendly casualties. Quick, violent, and simultaneous action throughout the depth of the defender's area of operations can hurt, confuse, and even paralyze an enemy force just as that enemy force is most exposed and vulnerable. Synchronization of decisive, shaping, and sustaining operations facilitates mission success.

F. Preparation
The defense has inherent strengths. The defender arrives in the area of operations before the attacker and uses the available time to prepare. Defenders study the ground and select positions that allow the massing of fires on likely approaches. They combine natural and manmade obstacles to canalize attacking forces into engagement areas. Defending forces coordinate and rehearse actions on the ground, gaining intimate familiarity with the terrain. They place security, intelligence, and reconnaissance forces throughout the area of operations. These preparations multiply the effectiveness of the defense. Commanders continue defensive preparations in depth, even as the close engagement begins.

G. Security
Commanders secure their forces principally through protection, military deception, inform and influence activities, and cyber electromagnetic activities. Security operations prevent enemy intelligence, surveillance, and reconnaissance assets from determining friendly locations, strengths, and weaknesses. Protection efforts preserve combat power. Military deception and cyber electromagnetic activities inaccurately portray friendly forces, mislead enemy commanders, and deny those same enemy commanders the ability to use cyberspace and the electromagnetic spectrum.

IV. Common Defensive Control Measures
Ref: ADRP 3-90, Offense and Defense (Aug '12), pp. 3-4 to 3-7.

The commander controls the defense by using control measures to provide the flexibility needed to respond to changes in the situation and allow the defending commander to rapidly concentrate combat power at the decisive point.

A. Battle Positions

A battle position is a defensive location oriented on a likely enemy avenue of approach. The battle position is an intent graphic that depicts the location and general orientation of the majority of the defending forces. A commander's use of a battle position does not direct the subordinate to position the subordinate's entire force within its bounds since it is not an area of operations. Units as large as battalion task forces and as small as squads or sections use battle positions. They may occupy the topographical crest of a hill, a forward slope, a reverse slope, or a combination of these areas. The commander selects positions based on terrain, enemy capabilities, and friendly capabilities. A commander can assign some or all subordinates battle positions within the area of operations. Multiple battle positions may be assigned to a single unit, which allows that unit to maneuver between battle positions. The commander specifies mission and engagement criteria to the unit assigned to a battle position. Security, support, and sustainment forces typically operate outside a unit's battle position.

There are five kinds of battle positions—primary, alternate, supplementary, subsequent, and strong point.

- The **primary position** is the position that covers the enemy's most likely avenue of approach into the area of operations. It is the best position from which to accomplish the assigned mission, such as cover an engagement area to prevent enemy penetration.

- An **alternate position** is a defensive position that the commander assigns to a unit or weapon system for occupation when the primary position becomes untenable or unsuitable for carrying out the assigned task. It covers the same area as the primary position. The commander locates alternate positions so the occupant can continue to fulfill the original task, such as covering the same avenue of approach or engagement area as the primary position. These positions increase the defender's survivability by allowing the defender to engage the enemy from multiple positions. For example, a unit moves to its alternate positions when the enemy brings suppressive fires on the primary position.

- A **supplementary position** is a defensive position located within a unit's assigned area of operations that provides the best sectors of fire and defensive terrain along an avenue of approach that is not the primary avenue where the enemy is expected to attack. For example, an avenue of approach into a unit's area of operations from one of its flanks normally requires establishing supplementary positions to allow a unit or weapon system to engage enemy forces traveling along that avenue.

- A **subsequent position** is a position that a unit expects to move to during the course of battle. A defending unit may have a series of subsequent positions. Subsequent positions can also have primary, alternate, and supplementary positions associated with them.

- A **strong point** is a heavily fortified battle position tied to a natural or reinforcing obstacle to create an anchor for the defense or to deny the enemy decisive or key terrain. The commander prepares a strong point for all-around defense. The commander positions strong points on key or decisive terrain. The unit occupying the strong point prepares positions for its weapon systems, vehicles, Soldiers, and supplies. The commander also establishes a strong point when anticipating that enemy actions will isolate a defending force retaining terrain critical to the defense.

B. Direct Control Measures

The commander engages the enemy force with all available fires when it enters the defending unit's engagement area. ADRP 1-02 defines direct fire control measures, such as target reference points, trigger lines, and engagement areas.

C. Disengagement Line

A disengagement line is a phase line located on identifiable terrain that, when crossed by the enemy, signals to defending elements that it is time to displace to their next position. The commander uses these lines in the delay and the defense when the commander does not want the defending unit to become decisively engaged. The commander establishes criteria for the disengagement, such as number of enemy vehicles by type, friendly losses, or enemy movement to flanking locations. Commanders may designate multiple disengagement lines, one for each system in the defense.

D. Fire Support Coordination Measures

The commander tries to engage the enemy at extended ranges and attrit the enemy force as the enemy's attack advances. To control indirect fires, the commander uses common FSCMs described in ADRP 1-02. The commander can also employ final protective fires. Final protective fire is an immediately available preplanned barrier of fires designed to impede enemy movement across defensive lines or areas (JP 1-02). Both direct- and indirect- fire weapons can provide final protective fires (FPFs). The commander can only assign each firing battery or platoon a single FPF. A FPF is a priority target for an element or system, and those fire units are laid on that target when they are not engaged in other fire missions. When the enemy force initiates its final assault into a defensive position, the defending unit initiates its FPFs to kill enemy infantry soldiers and suppress enemy armored vehicles.

E. Forward Edge of the Battle Area

The forward edge of the battle area is the foremost limit of a series of areas in which ground combat units are deployed, excluding the areas in which the covering or screening forces are operating, designated to coordinate fire support, the positioning of forces, or the maneuver of units (JP 3-09.3). The Army uses a forward edge of the battle area (FEBA) only during the defense. The FEBA is not a boundary, but it conveys the commander's intent. It marks the foremost limits of the areas in which the preponderance of ground combat units deploy, excluding the areas in which security forces are operating. MBA forces can temporarily move forward of the FEBA to expedite the retrograde operations of security forces. The commander designates a FEBA to coordinate fire support and to assist in the maneuver of subordinate forces. A phase line designating the forward-most point of the MBA indicates the FEBA. The FEBA shows the senior commander's planned limit for the effects of direct fires.

F. Main Battle Area

The main battle area is the area where the commander intends to deploy the bulk of the unit's combat power and conduct decisive operations to defeat an attacking enemy. The defending commander's major advantage is the ability to select the ground on which the battle takes place. The defender positions subordinate forces in mutually supporting positions in depth to absorb enemy penetrations or canalize them into prepared engagement areas, defeating the enemy's attack by concentrating the effects of overwhelming combat power. The natural defensive strength of the position determines the distribution of forces in relation to both frontage and depth. In addition, defending units typically employ field fortifications and obstacles to improve the terrain's natural defensive strength. The MBA includes the area where the defending force creates an opportunity to deliver a counterattack to defeat or destroy the enemy.

The MBA extends from the FEBA to the unit's rear boundary. The commander locates subordinate unit boundaries along identifiable terrain features and extends them out beyond the forward line of own troops (FLOT) by establishing forward boundaries. Unit boundaries should not split avenues of approach or key terrain.

V. Transition

Ref: ADRP 3-90, Offense and Defense (Aug '12), pp. 4-4 to 4-6.

If a defense is successful, the commander anticipates and attempts to transition to the offense. If the defense is unsuccessful, the commander transitions from a defensive posture into retrograde operations. Transition from one type of operation or task to another requires mental as well as physical agility on the part of involved commanders, staffs, and units as well as an accurate understanding of the situation.

The commander deliberately plans for sequential operations, assisting the transition process and allowing the setting of the conditions necessary for a successful transition.

A. Transition to Offense

A defending commander transitions to a focus on the offensive element of operations by anticipating when and where the enemy force will reach its culminating point or require an operational pause before it can continue. At those moments, the combat power ratios most favor the defending force. The enemy force will do everything it can to keep the friendly force from knowing when the enemy force is becoming overextended. Indicators that the enemy is becoming overextended include when—

- Enemy forces begin to transition to the defense—this defense may be by forces in or out of contact with friendly forces
- Enemy forces suffer heavy losses
- Enemy forces start to deploy before encountering friendly forces
- Enemy forces are defeated in most engagements
- Enemy forces are committed piecemeal in continued attacks
- Enemy reserve forces are identified among attacking forces
- Examination of captured or killed enemy soldiers and captured or destroyed enemy equipment and supplies shows that the enemy is unable to adequately sustain itself
- A noticeable reduction in the tempo of enemy operations
- Local counterattacks meet with unexpected success

B. Transition to the Retrograde

A retrograde usually involves a combination of delay, withdrawal, and retirement operations. These operations may occur simultaneously or sequentially. As in other operations, the commander's concept of operations and intent drive planning for the retrograde. Each form of retrograde has its unique planning considerations, but considerations common to all retrogrades are risk, the need for synchronization, and rear security. However, the following key considerations receive special emphasis during the transition from the defense to the retrograde.

The transition to the retrograde must be accompanied by efforts designed to—

- Reduce the enemy's strength and combat power
- Provide friendly reinforcements
- Concentrate forces elsewhere for the attack
- Prepare stronger defenses elsewhere within the area of operations
- Lure or force part or all of the enemy force into areas where it can be counterattacked

C. Transition to Stability Tasks

The transition to stability tasks is conditional, but it should be planned for in advance. A defending commander may transition to emphasize stability tasks, if the defense retained decisive terrain, denied vital areas to the enemy, and so successfully attrited the attacking enemy that offensive actions are superfluous. As in other operations, the commander's concept of operations and intent drive the design of and planning for stability tasks.

Chap 3

The Defense
I. Mobile Defense

Ref: FM 3-90 Tactics, chap 10; and FM 3-21.10 (FM 7-10) The Infantry Rifle Company, chap 5.

The mobile defense is a type of defensive operation that concentrates on the destruction or defeat of the enemy through a decisive attack by a striking force. It focuses on destroying the attacking force by permitting the enemy to advance into a position that exposes him to counterattack and envelopment. The commander holds the majority of his available combat power in a striking force for his decisive operation, a major counterattack. He commits the minimum possible combat power to his fixing force that conducts shaping operations to control the depth and breadth of the enemy's advance. The fixing force also retains the terrain required to conduct the striking force's decisive counterattack. The area defense, on the other hand, focuses on retaining terrain by absorbing the enemy into an interlocked series of positions, where he can be destroyed largely by fires.

The mobile defense focuses on defeating or destroying the enemy by allowing him to advance to a point where he is exposed to a decisive counterattack by the striking force. The decisive operation is a counterattack conducted by the striking force. (Photo by Jeong, Hae-jung).

Small units do not normally conduct a mobile defense because of their inability to fight multiple engagements throughout the width, depth, and height of the AO, while simultaneously resourcing striking, fixing, and reserve forces. Typically, the striking force in a mobile defense may consist of one-half to two-thirds of the defender's combat power. Smaller units generally conduct an area defense or a delay as part of the fixing force as the commander shapes the enemy's penetration or they attack as part of the striking force. Alternatively, they can constitute a portion of the reserve.

(The Defense) I. Mobile Defense 3-5

The factors of METT-TC may dictate that a unit conducts a mobile defense when defending against an enemy force with greater combat power but less mobility. A commander may also employ a mobile defense when defending a large area of operations (AO) without well-defined avenues of approach, such as flat, open terrain. The mobile defense is preferred in an environment where the enemy may employ weapons of mass destruction because this type of defense reduces the vulnerability of the force to attack and preserves its freedom of action.

I. Organization

The commander organizes his main body into two principal groups—the fixing force and the striking force. In the mobile defense, reconnaissance and security, reserve, and sustaining forces accomplish the same tasks as in an area defense. The commander completes any required adjustments in task organization before he commits his units to the fight.

A. The Fixing Force

Organized by the commander with the minimum combat power needed to accomplish its mission, the fixing force turns, blocks, and delays the attacking enemy force. It tries to shape the enemy penetration or contain his advance. Typically, it has most of the countermobility assets of the defending unit. The fixing force may conduct defensive actions over considerable depth within the main battle area (MBA). However, it must be prepared to stop and hold terrain on short notice to assist the striking force on its commitment. The operations of the fixing force establish the conditions for a decisive attack by the striking force at a favorable tactical location. The fixing force executes its portion of the battle essentially as a combination of an area defense and a delaying action. The actions of the fixing force are shaping operations.

B. The Striking Force

The striking force decisively engages the enemy as he becomes exposed in his attempts to overcome the fixing force. The term "striking force" is used rather than reserve because the term "reserve" indicates an uncommitted force. The striking force is a committed force and has the resources to conduct a decisive counterattack as part of the mobile defense. It is the commander's decisive operation.

The striking force contains the maximum combat power available to the commander at the time of its counterattack. The striking force is a combined arms force that has greater combat power and mobility than the force it seeks to defeat or destroy. The commander considers the effects of surprise when determining the relative combat power of the striking force and its targeted enemy unit. The striking force is normally fully task organized with all combat support (CS) and combat service support (CSS) assets before its actual commitment. The commander positions engineer mobility-enhancing assets with the lead elements of the striking force.

The striking force is the key to a successful mobile defense. All of its contingencies relate to its attack. If the opportunity does not exist to decisively commit the striking force, the defender repositions his forces to establish the conditions for success. The striking force must have mobility equal to or greater than that of its targeted enemy unit. It can obtain this mobility through proper task organization, countermobility operations to slow and disrupt enemy movements, and mobility operations to facilitate the rapid shifting of friendly formations. The striking force requires access to multiple routes because an attacking enemy normally goes to great length to deny the defending force freedom of action.

3-6 (The Defense) I. Mobile Defense

The mobile defense is dynamic in its approach. Standoff weaponry such as mines, IED, mortars, and rockets are used to disrupt enemy activity. (Dept. of Army photo by Yu, Hu Son).

II. Planning & Preparation

The key to successful mobile defensive operations is the integration and synchronization of all available assets to maximize the combat power of the defending unit, particularly the striking force. The commander achieves integration and synchronization when he can employ their combined effects at decisive times and places.

Preparations for conducting a mobile defense include developing the fixing force's defensive positions and Engagement Areas (EAs). The commander aggressively uses his reconnaissance assets to track enemy units as they approach. Engineers participate in conducting route and area reconnaissance to find and classify existing routes. They improve existing routes and open new routes for use during the battle

III. Conducting the Mobile Defense - A Small Unit Perspective

FM 3-90 Tactics divides the execution of a mobile defense into five phases for discussion purposes. The length and nature of each phase, if it occurs at all, varies from situation to situation according to the factors of METT-TC. The phases of defensive operations are gain and maintain enemy contact, disrupt the enemy, fix the enemy, maneuver, and follow through.

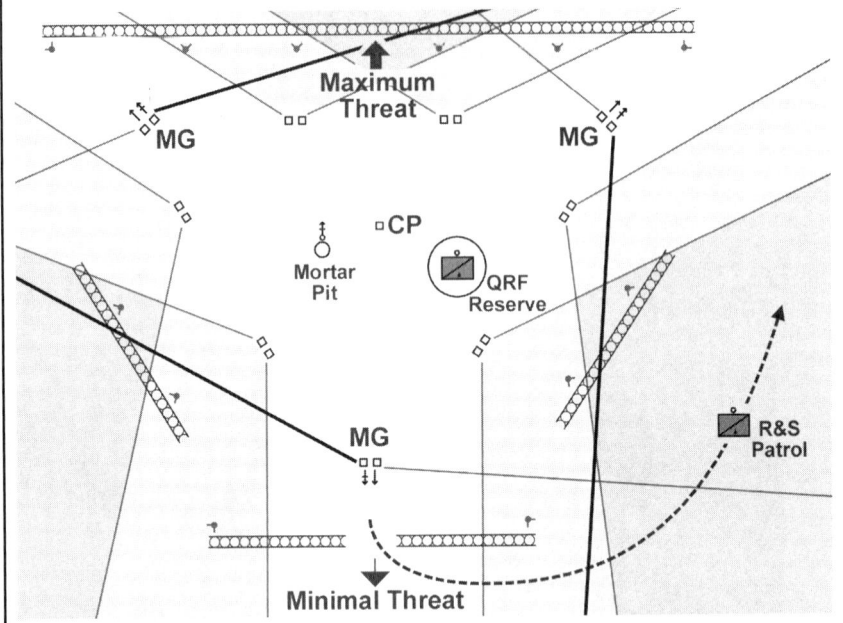

The mobile defense makes use of multiple defensive perimeters that are networked together for mutual support. All the principles of the defense are employed—obstacles, coordinated fires, R&S patrols, and a reserve force. (Ref: FM 7-10, chap 5, fig. 5-19).

1. Gain and Maintain Enemy Contact

The commander conducting a mobile defense focuses on discovering the exact location of the enemy and his strength to facilitate the effectiveness of the striking force. The security force (guard or cover) or the fixing force confirms the enemy's COA and main avenues of approach. The commander normally tasks other ISR assets to determine the location of enemy reserves and follow-on forces. Early detection of the enemy's decisive operation provides the commander with reaction time to adjust the fixing force's positions and shape the enemy penetration, which, in turn, provides the time necessary to commit the striking force. The striking force commander requires as close to real-time updates of the enemy situation as are possible to ensure that the striking force engages the enemy at the right location and time.

2. Disrupt the Enemy

In a mobile defense, the commander conducts shaping operations designed to shape the enemy's penetration into the MBA and disrupt the enemy's introduction of fresh forces into the fight. These shaping operations help establish the preconditions for committing the striking force by isolating the object of the striking force and destroying the enemy's key C2 nodes, logistics resupply units, and reserves. Whenever possible the commander sequences these shaping operations, to include offensive information operations, so that the impact of their effects coincides with the commitment of the striking force. To generate a tempo that temporarily paralyzes enemy C2, the intensity of these shaping operations may increase dramatically on the commitment of the striking force. The commander continues to conduct shaping operations once the striking force commits to prevent enemy forces from outside the objective area from interfering with executing the decisive counterattack.

3. Fix the Enemy

Fixing the enemy is the second half of shaping operations and results in establishing the conditions necessary for decisive operations by the striking force. Typically, the commander of the defending force allows the enemy force to penetrate into the defensive AO before the striking force attacks. The fixing force may employ a combination of area defense, delay, and strong point defensive techniques to shape the enemy penetration. The intent of the fixing force is not necessarily to defeat the enemy but to shape the penetration to facilitate a decisive counterattack by the striking force. The commander ensures that the missions and task organization of subordinate units within the fixing force are consistent with his concept for shaping the enemy penetration. Defensive positions within the fixing force may not be contiguous since the fixing force contains only the minimum-essential combat power to accomplish its mission.

4. Maneuver

The commander's situational understanding is critical in establishing the conditions that initiate the striking force's movement and in determining the general area that serves as a focus for the counterattack. Situational understanding includes identifying those points in time and space where the counterattack proves decisive. A force-oriented objective or an EA usually indicates the decisive point. The staff synchronizes the unit's activities in time and space to sufficiently mass the effects of the striking force at the right time and place.

The actions of the striking force are the echelon's decisive operation on its commitment. The commander's ISR systems focus entirely on tracking the enemy's advance. The striking force commander continuously receives intelligence and combat information updates that allow him to adjust his counterattack as necessary to defeat the targeted enemy. Once the enemy starts his attack, any forward-deployed elements of the striking force withdraw to AAs or attack positions and prepare for their commitment in counterattack.

The enemy attempts to discover the strength, composition, and location of the units that constitute the fixing force and the striking force. The commander uses security forces and information operations to deny the enemy this information and degrade the collection capabilities of enemy ISR assets. The commander routinely repositions to mislead the enemy and to protect his force. In addition, his plans and preparations incorporate defensive information operations. The commander normally tries to portray an area defense while hiding the existence and location of the striking force.

5. Follow Through

All defensive operations intend to create the opportunity to transition to the offense. In a mobile defense, that transitional opportunity generally results from the success of the striking force's attack. The commander exploits his success and attempts to establish conditions for a pursuit if his assessment of the striking force's attack is that there are opportunities for future offensive operations.

The striking force assembles in one or more areas depending on the width of the AO, the terrain, enemy capabilities, and the planned manner of employment. Before the enemy attack begins, the striking force may deploy all or some of its elements forward in the MBA to—
- Deceive the enemy regarding the purpose of the force
- Occupy dummy battle positions.
- Create a false impression of unit boundaries, which is important when operating with a mix of heavy and light forces or multinational forces.
- Conduct reconnaissance of routes between the striking force's AAs and potential EAs

On Point

At the small unit level, the mobile defensive perimeter is established much in the same way as a patrol base—using the triangle method. The difference between a mobile defense and a patrol base is that the mobile defensive perimeter will be used for a long period of time. While it will likely remain intentionally hidden from enemy detection, the development of advanced fighting positions and obstacles resembles more an area defense than the temporary a patrol base. However, unlike the area defense, the mobile defensive perimeter is formed in 360-degrees to fight from all angles.

1. **Obstacles**—preferably wire obstacles—are placed completely around the perimeter, in front of each of the lines of the triangle. Anti-personnel mines are carefully emplaced and marked on each squad's sector sketch.

2. **Communication**—preferably field phones—are established at each of the apexes and back to the CP in the center. The wire is appropriately hung and camouflaged in the trees or buried underground. Additionally, communication is established with adjacent defensive perimeters. These measures are essential in requesting or rendering aid in the event of an enemy offensive.

3. **Crew served weapons** such as mortars or recoilless rifles are located near the CP, while machine gun crews are located at each of the three apexes. Here the reserve force rests, eats and carries out other duties.

4. **OP/LP or R&S patrols** are sent out in irregular patterns to obtain intelligence on enemy movement in the immediate area. Additionally, these patrols are intended to thwart the enemy's effort to gain information regarding the size, activity, and location of the defensive perimeter.

5. Coordinated **secondary sectors of fire** are designated so that each of the fighting positions can also *fire in toward the center* of the defensive perimeter! This drastic measure will only take place if one of the defensive lines of the triangle is breeched and the CP is abandoned. A designated signal indicates that the CP has withdrawn, issuing the order to "commence fire" for the other two lengths of the defensive perimeter. The advantage of such a drastic measure is that it adds depth to the defense because the remaining two lines can fire upon the enemy attack force as they attempt to clear and reorganize on their objective. This measure allows the other two lines of the defensive triangle to take the enemy under fire from two directions.

6. In the event one line of the triangle is overrun, the **reserve force** also withdraws with the command team. This frees the reserve force to conduct counterattacks against the enemy's flank or rear—from directions the enemy is ill prepared to fight.

Chap 3

The Defense

II. Area Defense

Ref: FM 3-90 Tactics, chap 8 and 9; FM 3-21.10 (FM 7-10) The Infantry Rifle Company, chap 5; and FM 3-21.8 (FM 7-8) Infantry Rifle Platoon and Squad, chap 8.

The area defense is a type of defensive operation that concentrates on denying enemy forces access to designated terrain for a specific time rather than destroying the enemy outright. An area defense capitalizes on the strength inherent in closely integrated defensive organization on the ground.

A commander should conduct an area defense when the following conditions occur:

- When directed to defend or retain specified terrain
- When he cannot resource a striking force
- The forces available have less mobility than the enemy
- The terrain affords natural lines of resistance and limits the enemy to a few well-defined avenues of approach, thereby restricting the enemy's maneuver
- There is enough time to organize the position
- Terrain constraints and lack of friendly air superiority limit the striking force's options in a mobile defense to a few probable employment options

The area defense retains dominance over a given geographical location. It does this by employing a fortified defensive line, a screening force, and a reserve force. Particular care is taken in coordinating and synchronizing fire control measures in order to repel any enemy attack.

The reserve force is perhaps the single most decisive element of the defense. The reserve may reinforce a failing line, occupy a secondary defensive position, or counterattack. (Photo by Jeong, Hae-jung).

(The Defense) II. Area Defense 3-11

I. Area Defense - Organization

The platoon will normally defend in accordance with command orders using one of these basic techniques *(see pp. 3-23 to 3-28)*:
- Defend an area
- Defend a battle position
- Defend a strongpoint
- Defend a perimeter
- Defend a reverse slope

The commander conducting an area defense combines static and mobile actions to accomplish his assigned mission. Static actions usually consist of fires from prepared positions. Mobile actions include using the fires provided by units in prepared positions as a base for counterattacks and repositioning units between defensive positions. The commander can use his reserve and uncommitted forces to conduct counterattacks and spoiling attacks to desynchronize the enemy or prevent him from massing.

A well-conducted area defense is anything but static. It's actually quite active in that it continues to advance its own fighting position while it patrols forward to gather intelligence on the enemy. The area defense has great depth and will track and channel the enemy from considerable distances beyond the defense.

Primary Positions
In addition to establishing the platoon's primary positions, the platoon leader and subordinate leaders normally plan for preparation and occupation of alternate, supplementary, and subsequent positions. This is done IAW the company order. The platoon and/or company reserve need to know the location of these positions. The following are tactical considerations for these positions.

Alternate Positions
The following characteristics and considerations apply to an alternate position:
- Covers the same avenue of approach or sector of fire as the primary position
- Located slightly to the front, flank, or rear of the primary position
- Positioned forward of the primary defensive positions during limited visibility operations
- Normally employed to supplement or support positions with weapons of limited range, such as Infantry squad positions. They are also used as an alternate position to fall back to if the original position is rendered ineffective or as a position for Soldiers to rest or perform maintenance

Supplementary Positions
The following characteristics and considerations apply to a supplementary position:
- Covers an avenue of approach or sector of fire different from those covered by the primary position
- Occupied based on specific enemy actions

Subsequent Positions
The following characteristics and considerations apply to a subsequent position:
- Covers the same avenue of approach and or sector of fire as the primary position
- Located in depth through the defensive area
- Occupied based on specific enemy actions or conducted as part of the higher headquarters' scheme of maneuver

As part of a larger element, the platoon conducts defensive operations in a sequence of integrated and overlapping phases:
- Reconnaissance, security operations, and enemy preparatory fires
- Occupation
- Approach of the enemy main attack
- Enemy assault
- Counterattack
- Consolidation and reorganization

The defense makes use of obstacles to expose and/or slow the enemy advance in our prepared engagement areas. Troops use of wire obstacles and mines in defilades that cannot be covered by fire. (Photo by Jeong, Hae-jung).

At the small unit level, the area defense typically employs three teams and rotates those teams through the three tasked responsibilities—manning the line, manning the reserve, and patrolling forward. Each team takes its turn in conducting the three tasked responsibilities.

The line includes 2-man fighting positions and crew served weapon positions. It also includes the observation post/listening post (OP/LP) that is positioned just forward of our defensive line. Finally, while the command post (CP) is located behind our line near the reserve force, it is also part of the line force. The command team, however, does not rotate to the other assigned tasks, but rotates a rest plan amongst the command team.

This organization permits:
- About 50 percent of the force on the defensive line at all times
- A reserve force of 25 percent to add depth to our position
- A patrolling force of 25 percent forward to monitor enemy activity

The reserve and screening forces make up half of the total force for the area defense. The line force utilizes the other half of our troops. The troops are typically rotated in 2-hour intervals. That's two hours on the line, two hours patrolling, two hours on the line again, and two hours in reserve (where troops can implement a sleep plan).

(The Defense) II. Area Defense 3-13

II. Planning & Preparation

Platoons establish defensive positions IAW the platoon leader and commander's plan. They mark engagement areas using marking techniques prescribed by unit SOP. The platoon physically marks obstacles, TRPs, targets, and trigger lines in the engagement area. During limited visibility, the platoon can use infrared light sources to mark TRPs for the rifle squads. When possible, platoons should mark TRPs with both a thermal and an infrared source so the rifle squads can use the TRP.

Fire control measures are critical for the success of the area defense. These control measures help the defense to engage the attacking enemy at greater distances, synchronize the final defensive fires of the line, and minimize the possibility of fratricide. (Photo by Jeong, Hae-jung).

A. Range Card

A range card is a sketch of a sector that a direct fire weapons system is assigned to cover. Range cards aid in planning and controlling fires. They also assist crews in acquiring targets during limited visibility, and orient replacement personnel, platoons, or squads that are moving into position. During good visibility, the gunner should have no problems maintaining orientation in his sector. During poor visibility, he may not be able to detect lateral limits. If the gunner becomes disoriented and cannot find or locate reference points or sector limit markers, he can use the range card to locate the limits. The gunner should make the range card so he becomes more familiar with the terrain in his sector. He should continually assess the sector and, if necessary, update his range card.

B. Sector Sketch

The sector sketch illustrates how each fighting position is interlocked and how the left and right flanks of the defensive line are secured. The flanks are tied in with adjacent friendly units or with naturally occurring or man-made obstacles to stop the enemy approach. Additionally, the sector sketch identifies engagement areas, final protective fires, locations of our wire and landmine obstacles, the OP/LP, the CP, and assigned alternate fighting positions.

C. Sectors of Fire

Each fighting position must be placed so that it has an interlocking sector of fire. This means that every fighting position has supporting fire to its immediate front from the fighting positions to its left and right. This also ensures that no matter which section of the line the enemy attacks, the enemy will be engaged by a minimum of three fighting positions simultaneously.

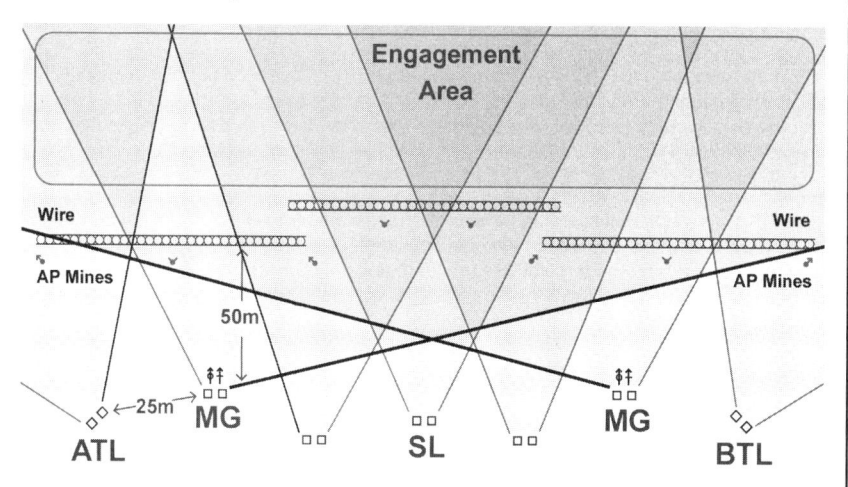

The effort is to use obstacles and terrain to slow or stop the enemy in the engagement area. Weapons are coordinated with interlocking fires to cover the obstacles and destroy the enemy. (Ref: FM 7-8, chap 2, section VI, fig. 2-42).

D. Engagement Areas

Engagement areas indicate the location the commander intends to defeat the enemy. The success of any engagement depends on how effectively the platoon leader can integrate the obstacle and indirect fire plans with his direct fire plan in the engagement area to achieve the platoon's purpose. At the platoon level, engagement area development remains a complex function that requires parallel planning and preparation if the platoon is to accomplish its assigned tasks. Despite this complexity, engagement area development resembles a drill. The platoon leader and his subordinate leaders use a standardized set of procedures. Beginning with an evaluation of the factors of METT-TC, the development process covers these steps:

- Identify likely enemy avenues of approach
- Identify the enemy scheme of maneuver
- Determine where to kill the enemy
- Plan and integrate obstacles
- Emplace weapons systems
- Plan and integrate indirect fires
- Conduct an engagement area rehearsal

III. Conducting the Area Defense - A Small Unit Perspective

The area defense forms the forward line of troops (FLOT). Typically, the FLOT defends against one direction in approximately a 120° frontal-fan that overlooks the designated engagement areas within the forward edge of battle area (FEBA). The defensive line interlocks with friendly units or impassable terrain to the left and right of the position.

1. Prior to moving up to the FEBA, the patrol leader (PL) moves the patrol into security halt at an appropriate distance. The PL conducts a leader's recon of the FEBA and selects the best location to establish a defensive line. The PL issues a five-point contingency plan with the assistant patrol leader (APL) before leaving on the recon.

2. The PL moves up to the FEBA with at least a security team. The security team is placed in over-watch at the designated release point. This allows the PL to move more freely about the terrain and determine how best to place the defensive line. The leader's recon may require moving the entire recon team across the FEBA to consider possible avenues for an enemy attack.

3. Once the PL confirms or adjusts the plan, the security team is given a five-point contingency plan and left at the release point to monitor the FEBA. The PL returns to the patrol to disseminate any changes to the original plan.

4. Upon returning, the rest of the patrol forms into the line team, the reserve force, and the screening force. The order of march depends on the OPORD, but typically the screening force is first in formation so that it may move forward for security as the line team establishes the FLOT. The PL leads the patrol and links up with the security team at the release point.

Mortars serve as combat force multipliers. They add significant punch against targets within the engagement area and break up the enemy's attack formations. Their placement and fires are carefully considered. (Dept. of Army photo by Gary L. Kieffer).

5. After linking up with the security team, the PL designates the location of fighting positions for crew-served weapons and indicates their primary direction of fire. The PL then designates a CP at an appropriate distance behind the line, typically towards the center of the entire line. The PL designates an OP/LP location just forward of any wire obstacles. The PL designates a 'fall back' position—an alternate position that can still cover the same engagement area within the FEBA.

6. The PL coordinates with units to his left and right to be certain he has adequately linked in with friendly defenses. Meanwhile the team leaders disseminate to every member of the patrol the location of the CP, the OP/LP, and the alternate position.

7. At this point, the line team leader assigns each member a fighting position and a sector of fire to fill in the lazy "W". He ensures that all positions interlock with at least the position to their left and right. Fighting positions are 2-man positions and are placed only close enough that they mutually support each other's immediate front. This distance varies according to different types of terrain. But at a minimum, the positions must be able to cover forward obstacles with fire. Landmine and concertina wire obstacles are placed out of hand grenade range (40 meters or more) and are in full view of friendly troops.

8. The reserve force locates behind the CP, out of sight from the line and out of the way of assigned indirect fire crews. The reserve force familiarizes with the defensive line and the alternative position.

Communication devices are established at key locations—the CP, flanks, and OP/LP. Field telephones are faster than runners, and more secure than radios. (Photo by Jeong, Hae-jung).

9. Communication is established between key positions. This typically involves the use of field phones and landlines wire. Field phones are more secure than radios and are not concerned with high traffic and frequency availability. Most field phones transmit a distance of several kilometers—ample distance for a defensive position.

10. At this point in the defense, the PL assumes the role of the forward unit commander. He collects the sector sketches from the team leaders to create his own master sketch and reports progress to higher command.

(The Defense) II. Area Defense 3-17

IV. Priorities of Work in the Defense

Occupation and preparation of the defense site is conducted concurrently with the TLP and the development of the engagement area (if required). The platoon occupies defensive positions IAW the company commander's plan and the results of the platoon's reconnaissance. To ensure an effective and efficient occupation, the reconnaissance element marks the friendly positions. These tentative positions are then entered on the operational graphics. Each squad moves in or is led in by a guide to its marker. Once in position, each squad leader checks his position location. As the platoon occupies its positions, the platoon leader manages the positioning of each squad to ensure they locate IAW the tentative plan. If the platoon leader notes discrepancies between actual positioning of the squads and his plan, he makes the corrections. Security is placed out in front of the platoon. The platoon leader must personally walk the fighting positions to ensure that everyone understands the plan and that the following are IAW the plan:

- Weapons orientation and general sectors of fire
- Crew served weapons positions
- Rifle squads' positions in relation to each other

Unit priorities of work are normally found in SOPs. However, the commander will dictate the priorities of work for the company based on the factors of METT-TC. Several actions may be accomplished at the same time. Leaders must constantly supervise the preparation of fighting positions, both for tactical usefulness and proper construction.

Leaders must ensure that Soldiers prepare for the defense quickly and efficiently. Work must be done in order of priority to accomplish the most in the least amount of time while maintaining security and the ability to respond to enemy action. Below are basic considerations for priorities of work.

- Emplace local security (all leaders)
- Position and assign sectors of fire for each squad (platoon leader)
- Position and assign sectors of fire for the CCMS and medium machine gun teams (platoon leader)
- Position and assign sectors of fire for M249 MG, grenadiers, and riflemen (squad leaders)
- Establish command post and wire communications
- Designate FPLs and FPFs
- Clear fields of fire and prepare range cards
- Prepare sector sketches (leaders)
- Dig fighting positions
- Establish communication and coordination with the company and adjacent units
- Coordinate with adjacent units
- Review sector sketches
- Emplace antitank and Claymore mines, then wire and other obstacles
- Mark or improve marking for TRPs and other fire control measures
- Improve primary fighting positions and add overhead cover
- Prepare supplementary and then alternate positions (same procedure as the primary position)
- Establish sleep and rest plans
- Distribute and stockpile ammunition, food, and water
- Dig trenches to connect positions
- Continue to improve positions—construct revetments, replace camouflage, and add to overhead cover

The Defense
III. Retrograde

Ref: FM 3-90 Tactics, chap 11 and FM 3-21.10 (FM 7-10) The Infantry Rifle Company, pp. 5-43 to 5-50.

The retrograde is a type of defensive operation that involves organized movement away from the enemy. The enemy may force these operations or a commander may execute them voluntarily. In either case, the higher commander of the force executing the operation must approve the retrograde. Retrograde operations are transitional operations; they are not considered in isolation. The commander executes retrogrades to—

- Disengage from operations
- Gain time without fighting a decisive engagement
- Resist, exhaust, and damage an enemy in situations that do not favor a defense
- Draw the enemy into an unfavorable situation or extend his lines of communication (LOCs)
- Preserve the force or avoid combat under undesirable conditions, such as continuing an operation that no longer promises success
- Reposition forces to more favorable locations or conform to movements of other friendly troops
- Position the force for use elsewhere in other missions
- Simplify the logistic sustainment of the force by shortening LOCs
- Position the force where it can safely conduct reconstitution
- Adjust the defensive scheme, such as secure more favorable terrain

The retrograde is a maneuver that places time and space between friendly forces and an attacking enemy in order to reconsolidate and coordinate an effective defense or counterattack. (Photo by Jeong, Hae-jung).

Conducting the Retrograde - A Small Unit Perspective

There are three forms of retrograde—delay, withdraw, and retirement. In all cases of retrograde, combat leaders must be very careful to maintain unit integrity and discipline.

I. Delay

This operation allows the unit to trade space for time, avoiding decisive engagement and safeguarding its elements. A delay is a series of defensive and offensive actions over subsequent positions in depth. It is an economy of force operation that trades space for time. While the enemy gains access to the vacated area (space), friendly elements have time to conduct necessary operations, while retaining freedom of action and maneuver. This allows friendly forces to influence the action; they can prevent decisive engagement or postpone action to occur at a more critical time or place on the battlefield. For either type of delay mission, the flow of the operation can be summarized as "hit hard, then move." A successful delay has three key components.

- The ability to stop or slow the enemy's momentum while avoiding decisive engagement
- The ability to degrade the enemy's combat power
- The ability to maintain a mobility advantage

There are two types of delaying operations at the small unit level:

A. Delay Within a Sector

The company might be assigned a mission to delay within a sector AO. The higher commander normally provides guidance regarding intent and desired effect on the enemy, but he minimizes restrictions regarding terrain, time, and coordination with adjacent forces. This form of a delay is normally assigned when force preservation is the highest priority and there is considerable depth to the AO.

B. Delay Forward of a Specified Line for a Specified Time

The company might be assigned a mission to delay forward of a specific control measure for a specific period. This mission is assigned when the battalion must control the enemy's attack and retain specified terrain to achieve some purpose relative to another element, such as setting the conditions for a counterattack, for completion of defensive preparations, or for the movement of other forces or civilians. The focus of this delay mission is clearly on time, terrain, and enemy destruction. It carries a much higher risk for the battalion, with the likelihood of all or part of the unit becoming decisively engaged. The timing of the operation is controlled graphically by a series of phase lines with associated dates and times to define the desired delay-until period.

Delay missions usually conclude in one of three ways: a defense, a withdrawal, or a counterattack. Planning options should address all three possibilities.

The delaying force always takes advantage of terrain where friendly forces can direct effective and high volume fires at the enemy as they advance. The more challenging the terrain is to maneuver across, the more likely that terrain offers appropriate avenues of escape with which the delaying force can displace. Open terrain also has advantages, namely that the enemy can be engaged at farther distances.

When displacing from one defensive position, the next delaying position must be able to take the enemy under fire. This action not only protects the forward-most friendly delaying force as it displaces, but also serves to distract the enemy force with violent fires. The enemy will soon find that their initial objective (the forward friendly delaying force) has withdrawn, and they must now deal with yet another threat. In this manner, the delaying force is split into multiple teams in order to complicate the situation for the attacking enemy.

II. Withdrawal

A withdrawal is a planned operation in which a force in contact disengages from an The commander uses this operation to break enemy contact, especially when he needs to free the unit for a new mission. Withdrawal is a planned operation in which a force in contact disengages from an enemy force. Withdrawals may or may not be conducted under enemy pressure. The two types of withdrawals are assisted and unassisted.

A. Assisted

The assisting force occupies positions to the rear of the withdrawing unit and prepares to accept control of the situation. It can also assist the withdrawing unit with route reconnaissance, route maintenance, fire support, and sustainment. Both forces closely coordinate the withdrawal. After coordination, the withdrawing unit delays to a battle handover line, conducts a passage of lines, and moves to its final destination.

B. Unassisted

The withdrawing unit establishes routes and develops plans for the withdrawal and then establishes a security force as the rear guard while the main body withdraws. Sustainment and CS elements normally withdraw first followed by combat forces. To deceive the enemy as to the friendly movement, battalion may establish a detachment left in contact (DLIC) if withdrawing under enemy pressure. As the unit withdraws, the DLIC disengages from the enemy and follows the main body to its final destination.

Phases

Withdrawals are accomplished in three overlapping phases, as follows.

- **Preparation.** The commander dispatches quartering parties, issues WARNOs, and initiates planning. Nonessential vehicles are moved to the rear.
- **Disengagement.** Designated elements begin movement to the rear. They break contact and conduct tactical movement to a designated assembly area or position.
- **Security.** In this phase, a security force protects and helps the other elements as they disengage or move to their new positions. This is done either by a DLIC, which the unit itself designates in an unassisted withdrawal, or by a security force provided by the higher headquarters in an assisted withdrawal. As necessary, the security force assumes responsibility for the sector, deceives the enemy, and protects the movement of disengaged elements by providing overwatch and suppressive fires. In an assisted withdrawal, the security phase ends when the security force has assumed responsibility for the fight and the withdrawing element has completed its movement. In an unassisted withdrawal, this phase ends when the DLIC completes its disengagement and movement to the rear.

III. Retirement

A retirement is conducted to move a force that is not in contact away from the enemy. Typically, the company conducts a retirement as part of a larger force while another unit's security force protects their movement. A retiring unit organizes for combat but does not anticipate interference by enemy ground forces. Triggers for a retirement may include the requirement to reposition forces for future operations or to accommodate other changes to the current CONOP. The retiring unit should move sustainment elements and supplies first, and then should move toward an assembly area that supports preparations for the next mission. Where speed and security are the most important considerations, units conduct retirements as tactical road marches.

The retiring unit generally moves toward an assembly area, which should support the preparations for the unit's next mission. When determining the routes the retiring force takes to the assembly area, the commander considers the unit's capability to support defensive actions if combat occurs during the retirement.

On Point

Retrogrades conducted under pressure of enemy contact are extremely dangerous combat operations. This is partly due to the fact that the enemy is almost always much larger than friendly forces, and almost always is on the offensive. But retrogrades are also dangerous because they often require friendly forces to break up into multiple elements, synchronize the fight, and then coordinate a link-up under austere conditions.

Retrograde operations are conducted to improve a tactical situation or to prevent a worse situation from developing. Companies normally conduct retrogrades as part of a larger force but may conduct independent retrogrades (withdrawal) as required such as on a raid. Retrograde operations accomplish the following:

- Resist, exhaust, and defeat enemy forces
- Draw the enemy into an unfavorable situation
- Avoid contact in undesirable conditions
- Gain time
- Disengage a force from battle for use elsewhere in other missions
- Reposition forces, shorten lines of communication, or conform to movements of other friendly units
- Secure more favorable terrain

The delay and withdraw retrograde operations require an enormous amount of coordinated fires. In turn this means that communication between maneuvering and displacing friendly elements becomes paramount. Without it, fratricide becomes certain.

Reconstitution

Combat leaders must recognize that troops do not like to retrograde. Retrogrades can negatively affect the participating soldiers' attitude more than any other type of operation because they may view the retrograde as a defeat. A commander must not allow retrograde operations to reduce or destroy unit morale. Leaders must maintain unit aggressiveness. By planning and efficiently executing the retrograde and ensuring that soldiers understand the purpose and duration of the operation, the commander can counter any negative effects of the operation on unit morale. After completing a retrograde operation, the commander may reconstitute the force.

Note: FM 4-100.9 establishes the basic principles of reconstitution.

IV. Small Unit Defensive Techniques

Ref: FM 3-21.8 (FM 7-8) The Infantry Rifle Platoon and Squad, pp. 8-23 to 8-33.

Though the outcome of decisive combat derives from offensive actions, leaders often find it is necessary, even advisable, to defend. The general task and purpose of all defensive operations is to defeat an enemy attack and gain the initiative for offensive operations. It is important to set conditions of the defense so friendly forces can destroy or fix the enemy while preparing to seize the initiative and return to the offense. The platoon may conduct the defense to gain time, retain key terrain, facilitate other operations, preoccupy the enemy in one area while friendly forces attack him in another, or erode enemy forces. A well coordinated defense can also set the conditions for follow-on forces and follow-on operations.

Platoon Defensive Techniques

A platoon will normally defend IAW using one of these basic techniques:

- **I** Defend an Area
- **II** Defend a Battle Position
- **III** Defend a Strongpoint
- **IV** Defend a Perimeter
- **V** Defend a Reverse Slope

Ref: FM 3-21.8 (FM 7-10), p. 8-23.

I. Defend An Area

Defending an area sector allows a unit to maintain flank contact and security while ensuring unity of effort in the scheme of maneuver. Areas afford depth in the platoon defense. They allow the platoon to achieve the platoon leader's desired end state while facilitating clearance of fires at the appropriate level of responsibility. The company commander normally orders a platoon to defend an area when flexibility is desired, when retention of specific terrain features is not necessary, or when the unit cannot concentrate fires because of any of the following factors:

- Extended frontages
- Intervening, or cross-compartmented, terrain features
- Multiple avenues of approach

The platoon is assigned an area defense mission to prevent a specific amount of enemy forces from penetrating the area of operations. To maintain the integrity of the area defense, the platoon must remain tied to adjacent units on the flanks. The platoon may be directed to conduct the defense in one of two ways.

He may specify a series of subsequent defensive positions within the area from where the platoon will defend to ensure that the fires of two platoons can be massed.

He may assign an area to the platoon. The platoon leader assumes responsibility for most tactical decisions and controlling maneuvers of his subordinate squads by assigning them a series of subsequent defensive positions. This is done IAW guidance from the company commander in the form of intent, specified tasks, and the concept of the operation. The company commander normally assigns an area to a platoon only when it is fighting in isolation.

II. Defend a Battle Position

The company commander assigns the defensive technique of defending a battle position to his platoons when he wants to mass the fires of two or more platoons in a company engagement area, or to position a platoon to execute a counterattack. A unit defends from a battle position to—

- Destroy an enemy force in the engagement area
- Block an enemy avenue of approach
- Control key or decisive terrain
- Fix the enemy force to allow another friendly unit to maneuver

The company commander designates engagement areas to allow each platoon to concentrate its fires or to place it in an advantageous position for the counterattack. Battle positions are developed in such a manner to provide the platoon the ability to place direct fire throughout the engagement area. The size of the platoon battle position can vary, but it should provide enough depth and maneuver space for subordinate squads to maneuver into alternate or supplementary positions and to counterattack. The battle position is a general position on the ground. The platoon leader places his squads on the most favorable terrain in the battle position based on the higher unit mission and commander's intent. The platoon then fights to retain the position unless ordered by the company commander to counterattack or displace. The following are basic methods of employing a platoon in a battle position:

1. Same battle position, same avenue of approach

Rifle squads are on the same battle position covering the same avenue of approach. The platoon can defend against mounted and dismounted attacks and move rapidly to another position.

All squads are in the same battle position when the terrain provides good observation, fields of fire, and cover and concealment.

Employing all the squads of the platoon on the same battle position covering the same avenue of approach is the most conservative use of the platoon. Its primary advantages are that it facilitates command and control functions because of the proximity of squad elements on the same approach and it provides increased security.

2. Same battle position, multiple avenues of approach
Rifle squads occupy the same battle position but cover multiple enemy avenues of approach.

3. Different battle positions, same avenue of approach
Rifle squads are on different battle positions covering the same avenue of approach. If positioned on separate battle positions, rifle squads must fight in relation to each other when covering the same avenues of approach. A weapons squad can provide supporting fires for the rifle squads from their primary, alternate, or supplementary positions. All squads are positioned to engage enemy forces on the same avenue of approach, but at different ranges.

4. Different battle positions, multiple avenues of approach
Squads may be employed on different battle positions and multiple avenues of approach to ensure that the squad battle positions cannot be fixed, isolated, or defeated by the enemy.

III. Defend a Strongpoint

Defending a strongpoint is not a common mission for an Infantry platoon. A strongpoint defense requires extensive engineer support (expertise, materials, and equipment), and takes a long time to complete. When the platoon is directed to defend a strongpoint, it must retain the position until ordered to withdraw. The success of the strong-point defense depends on how well the position is tied into the existing terrain. This defense is most effective when it is employed in terrain that provides cover and concealment to both the strongpoint and its supporting obstacles. Mountainous, forested, or urban terrain can be adapted easily to a strongpoint defense. Strongpoints placed in more open terrain require the use of reverse slopes or of extensive camouflage and deception efforts. This defensive mission may require the platoon to—

- Hold key or decisive terrain critical to the company or battalion scheme of maneuver
- Provide a pivot to maneuver friendly forces
- Block an avenue of approach
- Canalize the enemy into one or more engagement areas

A. Characteristics of the Strongpoint Defense

The prime characteristic of an effective strongpoint is that it cannot be easily overrun or bypassed. It must be positioned and constructed so the enemy knows he can reduce it only at the risk of heavy casualties and significant loss of materiel. He must be forced to employ massive artillery concentrations and dismounted Infantry assaults in his attack, so the strongpoint must be tied in with existing obstacles and positioned to afford 360-degree security in observation and fighting positions.

B. Techniques and Considerations

A variety of techniques and considerations are involved in establishing and executing the strongpoint defense, including considerations for displacement and withdrawal from the strongpoint.

The platoon leader begins by determining the projected size of the strongpoint. He does this through assessing the number of weapons systems and individual Soldiers available to conduct the assigned mission, and by assessing the terrain on which the platoon will fight. He must remember that although a strongpoint is usually tied into a company defense and flanked by other defensive positions, it must afford 360-degree observation and firing capability.

The platoon leader must ensure that the layout and organization of the strongpoint maximizes the capabilities of the platoon's personnel strength and weapons systems without sacrificing the security of the position. Platoon options range from positioning CCMS outside the strongpoint (with the rifle squads occupying fighting positions inside it), to placing all assets within the position. From the standpoint of planning and terrain management, placing everything in the strongpoint is the most difficult option and potentially the most dangerous because of the danger of enemy encirclement

Weapons Positions

In laying out the strongpoint, the platoon leader designates weapon positions that support the company defensive plan. Once these primary positions have been identified, he continues around the strongpoint, siting weapons on other possible enemy avenues of approach and engagement areas until he has the ability to orient effectively in any direction. The fighting positions facing the company engagement area may be along one line of defense or staggered in depth along multiple lines of defense (if the terrain supports positions in depth).

Reserve Force

The platoon's reserve may be comprised of a fire team, squad, or combination of the two. The platoon leader must know how to influence the strongpoint battle by employing his reserve. He has several employment options including reinforcing a portion of the defensive line or counterattacking along a portion of the perimeter against an identified enemy main effort.

Routes or Axes

The platoon leader should identify routes or axes that will allow the reserve to move to any area of the strongpoint. He should then designate positions the reserve can occupy once they arrive. These routes and positions should afford sufficient cover to allow the reserve to reach its destination without enemy interdiction. The platoon leader should give special consideration to developing a direct fire plan for each contingency involving the reserve. The key area of focus may be a plan for isolating an enemy penetration of the perimeter. Rehearsals cover actions the platoon takes if it has to fall back to a second defensive perimeter, including direct fire control measures necessary to accomplish the maneuver. FPF may be employed to assist in the displacement.

Obstacles

Engineers support strongpoint defense by reinforcing the existing obstacles. Priorities of work will vary depending on the factors of METT-TC, especially the enemy situation and time available. For example, the first 12 hours of the strongpoint construction effort may be critical for emplacing countermobility obstacles and survivability positions, and command and control bunkers. If the focus of engineer support is to make the terrain approaching the strongpoint impassable, the battalion engineer effort must be adjusted accordingly.

The battalion obstacle plan provides the foundation for the company strongpoint obstacle plan. The commander or platoon leader determines how he can integrate protective obstacles (designed to defeat dismounted enemy Infantry assaults) into the overall countermobility plan. If adequate time and resources are available, he should plan to reinforce existing obstacles using field-expedient demolitions.

Once the enemy has identified the strongpoint, he will mass all the fires he can spare against the position. To safeguard his rifle squads, the platoon leader must arrange for construction of overhead cover for individual fighting positions. If the strongpoint is in a more open position (such as on a reverse slope), he may also plan for interconnecting trenchlines. This will allow Soldiers to move between positions without exposure to direct and indirect fires. If time permits, these crawl trenches can be improved to fighting trenches or standard trenches.

IV. Defend a Perimeter

A perimeter defense allows the defending force to orient in all directions. In terms of weapons emplacement, direct and indirect fire integration, and reserve employment, a platoon leader conducting a perimeter defense should consider the same factors as for a strongpoint operation.

The perimeter defense allows only limited maneuver and limited depth. Therefore, the platoon may be called on to execute a perimeter defense under the following conditions:

- Holding critical terrain in areas where the defense is not tied in with adjacent units
- Defending in place when it has been bypassed and isolated by the enemy
- Conducting occupation of an independent assembly area or reserve position
- Preparing a strongpoint
- Concentrating fires in two or more adjacent avenues of approach
- Defending fire support or engineer assets
- Occupying a patrol base

The major advantage of the perimeter defense is the platoon's ability to defend against an enemy avenue of approach. A perimeter defense differs from other defenses in that—

- The trace of the platoon is circular or triangular rather than linear
- Unoccupied areas between squads are smaller
- Flanks of squads are bent back to conform to the plan
- The bulk of combat power is on the perimeter
- The reserve is centrally located

V. Defend a Reverse Slope

The platoon leader's analysis of the factors of METT-TC often leads him to employ his forces on the reverse slope. If the rifle squads are on a mounted avenue of approach, they must be concealed from enemy direct fire systems. This means rifle squads should be protected from enemy tanks and observed artillery fire.

The majority of a rifle squad's weapons are not effective beyond 600 meters. To reduce or prevent destruction from enemy direct and indirect fires beyond that range, a reverse-slope defense should be considered. Using this defense conflicts to some extent with the need for maximum observation forward to adjust fire on the enemy, and the need for long-range fields of fire for CCMS. In some cases it may be necessary for these weapons systems to be deployed forward while the rifle squads remain on the reverse slope. CCMS gunners withdraw from their forward positions as the battle closes. Their new positions should be selected to take advantage of their long-range fires, and to get enfilade shots from the depth and flanks of the reverse slope.

The nature of the enemy may change at night, and the rifle squads may occupy the forward slope or crest to deny it to the enemy. In these circumstances, it is feasible for a rifle squad to have an alternate night position forward. The area forward of the topographical crest must be controlled by friendly forces through aggressive patrolling and both active and passive reconnaissance measures. The platoon should use all of its night vision devices to deny the enemy undetected entry into the platoon's defensive area. CCMS are key parts of the platoon's surveillance plan and should be positioned to take advantage of their thermal sights. The enemy must not be allowed to take advantage of reduced visibility to advance to a position of advantage without being taken under fire.

The company commander normally makes the decision to position platoons on a reverse slope. He does so when—

- He wishes to surprise or deceive the enemy about the location of his defensive position
- Forward slope positions might be made weak by direct enemy fire
- Occupation of the forward slope is not essential to achieve depth and mutual support
- Fields of fire on the reverse slope are better or at least sufficient to accomplish the mission
- Forward slope positions are likely to be the target of concentrated enemy artillery fires

Advantages

The following are advantages of a reverse-slope defense:

- Enemy observation of the position, including the use of surveillance devices and radar, is masked
- Enemy cannot engage the position with direct fire without coming within range of the defender's weapons
- Enemy indirect fire will be less effective because of the lack of observation
- Enemy may be deceived about the strength and location of positions
- Defenders have more freedom of movement out of sight of the enemy

Disadvantages

Disadvantages of a reverse-slope defense include the following:

- Observation to the front is limited
- Fields of fire to the front are reduced
- Enemy can begin his assault from a closer range

Obstacles

Obstacles are necessary in a reverse-slope defense. Because the enemy will be engaged at close range, obstacles should prevent the enemy from closing too quickly and overrunning the positions. Obstacles on the reverse slope can halt, disrupt, and expose enemy vehicles to flank antitank fires. Obstacles should also block the enemy to facilitate the platoon's disengagement.

Train, Advise & Assist

Chap 4

Ref: ADRP 3-07, Stability (Sept '12), chap. 3 and adaptations from FM 3-07.1, Security Force Assistance (May '09, rescinded), chap. 1-3.

In the complex, dynamic operational environments of the 21st century, significant challenges to sustainable peace and security exist. Sources of instability that push parties toward violence include religious fanaticism, global competition for resources, climate change, residual territorial claims, ideology, ethnic tension, elitism, greed, and the desire for power. These factors create belts of state fragility and instability that threaten U.S. national security.

Throughout U.S. history, U.S. forces have learned that military force alone cannot secure sustainable peace. U.S. forces can only achieve sustainable peace through a comprehensive approach in which military objectives nest in a larger cooperative effort of the departments and agencies of the U.S. Government, intergovernmental and nongovernmental organizations, multinational partners, the private sector, and the host nation.

The recent announcement of "Advisory Brigades" by Army Chief of Staff Gen. Mark A. Milley focuses on the need for specialized units to train and advise foreign forces. Called train, advise and assist brigades, the units would deploy to different combatant command areas to help train allies and partners, similar to what units have been doing in Iraq and Afghanistan. "I look at it as if you get a two-for," Milley said. "You get a day-to-day engagement that the combatant commanders want in order to train, advise and assist. And then in time of national emergency, you have at least four or five brigades with standing chains of command that can marry Soldiers up like the old COHORT units."

Military Engagement, Security Cooperation, and Stability

Military engagement, security cooperation, and stability missions, tasks, and actions encompass a wide range of actions where the military instrument of national power is tasked to support OGAs and cooperate with IGOs (e.g., UN, NATO) and other countries to protect and enhance national security interests, deter conflict, and set conditions for future contingency operations. Use of joint capabilities in these and related activities such as Security Force Assistance and Foreign Internal Defense helps shape the operational environment and keep the day-to-day tensions between nations or groups below the threshold of armed conflict while maintaining US global influence.

See pp. 4-5 to 4-6 for discussion of military engagement, security cooperation and deterrence. See pp. 4-7 to 4-16 discussion of stablity operations.

Refer to TAA2: Military Engagement, Security Cooperation & Stability SMARTbook (Foreign Train, Advise, & Assist) chap 2 for further discussion. Topics include the Range of Military Operations (JP 3-0), Security Cooperation & Security Assistance (Train, Advise, & Assist), Stability Operations (ADRP 3-07), Peace Operations (JP 3-07.3), Counterinsurgency Operations (JP & FM 3-24), Civil-Military Operations (JP 3-57), Multinational Operations (JP 3-16), Interorganizational Coordination (JP 3-08), and more.

(Train, Advise & Assist) Overview 4-1*

Security Force Assistance (SFA)

Ref: JP 3-22, Foreign Internal Defense (Jul 2010), chap. 1.

SFA aims to establish conditions that support the partner's end state, which includes legitimate, credible, competent, capable, committed, and confident security forces. This requires a force capable of securing borders, protecting the population, holding individuals accountable for criminal activities, regulating the behavior of individuals or groups that pose a security risk, and setting conditions in the operational area that enable the success of other actors.

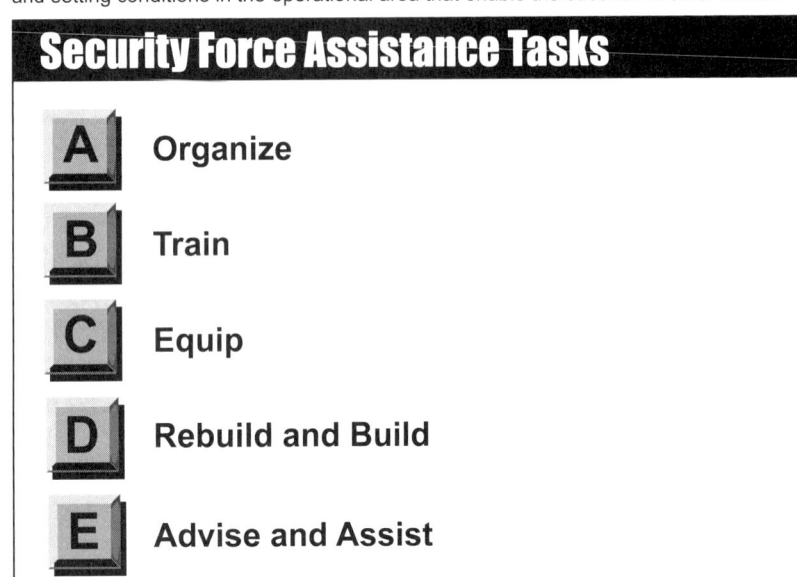

Security Force Assistance Tasks

A Organize

B Train

C Equip

D Rebuild and Build

E Advise and Assist

Ref: FM 3-07.1, Security Force Assistance, pp. 2-2 to 2-8.

Organize is a SFA task that encompasses all measures taken to assist FSF in improving its organizational structure, processes, institutions, and infrastructure. U.S. forces must understand the existing security organizations of FSF to better assist them. **Train** is a SFA task to assist FSF by developing programs and institutions to train and educate. These efforts must fit the nature and requirements of their security environment. **Equip** is a SFA task encompassing all efforts to assess and assist FSF with the procurement, fielding, and sustainment of equipment. All equipment must fit the nature of the operational environment. The SFA principle of ensuring long-term sustainment is a vital consideration for the equip task. **Rebuild and build** is a SFA task to assess, rebuild, and build the existing capabilities and capacities of FSF and their supporting infrastructure. This task requires an in-depth analysis of the capability, capacity, and structures required to meet the desired end state and operational environment. Some FSF may require assistance in building and rebuilding, while other FSF may only need assistance in building. **Advise and assist** is a SFA task in which U.S. personnel work with FSF to improve their capability and capacity. Advising establishes a personal and a professional relationship where trust and confidence define how well the advisor will be able to influence the foreign security force. Assisting is providing the required supporting or sustaining capabilities so FSF can meet objectives and the end state.

Refer to TAA2: Military Engagement, Security Cooperation & Stability SMARTbook (Foreign Train, Advise, & Assist) chap. 2 for further discussion.

Security Force Assistance Imperatives
Ref: FM 3-07.1, Security Force Assistance (May '09), pp. 2-1 to 2-2.

The imperatives of SFA come from the historical record and recent experience. These imperatives do not replace the principles of war or the principles of joint operations. Rather, they provide focus on how to successfully conduct SFA. The six imperatives apply to SFA at every level of war, for any echelon, and for any Soldier.

Since there is a close relationship between SFA and Army special forces conducting foreign internal defense, SFA planners should also consider the twelve Army special operations forces imperatives, especially at the tactical level (see FM 3-07.1, app. A).

Understand the Operational Environment
An in-depth understanding of the operational environment—including available FSF, opposing threats, and civil considerations—is critical to planning and conducting effective SFA. Units and Soldiers conducting SFA must clearly understand the theater, population, and FSF with which they are working, especially FSF capabilities. Diplomatic, informational, military, economic, sociological, psychological, and geographic research and understanding are essential prerequisites for successful SFA. Tactically, successful SFA requires identifying the friendly and hostile decisionmakers, their objectives and strategies, and the ways they interact.

Provide Effective Leadership
Leadership, a critical aspect of any application of combat power, proves important in the dynamic and complex environments associated with SFA. The operational environment in which SFA occurs places a premium on effective leadership at all levels, from the most junior to the most senior general officer and agency director.

Build Legitimacy
Ultimately, SFA aims to develop security forces that contribute to the legitimate governance of the local populace. Significant policy and legal considerations may apply to SFA activities. Legitimacy is the most crucial factor in developing and maintaining internal and international support. Legitimacy is a concept that goes beyond a strict legal definition; it includes the moral and political legitimacy of a host-nation government or partner organization.

Manage Information
Successful SFA disseminates timely and protected relevant information, integrates it during planning, and leverages it appropriately during execution. Effective and efficient information management supports decisionmaking throughout capacity building. Managing information encompasses the collection, analysis, management, application, and preparation of information. To meet expectations and success in general, units conducting SFA establish and integrate lessons learned.

Ensure Unity of Effort
SFA often includes many actors, making unity of effort essential for success. SFA will include U.S. and foreign security forces, including conventional forces, special forces, or a combination. Other civilian and military joint and military organizations are often involved in SFA. Planners integrate them into one cohesive effort.

Sustain the Effort
Sustainability consists of two major components: the ability to sustain SFA effort throughout the operation and the ability of the FSF to sustain their operations independently. While each situation will vary, Army personnel conducting SFA must avoid assisting FSF in techniques and procedures beyond the FSF's capability to sustain. U.S. tactics, techniques, and procedures must be modified to fit the culture, educational level, and technological capability of the FSF. Those involved in SFA must recognize the need for programs that are durable, consistent, and sustainable by both the U.S. and FSF.

A complex relationship exists among security cooperation, security assistance, and the military instrument of foreign internal defense.

Security Cooperation

Security cooperation is all Department of Defense interactions with foreign defense establishments to build defense relationships that promote specific United States security interests, develop allied and friendly military capabilities for self-defense and multinational operations, and provide United States forces with peacetime and contingency access to a host nation (JP 3-22). Security cooperation—usually coordinated by the U.S. military's security cooperation organization in a country—includes all Department of Defense (DOD) interactions with foreign defense and security establishments. These interactions include all DOD-administered security assistance programs that build defense and security relationships promoting specific U.S. security interests.

Such interests include all international armaments cooperation activities and security assistance activities to—

- Develop friendly, partner, and allied military capabilities for self-defense and multinational operations
- Build partnership capacity and enhance or establish relationships with regional national militaries that promote bilateral and coalition interoperability, strategic access, and regional stability

Security cooperation aims to promote stability, develop alliances, and gain and maintain access through security relationships that build both partner capacities and capabilities. The capacities and capabilities of partners directly correlate to the type of activities undertaken. Goals range from creating a positive relationship that allows freedom of movement to creating global security interoperability with core partners to addressing regional security organizations and alliance organizations. A broad range of interconnected and integrated security cooperation activities accomplishes security cooperation.

The Army supports security cooperation through security assistance, security force assistance, foreign internal defense, and security sector reform.

Security Assistance

Security assistance is a group of programs authorized by the Foreign Assistance Act of 1961, the Arms Export Control Act of 1976, or other related statutes. These programs permit the United States to provide defense articles, military training, and other defense-related services by grant, loan, credit, or cash sales in furtherance of national policies and objectives.

Security Force Assistance (SFA)

Security force assistance is Department of Defense activities that contribute to unified action by the United States Government to support the development of the capacity and capability of foreign security forces and their supporting institutions (JP 3-22). Military forces conduct these activities facilitating host nations to deter and defend against transnational internal threats to stability. The DOD also conducts security force assistance to assist host nations to defend against external threats; contribute to coalition operations; or organize, train, equip, and advise another country's security forces or supporting institutions. The only security force assistance activity conducted under combat conditions is combat advising.

Security Sector Reform (SSR)

Security sector reform (SSR) is an umbrella term that discusses reforming the security of an area. SSR includes integrated activities in support of defense and armed forces reform; civilian management and oversight; justice, police, corrections, and intelligence reform; national security planning and strategy support; border management; disarmament, demobilization, and reintegration; or reduction of armed violence.

I. Military Engagement, Security Coop & Deterrence

Ref: JP 3-0, Joint Operations (Aug '11), chap. V, section C (pp. V-9 to V-18).

The range of military operations (ROMO) is a fundamental construct that provides context. Military operations vary in scope, purpose, and conflict intensity across a range that extends from military engagement, security cooperation, and deterrence activities to crisis response and limited contingency operations and, if necessary, to major operations and campaigns.

Military engagement, security cooperation, and deterrence missions, tasks, and actions encompass a wide range of actions where the military instrument of national power is tasked to support OGAs and cooperate with IGOs (e.g., UN, NATO) and other countries to protect and enhance national security interests, deter conflict, and set conditions for future contingency operations. Use of joint capabilities in military engagement, security cooperation, and deterrence activities helps shape the operational environment and keep the day-to-day tensions between nations or groups below the threshold of armed conflict while maintaining US global influence.

(A Mongolian Armed Forces officer teaches U.S. Marines about the AK-47 Assault rifle as part of Khaan Quest in Mongolia. USMC photo by Lance Cpl. Nathan McCord.)

These activities generally occur continuously in all GCCs' AORs regardless of other ongoing contingencies, major operations, or campaigns. They usually involve a combination of military forces and capabilities separate from but integrated with the efforts of inter-organizational partners. Because DOS is frequently the major player in these activities, JFCs should maintain a working relationship with the chiefs of the US diplomatic missions in their area. Commanders and their staffs should establish and maintain dialogue with pertinent inter-organizational partners to share information and facilitate future operations.

(Train, Advise, Assist) I. Military Engagement, Security Coop & Deterrence 4-5*

A. Military Engagement

Military engagement is the routine contact and interaction between individuals or elements of the Armed Forces of the United States and those of another nation's armed forces, or foreign and domestic civilian authorities or agencies to build trust and confidence, share information, coordinate mutual activities, and maintain influence. Military engagement occurs as part of security cooperation, but also extends to interaction with domestic civilian authorities. Support to military engagement may include specific mission areas such as religious affairs and medical support.

B. Security Cooperation

Security cooperation involves all DOD interactions with foreign defense and security establishments to build defense relationships that promote specific US security interests, develop allied and friendly military and security capabilities for internal and external defense, and provide US forces with peacetime and contingency access to the HN. Developmental actions enhance a host government's willingness and ability to care for its people. Security cooperation is a key element of global and theater shaping operations. GCCs shape their AORs through security cooperation activities by continually employing military forces to complement and reinforce other instruments of national power.

C. Deterrence

Deterrence prevents adversary action through the presentation of a credible threat of counteraction. In both peace and war, the Armed Forces of the United States help to deter adversaries from using violence to reach their aims. Deterrence stems from an adversary's belief that a credible threat of retaliation exists, the contemplated action cannot succeed, or the costs outweigh the perceived benefits of acting. Thus, a potential aggressor chooses not to act for fear of failure, cost, or consequences. Ideally, deterrent forces should be able to conduct decisive operations immediately. However, if committed forces lack the combat power to conduct decisive operations, they conduct defensive operations while additional forces deploy. Effective deterrence requires a security cooperation plan that emphasizes the willingness of the US to employ forces in defense of its interests. Various joint operations (such as show of force and enforcement of sanctions) support deterrence by demonstrating national resolve and willingness to use force when necessary. Other operations (such as nation assistance and FHA) support deterrence by enhancing a climate of peaceful cooperation, thus promoting stability. Joint actions such as nation assistance, antiterrorism, DOD support to counterdrug (CD) operations, show of force operations, and arms control are applied to meet military engagement, security cooperation, and deterrence objectives.

Sustained presence contributes to deterrence and promotes a secure environment in which diplomatic, economic, and informational programs designed to reduce the causes of instability can perform as designed. Presence can take the form of forward basing, forward deploying, or pre-positioning assets. Forward presence activities demonstrate our commitment, lend credibility to our alliances, enhance regional stability, and provide a crisis response capability while promoting US influence and access. Joint force presence often keeps unstable situations from escalating into larger conflicts. The sustained presence of strong, capable forces is the most visible sign of US commitment to allies and adversaries alike. However, if sustained forward presence fails to deter an adversary, committed forces must be agile enough to transition rapidly to combat operations.

Refer to TAA2: Military Engagement, Security Cooperation & Stability SMARTbook (Foreign Train, Advise, & Assist) chap 1 for further discussion. Topics include the Range of Military Operations (JP 3-0), Security Cooperation & Security Assistance (Train, Advise, & Assist), Stability Operations (ADRP 3-07), Peace Operations (JP 3-07.3), Counterinsurgency Operations (JP & FM 3-24), Civil-Military Operations (JP 3-57), Multinational Operations (JP 3-16), Interorganizational Coordination (JP 3-08), and more.

II. Stability Operations

Ref: JP 3-07, Stability Operations (Sept '11), chap. 1.

1. Nature of Stability Operations

The Department of Defense (DOD) has learned through the difficult experiences of both Iraq and Afghanistan that success is not only defined in military terms; it also involves rebuilding infrastructure, supporting economic development, establishing the rule of law, building accountable governance, establishing essential services, building a capable host nation (HN) military responsible to civilian authority. In short, we must employ multiple instruments of national power to build a foreign nation's (FN's) internal capacity in a preventive mode to help them to defend themselves and maintain stability, or to enable the transition of responsibility back to the host country after defeat of an active insurgency. The US also expends resources to bring stability to areas and peoples affected by natural or man-made disasters.

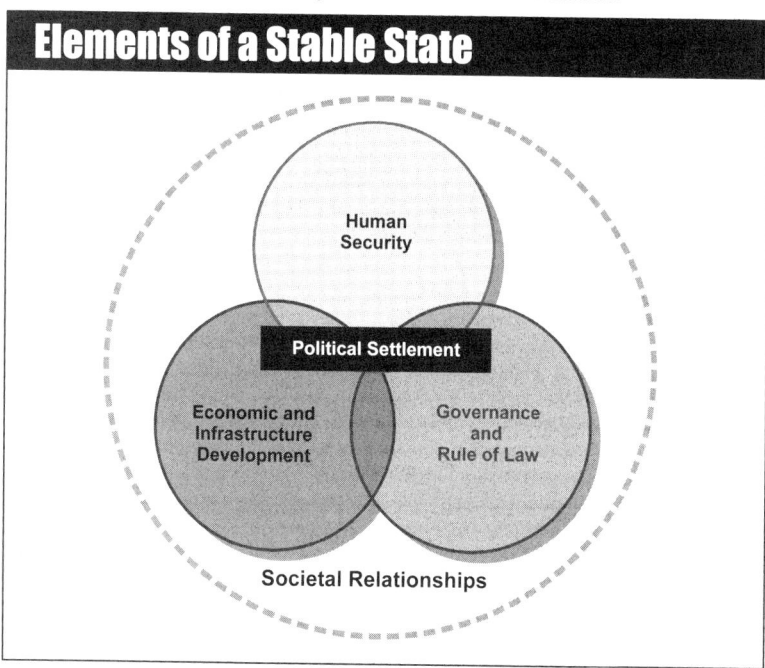

Elements of a Stable State

- Human Security
- Political Settlement
- Economic and Infrastructure Development
- Governance and Rule of Law
- Societal Relationships

Ref: JP 3-07, Stability Operations, fig I-1, p. I-3.

Refer to TAA2: Military Engagement, Security Cooperation & Stability SMARTbook (Foreign Train, Advise, & Assist) chap 3 for further discussion. Topics include the Range of Military Operations (JP 3-0), Security Cooperation & Security Assistance (Train, Advise, & Assist), Stability Operations (ADRP 3-07), Peace Operations (JP 3-07.3), Counterinsurgency Operations (JP & FM 3-24), Civil-Military Operations (JP 3-57), Multinational Operations (JP 3-16), Interorganizational Coordination (JP 3-08), and more.

End State Conditions for Stability

Ref: ADRP 3-07, Stability (Aug '12), pp. 1-13 to 1-16

To achieve conditions that ensure a stable and lasting peace, stability tasks in operations capitalize on coordination, cooperation, integration, and synchronization among military and nonmilitary organizations. These complementary civil-military efforts aim to strengthen legitimate governance, restore or maintain rule of law, support economic and infrastructure development, and foster a sense of national unity. These complementary efforts also seek to reform institutions to achieve sustainable peace and security and create the conditions that enable the host-nation government to assume responsibility for civil administration.

Successful efforts require an overarching framework that serves as a guide to develop strategy in pursuit of broader national or international policy goals. The following purpose-based framework—derived from work within the USG and led by the United States Institute of Peace with the Army's Peacekeeping and Stability Operations Institute—is founded on five broad conditions that describe the desired end state of a successful stability operation. In turn, a series of objectives link the execution of tactical tasks to that end state.

This framework provides the underpinnings for strategic, whole-of-government planning, yet also serves as a focal point for integrating operational- and tactical-level tasks. It is flexible and adaptive enough to support activities across the range of military operations but relies on concrete principles and fundamentals in application.

1. Safe and Secure Environment

Security is the most immediate concern of the military force, a concern typically shared by the local populace. A safe and secure environment is one in which these civilians can live their day-to-day lives without fear of being drawn into violent conflict and being victimized by criminals or by the forces there to protect them. Achieving security requires extensive collaboration with civil authorities, the trust and confidence of the people, immediate attention to any reported civilian harm as a result of operations, and strength of perseverance.

In the aftermath of conflict or disaster, conditions often create a significant security vacuum within the state. The government institutions are either unwilling or unable to provide security. In many cases, these institutions do not operate within internationally accepted norms. They are rife with corruption, abusing the power entrusted to them by the state. Sometimes these institutions actually embody the greatest threat to the populace. These conditions only serve to ebb away at the very foundation of the host nation's stability.

Many challenges threaten a safe and secure environment. Generally, the immediate threat to a safe and secure environment is a return to fighting by former warring parties. However, insurgent forces, criminal elements, and terrorists also significantly threaten the safety and security of the local populace.

2. Established Rule of Law

While military forces aim to establish a safe and secure environment, the rule of law requires much more—security of individuals and accountability for crimes committed against them. These basic elements are critical for a broader culture of rule of law to take hold in a society emerging from conflict. A broad effort integrates activities of many actors, focusing civilian and military law and order capabilities to support host-nation civil institutions in establishing and supporting the rule of law. These activities come from a shared sense of confidence among the population that the justice sector focuses on serving the public rather than pursuing narrow interests. Planning, preparing, and executing the transfer of responsibility from military to host-nation control for rule of law—although critical for building public confidence—often proves the most difficult and complex transition conducted in a stability operation. Failure to ensure continuity of rule of law through this transition threatens the safety and security of the local populace, erodes the legitimacy of the host nation, and impedes long-term development and achieving the desired end state.

*4-8 (Train, Advise, Assist) II. Stability Operations

3. Social Well-Being

The immediate needs of a host-nation population emerging from conflict or disaster generally consist of food, water, shelter, basic sanitation, and health care. International aid typically responds quickly, often due to their presence in, or proximity to, the affected area. If allowed, and once forces stabilize and secure the situation, local and international aid organizations provide for the immediate humanitarian needs of the people, establish sustainable assistance programs, and assist with displaced civilians.

However, forces also must attend to long-term requirements: developing educational systems, avoiding inadvertent civilian harm, addressing past abuses, and promoting peaceful coexistence among the host-nation people. These requirements most appropriately get supported from civilian actors, including other government agencies, intergovernmental organizations, and NGOs. Resolving issues of truth and justice are paramount to this process, and systems of amends, compensation, and reconciliation are essential.

4. Stable Governance

Since the end of the Cold War, all international interventions have aimed to establish stable governments with legitimate systems of political representation at the national, regional, and local levels. In a stable government, the host-nation populace regularly elects a representative legislature according to established rules and in a manner generally recognized as free and fair. Legislatures must be designed consistently with a legal framework and legitimate constitution. Officials must be trained, processes created, and rules established.

Typically, early elections in a highly polarized society empower elites, senior military leaders, and organized criminal elements. However, the local populace often seeks early and visible signs of progress. Effective reform processes begin with elections at the provincial or local level to minimize the likelihood of national polarization and reemergence of violent divisions in society. Popular leaders—capable of delivering services and meeting the demands of their constituents—and effective processes can emerge. Since elections can also become flashpoints for violence and instability between groups, U.S. forces consider security measures as part of the election process.

Successful, stable governments also require effective executive institutions. Such capacity building generally requires a long-term commitment of effort from the international community to reestablish effective ministries and a functional civil service at all levels of government. Stable governments also require free and responsible media, multiple political parties, and a robust civil society. Further, in many countries, formal systems of governance exist alongside informal governance systems, such as tribal elders. Such informal systems can play an important stabilization role, acting as an enduring and effective alternative to formal structures, which may have limited reaches within a country.

5. Sustainable Economy

Following conflict or a major disaster, economies tend toward a precarious state. They often suffer from serious structural problems that need immediate attention. However, they also possess significant growth potential. Commerce—legitimate and illicit—previously inhibited by circumstances emerges quickly to fill market voids and entrepreneurial opportunities. International aid and the requirements of intervening military forces often infuse the economy with abundant resources, stimulating rapid growth across the economic sector. However, much of this growth is temporary. It tends to highlight increasing income inequalities, the host-nation government's lagging capacity to manage and sustain growth, and expanding opportunities for corruption.

Rather than focus efforts toward immediately achieving economic growth, intervening elements aim to build on those aspects of the economic sector that enable the economy to become self-sustaining. These aspects include physical infrastructure, a sound fiscal and economic policy, an effective and predictable regulatory and legal environment, a viable workforce, business development and increased access to capital, and effective management of natural resources

II. Stability Operations Across the Range of Military Operations

The missions, tasks, and activities that make up stability operations fall into three broad categories: initial response activities, transformational activities, and sustainment activities.

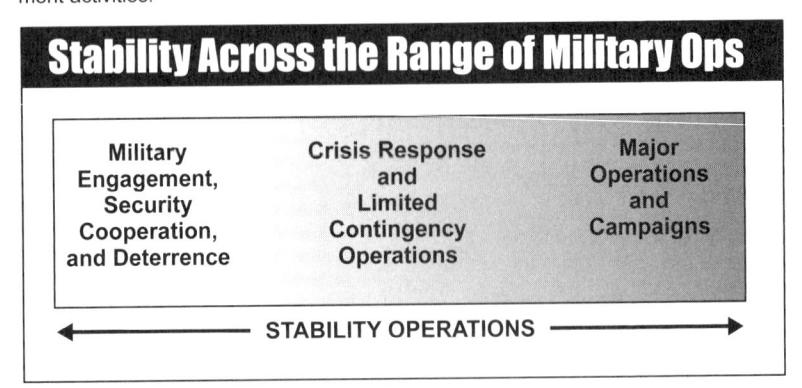

Stability Across the Range of Military Ops

Military Engagement, Security Cooperation, and Deterrence	Crisis Response and Limited Contingency Operations	Major Operations and Campaigns

◄──────────── STABILITY OPERATIONS ────────────►

Ref: JP 3-07, Stability Operations, fig I-2, p. I-4.

Initial response activities generally are tasks executed to stabilize the operational environment in an area in crisis, for instance during or immediately following conflict or a natural disaster. Initial response activities aim to provide a safe, secure environment and attend to the immediate humanitarian needs of a population. They support efforts to reduce the level of violence or human suffering while creating conditions that enable other organizations to participate safely in ongoing efforts.

Transformational activities are generally a broad range of security, reconstruction, and capacity building efforts. These activities aim to build HN capacity across multiple sectors. While establishing conditions that facilitate unified action to rebuild the HN and its supporting institutions, these activities are essential to instilling conditions that enable sustainable development.

Activities that foster sustainability encompass long-term efforts that capitalize on capacity building and reconstruction activities to establish conditions that enable sustainable development.

Military operations vary in size, purpose, and combat intensity within a range that extends from military engagement, security cooperation, and deterrence activities to crisis response and limited contingency operations, and, if necessary, major operations and campaigns, as illustrated in Figure I-2 above. The nature of the security environment may require US military forces to engage in several types of joint operations simultaneously across the range of military operations. Whether the prevailing context for the operation is one of traditional warfare or irregular warfare (IW), or even if the operation takes place outside of war, combat and stabilization are never sequential or alternative operations; the joint force commander (JFC) must integrate and synchronize stability operations with other operations (offense and defense) within each phase of any joint operation. The commander (CDR) for a particular operation determines the emphasis to be placed on each type of mission or activity in each phase of the operation.

III. Small Unit Stability Tasks

Stability operations are complex and demanding. A small unit in a stability operation -- an Infantry company in this example -- must master skills from negotiating to establishing OPs and checkpoints to escorting a convoy. The tasks and techniques in this section come from FM 3-21.10 (FM 7-10) The Infantry Rifle Company, and include lessons learned and should help the Infantry company commander implement these and other tasks.

A. Establish and Occupy a Lodgement Area or a Forward Operating Base (FOB)

A lodgment area (base camp) or forward operating base (FOB) is a well-prepared position used as a base of operations and staging area for the occupying unit. Like an assembly area or defensive strongpoint, the lodgment area also provides some force protection because it requires all-round security. However, several other factors distinguish a lodgment area from a less permanent position.

Due to the probability of long-term occupation, the lodgment requires a lot of preparation and logistical support. It needs shelters and facilities that can support the force and its attachments the whole time. Also, the area must be positioned and developed so the unit can effectively conduct its primary missions, such as PEO and counterterrorism, throughout its area of responsibility.

In establishing a lodgment, the Infantry company can either use existing facilities or request construction of new ones. Existing structures are immediately available, and require little or no construction support from engineers and members of the company. However, they might fall short of meeting the company's operational needs, and their proximity to other structures can pose security problems.

The company can establish and occupy a lodgment area as part of a battalion or, given enough support from battalion, as a separate element.

B. Monitor Compliance with an Agreement

Compliance monitoring involves observing belligerents and working with them to ensure they meet the conditions of one or more applicable agreements. Examples of the process include overseeing the separation of opposing combat elements, the withdrawal of heavy weapons from a sector, or the clearance of a minefield. Planning for compliance monitoring should cover, but is not limited to, the following considerations.

- Liaison teams, with suitable communications and transportation assets, are assigned to the headquarters of the opposing sides. Liaison personnel maintain communications with the leaders of their assigned element and talk directly to each other and to their mutual commander (the Infantry company or battalion commander).

- The commander positions himself at the point where violations are most likely to occur

- He positions platoons and squads where they can observe the opposing parties, instructing them to assess compliance and report any violations

- As directed, the commander keeps higher headquarters informed of all developments, including his assessment of compliance and noncompliance

C. Establishing Observation Posts and Checkpoints

Ref: FM 3-21.10 (FM 7-10) The Infantry Company, pp. 6-13 to 6-19

Observation Posts

Constructing and operating OPs is a high-frequency task for Infantry companies and subordinate elements whenever they must establish area security. Each OP is established for a specified time and purpose. Some OPs are overt (clearly visible) and deliberately constructed. Others are covert and designed to observe an area or target without the knowledge of the local population. Each type of OP must be integrated into supporting direct and indirect fire plans and into the overall observation plan.

An OP is similar in construction to a bunker and it is supported by fighting positions, barriers, and patrols. Covert operations may include sniper or designated marksmen positions over-watching TAIs. The Infantry company or a subordinate element might be directed to establish a checkpoint to achieve one or more of the following purposes:

- Obtain intelligence
- Identify enemy combatants or seize illegal weapons
- Disrupt enemy movement or actions
- Deter illegal movement
- Create an instant or temporary roadblock
- Control movement into the area of operations or onto a specific route
- Demonstrate the presence of US or peace forces
- Prevent smuggling of contraband
- Enforce the terms of peace agreements
- Serve as an OP, patrol base, or both

Checkpoints

One of the main missions conducted during OIF was the vehicle or traffic checkpoint. Units considered these standard steady-state operations and through repetitive execution could perform them virtually like battle drills; clearly beneficial given the often constrained planning and preparation time at company and platoon level. Checkpoint layout, construction, and operating should reflect METT-TC factors, including the amount of time available for emplacing it.

The Infantry company or a subordinate element might be directed to establish a checkpoint to achieve one or more of the following purposes.

- Obtain intelligence
- Identify enemy combatants or seize illegal weapons
- Disrupt enemy movement or actions
- Deter illegal movement
- Create an instant or temporary roadblock
- Control movement into the area of operations or onto a specific route
- Demonstrate the presence of US or peace forces
- Prevent smuggling of contraband
- Enforce the terms of peace agreements
- Serve as an OP, patrol base, or both

Some common types of checkpoints are discussed below.

1. Deliberate Checkpoints

These might be permanent or semi-permanent. They are typically constructed and employed to protect an operating base or well-established MSRs. Deliberate checkpoints are often used to secure the entrances to lodgment areas or base camps. They may also be used at critical intersections or along heavily traveled routes to monitor traffic and pedestrian flow. Deliberate checkpoints can be constructed so that all vehicles and personnel are checked or where only random searches occur (ROE and METT-TC dependent).

- They are useful deterrents and send a strong law and order or US presence message
- Deliberate checkpoints and their locations are known to terrorists and insurgents. Commanders must weigh the costs to the benefits of operating deliberate checkpoints.
- Commanders must consider that deliberate checkpoints may quickly become enemy targets and US Soldiers operating deliberate checkpoints are highly visible and viable targets for enemy attack

2. Hasty Checkpoints

Such checkpoints are planned and used only for a short, set period. Hasty checkpoints are normally employed during the conduct of a vehicle or foot patrol. The hasty checkpoint is similar in nature to the deliberate checkpoint but only uses transportable materials.

- The hasty checkpoint is mobile and can be quickly positioned
- While more adaptable, the hasty checkpoint does not send the constant visual reminder of US presence to the local population that the deliberate checkpoint does
- Because they can be quickly established and removed, hasty checkpoints are likely to be more effective in disrupting enemy actions. They are also less likely to be deliberately targeted by enemy forces.

3. Snap Checkpoints

Such checkpoints are conducted when specific intelligence indicates that a checkpoint hinders the enemy's freedom of movement at a specific time and place. Snap checkpoints are very similar to hasty checkpoints. The major difference is that hasty checkpoints are often random actions conducted as part of a patrol, whereas snap checkpoints are deliberate and based on either enemy analysis or quickly developed actionable intelligence. Snap checkpoints are normally conducted immediately and often with little to no deliberate planning.

4. Vehicular Traffic Stop Checkpoints

Such checkpoints are conducted by multiple sections of vehicle-equipped Infantrymen. This type of operation involves two or three sections of vehicles that patrol an area looking for a specific type of vehicle or specific personnel such as a particular model and color of car. Once this vehicle or person is identified, the vehicle or person is forced to stop and then searched. Normally the vehicle sections move single file with enough distance between the first two sections to allow civilian traffic to move between the sections (50 to 500 meters based on visibility, road conditions, and METT-TC.) If either section spots a targeted vehicle or person in a static or parked position, then the patrol cordons and searches the area, again based on METT-TC, or requests additional assistance. The patrol should move slightly slower than normal civilian traffic so that civilian traffic will pass the rear section. As civilian traffic passes the rear section, the patrol radio to the lead section if it spots a targeted vehicle. Once a targeted vehicle has moved between the two sections, both sections move abreast to effectively block the road and close the distance between themselves. They block in the targeted vehicle. The sections slowly force the targeted vehicle to pull to the side of the road and stop, and then they use normal vehicle search techniques. A third section can be employed as a reserve, as additional security, or simply as additional Soldiers.

(Train, Advise, Assist) II. Stability Operations 4-13*

D. Search

Searches are an important aspect of populace and resource control. The need to conduct search operations or to employ search procedures is a continuous requirement. A search can orient on people, materiel, buildings, or terrain. A search usually involves both civil police and Soldiers but may involve only Soldiers. Misuse of search authority can adversely affect the outcome of operations. Soldiers must conduct and lawfully record the seizure of contraband, evidence, intelligence material, supplies, or other minor items for their seizure to be of future legal value. Proper use of authority during searches gains the respect and support of the people.

E. Patrol

Note: See chap. 8 for information on patrols and patrolling.

Patrolling is also a high-frequency task during stability operations. The primary advantage of the dismounted patrol is that they provide a strong presence and enable regular interface with the local population. This procedure greatly helps in gathering vital information as well as in developing the base of knowledge of the unit's AO. Planning and execution of an area security patrol and presence patrol are similar to procedures for other tactical patrols except that the patrol usually occurs in urban areas and patrol leaders must consider political implications and ROE.

- **Presence Patrols.** US forces are deployed increasingly in combat operations in urban areas and in support of stability operations missions all around the world. The Infantry company and platoons conduct a presence patrol much the same as a combat patrol, and the planning considerations are similar. The main difference is that the patrol wants to both show force and lend confidence and stability to the local population of the host nation (HN).
- **Vehicle-Supported Patrols.** Infantry units might find themselves conducting frequent vehicle-assisted or vehicle-mounted patrols. The same considerations that apply to any dismounted patrol apply to vehicle-mounted patrols.

F. Escort a Convoy

This mission requires the Infantry company to provide a convoy with security and close-in protection from direct fire while on the move. Infantry forces must be augmented with additional transportation assets to carry out this mission.

G. Open and Secure Routes

This task is a mobility operation normally conducted by the engineers. The Infantry company might be tasked to assist in route clearance and to provide overwatch support. Route clearance may achieve one of several tactical purposes.

H. Conduct Reserve Operations

Reserve operations in the stability environment are similar to those in other tactical operations. They too allow the Infantry company commander to plan for a variety of contingencies based on the higher unit's mission.

I. Control Crowds

Large crowds or unlawful civil gatherings or disturbances pose a serious threat to US troops. Commanders must consider the effects of mob mentality, the willingness of enemies to manipulate media, and the ease with which a small, isolated group of Soldiers can be overwhelmed by masses of people. The police forces of each state and territory are normally responsible for controlling crowds involved in mass demonstrations, industrial, political and social disturbances, riots, and other civil disturbances. The prime role of US troops in the control of unlawful assemblies or demonstrations is to support and protect the police, innocent bystanders, and property.

*4-14 (Train, Advise, Assist) II. Stability Operations

III. Peace Operations (PO)

Ref: JP 3-07.3, Peace Operations (Aug '12), chap. I, and ATP 3-07.31, Peace Operations (Nov '14), chap. I.

Peace operations (PO) are crisis response and limited contingency operations, and normally include international efforts and military missions to contain conflict, redress the peace, and shape the environment to support reconciliation and rebuilding and to facilitate the transition to legitimate governance. PO may be conducted under the sponsorship of the United Nations (UN), another intergovernmental organization (IGO), within a coalition of agreeing nations, or unilaterally.

Peace operations (PO) are crisis response and limited contingency operations, including international efforts and military missions to contain conflict, redress the peace, and shape the environment in support of reconciliation and rebuilding and facilitating the transition to legitimate governance. (United Nations photo.)

Refer to TAA2: Military Engagement, Security Cooperation & Stability SMARTbook (Foreign Train, Advise, & Assist) chap 4 for further discussion. Topics include the Range of Military Operations (JP 3-0), Security Cooperation & Security Assistance (Train, Advise, & Assist), Stability Operations (ADRP 3-07), Peace Operations (JP 3-07.3), Counterinsurgency Operations (JP & FM 3-24), Civil-Military Operations (JP 3-57), Multinational Operations (JP 3-16), Interorganizational Coordination (JP 3-08), and more.

Types of Peace Operations

Ref: JP 3-07.3, Peace Operations (Aug '12), chap. I.

PO includes the five types of operations. *Note: The US adopted the term peace operations while others such as the North Atlantic Treaty Organization (NATO) adopted the term peace-support operations.*

1. Peacekeeping Operations (PKO)

PKO consist of military support to diplomatic, informational, and economic efforts to establish or maintain peace in areas of potential or actual conflict. PKO take place following diplomatic negotiation and agreement among the parties to a dispute, the sponsoring organization, and potential force-contributing nations. Before PKO begin, a credible truce or cease fire is in effect, and the parties to the dispute must consent to the operation. A main function of the PKO force is to establish a presence that inhibits hostile actions by the disputing parties and bolsters confidence in the peace process. Agreements often specify which nations' forces are acceptable, as well as the size and type of forces each will contribute.

2. Peace Enforcement Operations (PEO)

Peace enforcement operations (PEO) enforce the provisions of a mandate designed to maintain or restore peace and order. PEO may include the enforcement of sanctions and exclusion zones, protection of personnel providing FHA, restoration of order, and forcible separation of belligerent parties. PEO may be conducted pursuant to a lawful mandate or in accordance with international law and do not require the consent of the HN or the parties to the conflict, although broad based consent is preferred. Forces conducting PEO use force or the threat of force to coerce or compel compliance with resolutions or sanctions. Forces conducting PEO generally have full combat capabilities, although there may be some restrictions on weapons and targeting.

3. Peace Building (PB)

PB consists of actions that support political, economic, social, and security aspects of society. Although the major responsibility for PB is with the civil sector, early in PO, when critical and immediate tasks normally carried out by civilian organizations temporarily exceed their capabilities, PO forces should assist and cooperate with the HN civil sector, nongovernmental organizations (NGOs), and IGOs, to ensure that those tasks are accomplished.

4. Peacemaking (PM)

Peacemaking is the process of diplomacy, mediation, negotiation, or other forms of peaceful settlement that arranges an end to a dispute and resolves the issues that led to the conflict. Military support to the peacemaking process includes military-to-military relations, security assistance, or other activities, which influence disputing parties to seek a diplomatic settlement. An example of military support to peacemaking was the involvement of the Supreme Allied Commander Europe and the Joint Staff plans directorate during the development of the Dayton Accords by the presidents of Bosnia, Croatia, and Serbia outlining a General Framework Agreement for Peace in Bosnia and Herzegovina.

5. Conflict Prevention

Conflict prevention employs complementary diplomatic, civil, and military means to monitor and identify the causes of a conflict, and takes timely action to prevent the occurrence, escalation, or resumption of hostilities. Chapter VI of the United Nations (UN) Charter covers activities aimed at conflict prevention. Conflict prevention includes fact-finding missions, consultations, warnings, inspections, and monitoring.

Refer to TAA2: Military Engagement, Security Cooperation & Stability SMARTbook (Foreign Train, Advise, & Assist) chap. 4 for further discussion.

Chap 4
IV. Counterinsurgency Operations (COIN)

Ref: JP 3-24, Counterinsurgency (Nov '13), chap. 1 and FM 3-24, Insurgencies and Countering Insurgencies (May '14), chap. 1.

I. Insurgency

Insurgency is the organized use of subversion and violence to seize, nullify, or challenge political control of a region. Insurgency uses a mixture of subversion, sabotage, political, economic, psychological actions, and armed conflict to achieve its political aims. It is a protracted politico-military struggle designed to weaken the control and legitimacy of an established government, a military occupation government, an interim civil administration, or a peace process while increasing insurgent control and legitimacy—the central issues in an insurgency. Each insurgency has its own unique characteristics but they have the following aspects: a strategy, an ideology, an organization, a support structure, the ability to manage information, and a supportive environment.

Insurgency and counterinsurgency (COIN) are complex subsets of warfare. Globalization, technological advancement, urbanization, and extremists who conduct suicide attacks for their cause have certainly influenced contemporary conflict; however, warfare in the 21st century retains many of the characteristics it has exhibited since ancient times. (Dept. of Army photo by Staff Sgt. Jason T. Bailey).

Refer to TAA2: Military Engagement, Security Cooperation & Stability SMARTbook (Foreign Train, Advise, & Assist) chap 5 for further discussion. Topics include the Range of Military Operations (JP 3-0), Security Cooperation & Security Assistance (Train, Advise, & Assist), Stability Operations (ADRP 3-07), Peace Operations (JP 3-07.3), Counterinsurgency Operations (JP & FM 3-24), Civil-Military Operations (JP 3-57), Multinational Operations (JP 3-16), Interorganizational Coordination (JP 3-08), and more.

(Train, Advise, Assist) IV. Counterinsurgency Operations 4-17*

Insurgencies will continue to challenge security and stability around the globe in the 21st century. While the possibility of large scale warfare remains, few nations are likely to engage the US, allies, and partner nations. Globalization, numerous weak nation-state governments, demographics, radical ideologies, environmental concerns, and economic pressures are exacerbated by the ease of interaction among insurgent groups, terrorists, and criminals; and all put both weak and moderately governed states at risk. Today, a state's failure can quickly become not only a misfortune for its local communities, but a threat to global stability and US national interests.

Long-standing external and internal tensions tend to exacerbate or create core grievances within some countries, which can result in political strife, instability, or, if exploited by some groups to gain political advantage, even insurgency. Moreover, some transnational terrorists with radical political and religious ideologies may intrude in weak or poorly governed states to form a wider, more networked threat

The United States has supported numerous allies and partner nations to prevent or disrupt threats to their stability and security through foreign assistance and security cooperation (SC) activities as part of geographic combatant commanders' (GCCs') theater campaign plans in conjunction with other USG efforts. DoD efforts can include counterterrorism (CT) operations and foreign internal defense (FID) programs supported by stability operations tasks. If a friendly nation appears vulnerable to an insurgency, and it is in the best interest of the USG to help the host nation (HN) mitigate that insurgency, the USG would support the affected nation's internal defense and development (IDAD) strategy and program through a FID program. When an HN government supported by a FID program appears to be overwhelmed by internal threats, and if it is in the national security interests of the USG, then the third category of FID, US combat operations, may be directed by the President.

COIN is a comprehensive civilian and military effort designed to simultaneously defeat and contain insurgency and address its root causes. COIN is primarily a political struggle and incorporates a wide range of activities by the HN government of which security is only one, albeit an important one. Unified action is required to successfully conduct COIN operations and should include all HN, US, and multinational partners. The HN government in coordination with the chief of mission (COM) should lead the COIN efforts. When the operational environment (OE) is not conducive to a civilian agency lead for the COIN effort within a specific area, the joint force commander (JFC) must be cognizant of and able to lead the unified action required for effective COIN.

II. Governance and Legitimacy

A. Governance

Governance is the ability to serve the population through the rules, processes, and behavior by which interests are articulated, resources are managed, and power is exercised in a society. A state's ability to provide effective governance rests on its political and bureaucratic willingness, capability, and capacity to establish rules and procedures for decision making and on its ability to provide public services in a manner that is predictable and tolerable to the local population. In an ungoverned area (UGA), the state or the central government is unable or unwilling to extend control, effectively govern, or influence the local population. A UGA can also indicate where a provincial, local, tribal, or otherwise autonomous government does not fully or effectively govern. UGA is a broad term that encompasses under-governed, misgoverned, contested, and exploitable areas, characterized by the traits of inadequate governance capacity, insufficient political will, gaps in legitimacy, the presence or recent presence of conflict, or restrictive norms of behavior.

Insurgent Elements

Ref: JP 3-24, Counterinsurgency (Nov '13), pp. II-16 to II-18.

Insurgent organizations are often composed of different elements that perform complementary but distinct roles. Some elements openly challenge the government through public actions and guerrilla and terrorist attacks. Other elements operate through covert or clandestine methods, subverting existing political and civil institutions to support the insurgency or damage the legitimacy of the HN government. The proportion or presence of each element relative to the larger organization depends on the strategic approach the insurgency uses and the opportunity, motive, means factors. In many cases, these categories overlap and individuals may shift between them as the conflict and the insurgency evolve. This is especially true where insurgencies are based on existing social networks such as tribes and clans. The following categories should be regarded as illustrative; each insurgency should be carefully analyzed to identify the overt and covert elements within its organizational structure.

Political and Military Leadership. Leaders provide overall direction in more organized insurgencies. These leaders are the "key idea people" or strategic planners and are responsible for developing the insurgent narrative. They usually exercise leadership through some mixture of force of personality, the power of ideology, public esteem, or personal charisma. In some insurgencies, they may hold their position through religious, clan, or tribal authority. The leaders of movements based on religious extremism may also be religious figures. In loosely organized insurgencies, authority may be distributed across the leaders of multiple smaller groups that share similar or overlapping goals, such as expelling an occupier.

Underground. The underground is that element of the insurgent organization that conducts operations in areas normally denied to the auxiliary and the guerrilla force. The underground is a cellular organization within the insurgency that conducts covert or clandestine activities that are compartmentalized. This secrecy may be by necessity, by design, or both, depending on the situation. Most underground operations are required to take place in and around population centers that are held by counterinsurgent forces. Underground members often fill leadership positions, overseeing specific functions that are carried out by the auxiliary.

Guerrillas. Guerrillas conduct the actual fighting and provide security. They support the insurgency's broader agenda and maintain local control. Guerrillas protect and expand the counter state, if the insurgency establishes such an institution. They also protect training camps and networks that facilitate the flow of money, instructions, and foreign and local fighters. Guerrillas include any individual member of the insurgency who commits or attempts an act of overt violence or terrorism in support of insurgent goals. Guerrilla leaders are considered part of the combatant element for analyzing insurgencies.

Support Base. Sometimes referred to as the auxiliary, the insurgency's support base typically conceals its involvement with the movement. Ranging from sympathetic individuals who store weapons or warn of COIN force activities to major providers of finances or materiel, these supporters are critical to the insurgency but generally do not participate in combat operations. Typical activities include running safe houses; storing weapons and supplies; acting as couriers; providing intelligence collection; giving early warning of counterinsurgent movements; providing funding from lawful and unlawful sources; and providing forged or stolen documents and access or introductions to potential supporters. COIN forces face key challenges in distinguishing between voluntary supporters and those who have been coerced into cooperating with insurgency; understanding the complex motives of supporters; and neutralizing or co-opting them without appearing oppressive to the broader population that is unaware of their activities.

The authority to govern is dependent upon the successful amalgamation and interplay of four factors: mandate, manner, support and consent, and expectations. When the relationship between the government and those governed breaks down, challenges to authority may result. If a significant section of the population, or just an extreme faction, believes it cannot achieve a remedy through established political discourse, it may resort to insurgency.

- **Mandate**. The perceived legitimacy of the mandate that establishes a state authority, whether through the principles of universal suffrage, a recognized or accepted caste/tribal model, or authoritarian rule.
- **Manner**. The way in which those exercising that mandate conduct themselves, both individually and collectively in meeting the expectations of the local population(s).
- **Support and Consent**. The extent to which local populations consent to, or comply with, the manner/authority of those exercising the mandate. Consent may range from active support, passive support, or indifference, through unwilling compliance.
- **Expectations**. The relative quality or amount of support that local populations expect from their government.

Support to HN Government

Successful COIN operations require an HN government that is capable and willing to counter the insurgency and address its root causes. Typically this involves a mix of political reform, improved governance, and/or targeted economic development initiatives. COIN involves a careful balance between constructive dimensions (enhancing the capacity of the HN government to address the root causes of insurgency) and destructive dimensions (destroying and marginalizing the insurgency's political and military capabilities). In some situations the USG may need to take the lead for the HN government, especially in the early stages of a COIN effort. However, COIN activities should be transitioned back to an HN-led effort as soon as possible.

B. Legitimacy

Many governments rule through a combination of consent and influence, and in some cases, coercion. Legitimacy is a significant indicator of the extent to which systems of authority, decisions, and conduct are accepted by the local population. Political legitimacy of a government determines the degree to which the population will voluntarily or passively comply with the decisions and rules issued by a governing authority. Governments described as legitimate rule primarily with the consent of the governed; those described as illegitimate tend to rely heavily on coercion. Citizens obey illegitimate governments because they fear retribution, rather than because they voluntarily accept its rule. While a legitimate government may employ limited coercion to enforce the rule of law, most of its citizens voluntarily accept its authority.

Legitimacy in COIN

The struggle for legitimacy with the relevant population typically is a central theme of the conflict between the insurgency and the HN government. The HN government generally needs some level of legitimacy among the population in order to retain confidence of the populace and an acknowledgment of governing power. The insurgency will attack the legitimacy of the HN government while attempting to develop its own legitimacy with the population. The COIN effort must reduce the credibility of the insurgency while strengthening the legitimacy of the HN government. In a COIN environment high legitimacy of the HN government magnifies the resources/capabilities of the COIN effort (through such means as a populace willing to report on insurgents) and allows the HN to concentrate finite resources on targeting the insurgency. In dealing with an enemy like the insurgents, who are drawn from segments of the population, it is often a particular challenge for the HN to be seen as legitimate in public opinion.

Chap 5
Tactical Enabling Tasks
I. Security Operations

Ref: ADRP 3-90, Offense and Defense (Aug '12), chap 5, pp. 5-3 to 5-4.

Security operations are those operations undertaken by a commander to provide early and accurate warning of enemy operations, to provide the force being protected with time and maneuver space within which to react to the enemy, and to develop the situation to allow the commander to effectively use the protected force.

Security operations must provide information regarding enemy movement and capacity while giving the commander enough time and space with which to form an effective response. (Photo by Jeong, Hae-jung).

I. Forms of Security Operations

There are five forms of security operations -- screen, guard, cover, area security and local security.

A. Screen

Screen is a form of security operations that primarily provides early warning to the protected force. A unit performing a screen observes, identifies, and reports enemy actions. Generally, a screening force engages and destroys enemy reconnaissance elements within its capabilities—augmented by indirect fires—but otherwise fights only in self-defense. The screen has the minimum combat power necessary to provide the desired early warning, which allows the commander to retain the bulk of his combat power for commitment at the decisive place and time. A screen provides the least amount of protection of any security mission; it does not have the combat power to develop the situation.

A screen is appropriate to cover gaps between forces, exposed flanks, or the rear of stationary and moving forces. The commander can place a screen in front of a stationary formation when the likelihood of enemy action is small, the expected enemy force is small, or the main body needs only limited time, once it is warned, to react effectively. Designed to provide minimum security with minimum forces, a screen is usually an economy-of-force operation based on calculated risk. If a significant enemy force is expected or a significant amount of time and space is needed to provide the required degree of protection, the commander should assign and resource a guard or cover mission instead of a screen. The security element forward of a moving force must conduct a guard or cover because a screen lacks the combat power to defeat or contain the lead elements of an enemy force.

Security Fundamentals

Principles of Security Operations	Techniques Used to Perform Security Operations	Information Required from Controlling Headquarters
• Three General Orders • Provide early and accurate warning • Provide reaction time and maneuver space • Orient on the force / facility being secured • Perform continuous reconnaissance • Maintain enemy contact	• Observation post • Combat outpost • Battle position • Patrols • Combat formations • Movement techniques • Infiltration • Movement to contact • Dismounted, mounted, and air insertion • Roadblocks • Checkpoints • Convoy and route security • Searches	• Trace of the security area (front, sides, and rear boundaries), and initial position within the area • Time security is to be established • Main body size and location • Mission, purpose and commander's intent of the controlling headquarters • Counterreconnaissance and engagement criteria • Method of movement to occupy the area (zone reconnaissance, infiltration, tactical road march, movement to contact; mounted, dismounted, or air insertion) • Trigger for displacement and method of control when displacing. • Possible follow-on missions

Ref: FM 3-21.8 (FM 7-8), table H-1, p. H-2.

Critical Tasks for a Screen

Unless the commander orders otherwise, a security force conducting a screen performs certain tasks within the limits of its capabilities. A unit can normally screen an avenue of approach two echelons larger than itself. If a security force does not have the time or other resources to complete all of these tasks, the security force commander must inform the commander assigning the mission of the shortfall and request guidance on which tasks must be completed and their priority. After starting the screen, if the security unit commander determines that he cannot complete an assigned task, such as maintain continuous surveillance on all avenues of approach into an AO, he reports and awaits further instructions. Normally, the main force commander does not place a time limit on the duration of the screen, as doing so may force the screening force to accept decisive engagement. Screen tasks are to:

- Allow no enemy ground element to pass through the screen undetected and unreported
- Maintain continuous surveillance of all avenues of approach larger than a designated size into the area under all visibility conditions
- Destroy or repel all enemy reconnaissance patrols within its capabilities.
- Locate the lead elements of each enemy advance guard and determine its direction of movement in a defensive screen
- Maintain contact with enemy forces and report any activity in the AO
- Maintain contact with the main body and any security forces operating on its flanks
- Impede and harass the enemy within its capabilities while displacing

II. Fundamentals of Security Ops

Ref: Adapted from FM 3-90 Tactics, pp. 12-2 to 12-3.

1. Provide Early and Accurate Warning

The security force provides early warning by detecting the enemy force quickly and reporting information accurately to the main body commander. The security force operates at varying distances from the main body based on the factors of METT-TC. As a minimum, it should operate far enough from the main body to prevent enemy ground forces from observing or engaging the main body with direct fires. The earlier the security force detects the enemy, the more time the main body has to assess the changing situation and react. The commander positions ground security and aeroscouts to provide long-range observation of expected enemy avenues of approach, and he reinforces and integrates them with available intelligence collection systems to maximize warning time.

2. Provide Reaction Time and Maneuver Space

The security force provides the main body with enough reaction time and maneuver space to effectively respond to likely enemy actions by operating at a distance from the main body and by offering resistance to enemy forces. The commander determines the amount of time and space required to effectively respond from information provided by the intelligence preparation of the battlefield (IPB) process and the main body commander's guidance regarding time to react to enemy courses of action (COA) based on the factors of METT-TC. The security force that operates farthest from the main body and offers more resistance provides more time and space to the main body. It attempts to hinder the enemy's advance by acting within its capabilities and mission constraints.

3. Orient on the Force or Facility to Be Secured

The security force focuses all its actions on protecting and providing early warning to the secured force or facility. It operates between the main body and known or suspected enemy units. The security force must move as the main body moves and orient on its movement. The security force commander must know the main body's scheme of maneuver to maneuver his force to remain between the main body and the enemy. The value of terrain occupied by the security force hinges on the protection it provides to the main body commander.

4. Perform Continuous Reconnaissance

The security force aggressively and continuously seeks the enemy and reconnoiters key terrain. It conducts active area or zone reconnaissance to detect enemy movement or enemy preparations for action and to learn as much as possible about the terrain. The ultimate goal is to determine the enemy's COA and assist the main body in countering it. Terrain information focuses on its possible use by the enemy or the friendly force, either for offensive or defensive operations. Stationary security forces use combinations of OPs, aviation, patrols, intelligence collection assets, and battle positions (BPs) to perform reconnaissance. Moving security forces perform zone, area, or route reconnaissance along with using OPs and BPs, to accomplish this fundamental.

5. Maintain Enemy Contact

Once the security force makes enemy contact, it does not break contact unless specifically directed by the main force commander. The security asset that first makes contact does not have to maintain that contact if the entire security force maintains contact with the enemy. The security force commander ensures that his subordinate security assets hand off contact with the enemy from one security asset to another in this case. The security force must continuously collect information on the enemy's activities to assist the main body in determining potential and actual enemy COAs and to prevent the enemy from surprising the main body. This requires continuous visual contact, the ability to use direct and indirect fires, freedom to maneuver, and depth in space and time.

Screen Movement Methods

Method	Characteristics	Advantages	Disadvantages
Alternate Bounds by OPs	■ Main body moves faster ■ Conducted by platoon or company/troop ■ Contact is possible ■ Conducted rear to front	■ Very secure method ■ Maintains maximum surveillance over the security area	■ Execution takes time ■ Disrupts unit integrity
Alternate Bounds by Units	■ Main body moves faster ■ Conducted by platoon or company/troop ■ Contact is possible ■ Conducted rear to front	■ Execution does not take a great deal of time ■ Maintains good surveillance over the security area ■ Maintains unit integrity	■ May leave temporary gaps in coverage
Successive Bounds	■ Main body moving slowly ■ Conducted by platoon or company/troop ■ Contact is possible ■ Conducted simultaneously or in succession ■ Units should maintain an air screen during ground movement	■ Most secure method ■ Maintains maximum surveillance ■ Maintains unit integrity	■ Execution takes the most time ■ Unit is less secure when all elements are moving simultaneously ■ Simultaneous movement may leave temporary gaps
Continuous Marching	■ Main body is moving relatively quickly ■ Performed as a route reconnaissance ■ Enemy contact not likely ■ Unit should maintain an air screen on the flank	■ OPs displace quickly ■ Maintains unit integrity	■ Least secure method

Ref: FM 3-90, table 12-1, p. 12-6.

B. Guard

Guard is a form of security operations whose primary task is to protect the main body by fighting to gain time while also observing and reporting information and preventing enemy ground observation of and direct fire against the main body. Units conducting a guard mission cannot operate independently because they rely upon fires and combat support assets of the main body.

A guard differs from a screen in that a guard force contains sufficient combat power to defeat, cause the withdrawal of, or fix the lead elements of an enemy ground force before it can engage the main body with direct fire. A guard force routinely engages enemy forces with direct and indirect fires. A screening force, however, primarily uses indirect fires or close air support to destroy enemy reconnaissance elements and slow the movement of other enemy forces. A guard force uses all means at its disposal, including decisive engagement, to prevent the enemy from penetrating to a position were it could observe and engage the main body. It operates within the range of the main body's fire support weapons, deploying over a narrower front than a comparable-size screening force to permit concentrating combat power.

Types of Guard Operations

1. Advance guard
2. Flank guard
3. Rear guard

Ref: FM 3-90, pp. 12-21 to 12-25.

The three types of guard operations are advance, flank, and rear guard. A commander can assign a guard mission to protect either a stationary or a moving force.

Guard tasks:
- Destroy the enemy advance guard
- Maintain contact with enemy forces and report activity in the AO
- Maintain continuous surveillance of avenues of approach into the AO under all visibility conditions
- Impede and harass the enemy within its capabilities while displacing
- Cause the enemy main body to deploy, and then report its direction of travel
- Allow no enemy ground element to pass through the security area undetected and unreported
- Destroy or cause the withdrawal of all enemy reconnaissance patrols
- Maintain contact with its main body and any other security forces operating on its flanks

C. Cover

Cover is a form of security operations whose primary task is to protect the main body by fighting to gain time while also observing and reporting information and preventing enemy ground observation of and direct fire against the main body.

The covering force's distance forward of the main body depends on the intentions and instructions of the main body commander, the terrain, the location and strength of the enemy, and the rates of march of both the main body and the covering force. The width of the covering force area is the same as the AO of the main body.

A *covering force* is a self-contained force capable of operating independently of the main body, unlike a screening or guard force. A covering force, or portions of it, often becomes decisively engaged with enemy forces. Therefore, the covering force must have substantial combat power to engage the enemy and accomplish its mission. A covering force develops the situation earlier than a screen or a guard force. It fights longer and more often and defeats larger enemy forces.

While a covering force provides more security than a screen or guard force, it also requires more resources. Before assigning a cover mission, the main body commander must ensure that he has sufficient combat power to resource a covering force and the decisive operation. When the commander lacks the resources to support both, he must assign his security force a less resource-intensive security mission, either a screen or a guard.

A covering force accomplishes all the tasks of screening and guard forces. A covering force for a stationary force performs a defensive mission, while a covering force for a moving force generally conducts offensive actions. A covering force normally operates forward of the main body in the offense or defense, or to the rear for a retrograde operation. Unusual circumstances could dictate a flank covering force, but this is normally a screen or guard mission.

1. Offensive Cover
An offensive covering force seizes the initiative early for the main body commander, allowing him to attack decisively. Some critical tasks include

- Performing zone reconnaissance along the main body's axis of advance or within the AO

- Clearing or bypassing enemy forces within the AO in accordance with bypass criteria
- Denying the enemy information about the strength, composition, and objective of the main body
- Penetrating the enemy's security area to locate enemy main defensive positions
- Determining enemy strengths and dispositions
- Locating gaps or weaknesses in the enemy's defensive scheme
- Defeating or repelling enemy forces as directed by the higher commander
- Deceiving the enemy into thinking the main body has been committed and causing him to launch counterattacks prematurely
- Fixing enemy forces to allow the main body to maneuver around enemy strengths or through weaknesses
- Destroying enemy reconnaissance, the advance guard, and the lead elements of the main body
- Determining the location of enemy assailable flanks
- Fixing enemy forces to allow the main body to maneuver around enemy strengths or through weaknesses

2. Defensive Cover

A defensive covering force prevents the enemy from attacking at the time, place, and combat strength of his choosing. Defensive cover gains time for the main body, enabling it to deploy, move, or prepare defenses in the MBA. It accomplishes this by disrupting the enemy's attack, destroying his initiative, and establishing the conditions for decisive operations. A defensive covering force emphasizes the following tasks:

- Prevent the main body from being surprised and becoming engaged by direct-fire weapons
- Defeat enemy advance guard formations
- Maintain continuous surveillance of high-speed avenues of approach into the security area
- Defeat all enemy reconnaissance formations before they can observe the main body
- Cause the deployment of the enemy main body
- Determine the size, strength, composition, and direction of the enemy's main effort
- Destroy, defeat, or attrit enemy forces within its capacity
- Deprive the enemy of his fire support and air defense umbrellas, or require him to displace them before he attacks the MBA
- Deceive the enemy regarding the location of main body and main defensive positions
- Avoid being bypassed

D. Area Security

Area security is a form of security operations conducted to protect friendly forces, installations, routes, and actions within a specific area.

Area security operations may be offensive or defensive in nature. They focus on the protected force, installation, route, or area. Forces to protect range from echelon headquarters through artillery and echelon reserves to the sustaining base. Protected installations can also be part of the sustaining base or they can constitute part of the area's infrastructure. Areas to secure range from specific points (bridges and defiles) and terrain features (ridge lines and hills) to large population centers and their adjacent areas.

Operations in noncontiguous AOs require commanders to emphasize area security. During offensive and retrograde operations, the speed at which the main body moves provides some measure of security. Rapidly moving units in open terrain can rely on technical assets to provide advance warning of enemy forces. In restrictive terrain, security forces focus on key terrain such as potential choke points.

E. Local Security

Local security consists of low-level security operations conducted near a unit to prevent surprise by the enemy.

Local security includes any local measure taken by units against enemy actions. It involves avoiding detection by the enemy or deceiving the enemy about friendly positions and intentions. It also includes finding any enemy forces in the immediate vicinity and knowing as much about their positions and intentions as possible. Local security prevents a unit from being surprised and is an important part of maintaining the initiative. The requirement for maintaining local security is an inherent part of all operations. Units perform local security when conducting full spectrum operations, including tactical enabling operations.

Units use both active and passive measures to provide local security.

1. Active Measures

- Using OPs and patrols
- Establishing specific levels of alert within the unit. The commander adjusts those levels based on the factors of METT-TC
- Establishing stand-to times. The unit SOP should detail the unit's activities during the conduct of stand-to

2. Passive Measures

Passive local security measures include using camouflage, movement control, noise and light discipline, and proper communications procedures. It also includes employing available ground sensors, night-vision devices, and daylight sights to maintain surveillance over the area immediately around the unit.

* Combat Outposts

A combat outpost is a reinforced OP capable of conducting limited combat operations. Using combat outposts is a technique for employing security forces in restrictive terrain that precludes mounted security forces from covering the area. They are also used when smaller OPs are in danger of being overrun by enemy forces infiltrating into and through the security area. The commander uses a combat outpost when he wants to extend the depth of his security area, when he wants his forward OPs to remain in place until they can observe the enemy's main body, or when he anticipates that his forward OPs will be encircled by enemy forces. Both mounted and dismounted forces can employ combat outposts.

On Point

Combat leaders require advanced warning of an enemy attack. Not every troop can remain at the ready 100 percent of the time. To protect the friendly force, security operations project patrols and targeting systems that can detect enemy activity.

Security operations conduct many of the same activities and use the same technologies as reconnaissance operations. The principal difference between the two is that while reconnaissance operations are dedicated to the enemy force and resources, security operations are dedicated to the friendly force and resources.

It may help to think of these two operations in simplistic terms. Reconnaissance operations tend to be offensive in nature, whereas security operations are more defensive in nature.

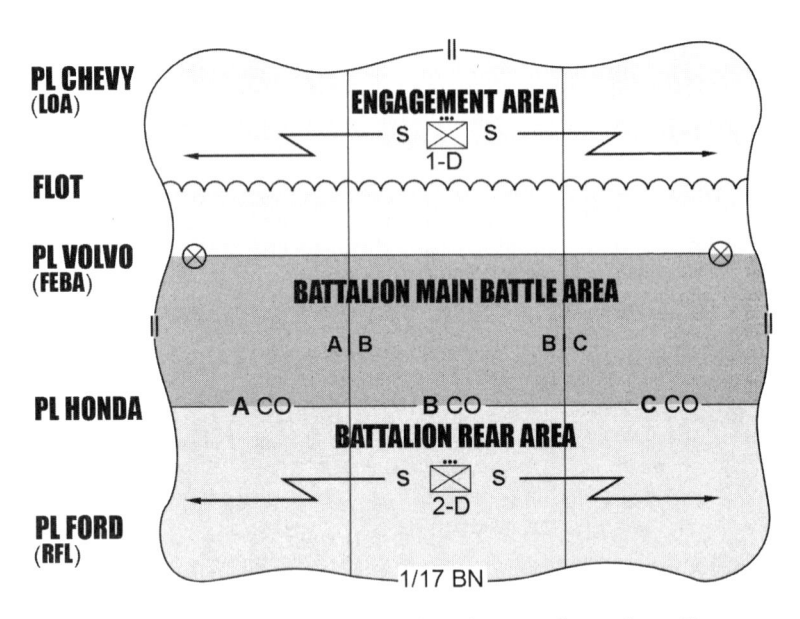

(Diagram: Battalion area showing PL CHEVY (LOA), FLOT, PL VOLVO (FEBA), PL HONDA, PL FORD (RFL); Engagement Area with S–1-D–S screen forward of FLOT; Battalion Main Battle Area with companies A|B, B|C and A CO, B CO, C CO; Battalion Rear Area with S–2-D–S screen; 1/17 BN.)

Security operations include activities throughout the area of operations. They are denoted by lightning bolts from the unit to the boundaries. In this case, the 'S' indicates screening operations in the engagement area forward of the FLOT, and in the battalion's rear area. (Ref: FM 3-90, chap 12, fig. 12-1).

The main body commander must designate the exact force to secure. This designation determines the limits of the security force's responsibilities. The security force must orient on the force it is securing. If the main body moves, the security force also moves to maintain its position in relation to the main body. The limited capabilities of most maneuver platoons prohibit them from having a mission separate from their parent company. Scout platoons are the exception to this rule.

The main body commander must determine when to establish the security force. He decides this based on the activity of the main body and expected enemy activity. He must allow enough time for the security force to move into and occupy the security area to prevent enemy forces from penetrating the security area undetected. The factors of METT-TC influence how the security force deploys to and occupies the screen line. If the security mission is the result of a current reconnaissance mission, the security force is already positioned to begin its mission. This occurs frequently when a reconnaissance mission halts at a designated PL. Analyzing the factors of METT-TC determines which deployment technique meets mission requirements.

Chap 5
Tactical Enabling Tasks
II. Reconnaissance

Ref: ADRP 3-90, Offense and Defense (Aug '12), chap 5, pp. 5-1 to 5-3.; and FM 7-92 Infantry Reconnaissance Platoon and Squad (Airborne, Air Assault, Light Infantry), chap 1 and 4; FM 3-21.10 (FM 7-10) The Infantry Rifle Company, pp. 8-1 to 8-4.

Reconnaissance is a mission undertaken to obtain, by visual observation or other detection methods, information about the activities and resources of an enemy or potential enemy, or to secure data concerning the meteorological, hydrographical or geographical characteristics and the indigenous population of a particular area. Reconnaissance primarily relies on the human dynamic rather than technical means. Reconnaissance is performed before, during, and after other operations to provide information used in the intelligence preparation of the battlefield (IPB) process, as well as by the commander in order to formulate, confirm, or modify his course of action (COA).

Reconnaissance is a process of gathering information to help the commander shape his understanding of the battlespace. Reconnaissance uses many techniques and technologies to collect this information, but it is still largely a human endeavor. (Photo by Jeong, Hae-jung).

Reconnaissance Objective

The commander orients his reconnaissance assets by identifying a reconnaissance objective within the area of operations (AO). The reconnaissance objective is a terrain feature, geographic area, or an enemy force about which the commander wants to obtain additional information. The reconnaissance objective clarifies the intent of the reconnaissance effort by specifying the most important result to obtain from the reconnaissance effort. The commander assigns a reconnaissance objective based on his priority information requirements (PIR) resulting from the IPB process and the reconnaissance asset's capabilities and limitations. The reconnaissance objective can be information about a specific geographical location, such as the cross-country trafficability, a specific enemy activity to be confirmed or denied, or a specific enemy unit to be located and tracked.

I. Reconnaissance Fundamentals

At this point the PL should know the specific recon mission and have received and developed the maneuver control measures. Before delving into the execution of each type of reconnaissance, it's important to delineate the fundamentals of recon patrols. There are seven rules to remember.

1. Ensure continuous recon
Reconnaissance happens before, during, and after an engagement. Before the engagement, the recon team develops the commander's picture of the battlefield. During the engagement, the recon team lets the commander know if the plan is having its intended effect upon the enemy force. After the engagement, the recon team helps the commander to determine the enemy's next move.

2. Don't keep recon assets in reserve
Of course, the recon team should not be run until it is exhausted. However, the recon team acts as the commander's eyes and ears forward. There is no reason to keep one ear or one eye in reserve! Recon assets must be managed to allow continuous reconnaissance. That includes a rest plan.

3. Orient on the objective
Don't just throw the recon team forward without a specified objective! Name the type of recon mission and name the objective within the AO. This helps the commander prioritize the recon assets and objectives, plus gives focus to the recon team for the most economical use of time.

4. Report information rapidly and accurately
Over time, information loses value because more often than not, the battlefield is rapidly changing. Recon teams give timely reports on exactly what they see (without exaggeration) and exactly what they do not see. A common mistake in reconnaissance is the failure to report when no enemy force or presence is detected. Failing to report does nothing for the commander. A report of negative activity gives the commander a better understanding of where the enemy *isn't* located, at least.

5. Retain freedom of maneuver
As stated earlier, a recon team that becomes engaged in a firefight with the enemy is fixed to a given location. Without the ability to maneuver, that recon team can only report what is to its immediate front. In short, it has become no more useful in developing the battlefield picture than any other line unit.

6. Gain and maintain contact
Recon teams seek to gain contact with the enemy. More often than not, the recon team uses a combination of stealth and surveillance to maintain contact with the enemy. The recon team maintains contact with the enemy until the commander orders them to withdraw. The recon patrol leader may also break contact if the recon team is decisively engaged, but then seeks to regain contact immediately.

7. Develop the situation rapidly
Once contact is gained, the recon team must quickly discern the threat. For an enemy force, that means identifying the approximate size of the enemy force, the activity and direction of movement, and possibly the enemy's disposition and capabilities. When evaluating an enemy obstacle, the recon team must discern the type of obstacle, the extent of the obstacle, and whether or not it is covered by enemy fire. Often, enemy obstacles tell the commander a fair amount of information regarding the capabilities and even the location of the enemy force.

II. Organization

The responsibility for conducting reconnaissance does not reside solely with specifically organized units. Every unit has an implied mission to report information about the terrain, civilian activities, and friendly and enemy dispositions, regardless of its battlefield location and primary function. Frontline troops and reconnaissance patrols of maneuver units at all echelons collect information on enemy units with which they are in contact. In rear areas, reserve maneuver forces, fire support assets, air defense, military police, host nation agencies, combat support, and combat service support elements observe and report civilian and enemy activity.

The small unit commander develops the enemy situation through active and passive reconnaissance. Passive reconnaissance includes techniques such as map, photographic and small, unmanned aerial systems (SUAS) reconnaissance and surveillance. Active methods include ground reconnaissance and reconnaissance by fire. Active reconnaissance operations are also classified as stealthy or aggressive:

- **Stealthy Reconnaissance.** Stealthy reconnaissance emphasizes procedures and techniques that allow the unit to avoid detection and engagement by the enemy. It is more time-consuming than aggressive reconnaissance. To be effective, stealthy reconnaissance must rely primarily on elements that make maximum use of covered and concealed terrain. The company's primary assets for stealthy reconnaissance are its Infantry squads or SUASs.

- **Aggressive Reconnaissance.** Aggressive reconnaissance is characterized by the speed and manner in which the reconnaissance element develops the situation once contact is made with an enemy force. A unit conducting aggressive reconnaissance uses both direct and indirect fires and movement to develop the situation. In conducting a patrol, the unit employs the principles of tactical movement to maintain security. The patrolling element maximizes the use of cover and concealment and conducts bounding overwatch as necessary to avoid detection.

III. Planning & Preparation

Reconnaissance planning starts with the company commander's identification of critical information requirements. The company commander then compares his CCIR list to that of the battalion commander. If the company commander identifies CCIR not covered on the battalion list, he shares them with the battalion commander and staff. The company commander requests that battalion or higher headquarter commit assets to confirm his CCIR. Based on the results of that request, the company commander can commit his forces to gather the information needed. This process begins while the unit is planning or preparing for an operation. It often continues during the conduct of the operation. Once the operation is under way, the commander continues to identify information requirements.

In planning for route, zone, or area reconnaissance, the company commander determines the objective of the mission, and identifies whether the reconnaissance will orient on the terrain or on the enemy force. He provides the company with clear guidance on the objective of the reconnaissance. The patrol leader (PL) typically identifies any additional maneuver control measures. This might include a direction of advance, anticipated listening halts, en route rally points (ERP), objective rally points (ORP) and target reference points (TRP).

III. Forms of the Reconnaissance

The four forms of reconnaissance operations—

- Route reconnaissance
- Zone reconnaissance
- Area reconnaissance
- Reconnaissance in force (RIF)

A. The Route Reconnaissance

Route reconnaissance is a form of reconnaissance that focuses along a specific line of communication, such as a road, railway, or cross-country mobility corridor. It provides new or updated information on route conditions, such as obstacles and bridge classifications, and enemy and civilian activity along the route. A route reconnaissance includes not only the route itself, but also all terrain along the route from which the enemy could influence the friendly force's movement.

The commander may assign a route reconnaissance as a separate mission or as a specified task for a unit conducting a zone or area reconnaissance. A scout platoon can conduct a route reconnaissance over only one route at a time. For larger organizations, the number of scout platoons available directly influences the number of routes that can be covered at one time. Integrating ground, air, and technical assets assures a faster and more complete route reconnaissance.

Depending on the length of route to be reconnoitered, the route recon may be conducted with great stealth, or it may be conducted with great mobility. The effort is to gather intelligence on the route and its conditions. (Dept. of Army photo by Arthur McQueen).

Route Reconnaissance Tasks
- Find, report, and clear within capabilities all enemy forces that can influence movement along the route
- Determine the trafficability of the route; can it support the friendly force?
- Reconnoiter all terrain that the enemy can use to dominate movement along the route, such as choke points, ambush sites, and pickup zones, landing zones, and drop zones
- Reconnoiter all built-up areas, contaminated areas, and lateral routes along the route
- Evaluate and classify all bridges, defiles, overpasses and underpasses, and culverts along the route

- Locate any fords, crossing sites, or bypasses for existing and reinforcing obstacles (including built-up areas) along the route
- Locate all obstacles and create lanes as specified in execution orders
- Report the above route information to the headquarters initiating the route reconnaissance mission, to include providing a sketch map or a route overlay

Note: See FM 3-34.212 and FM 3-20.95 for additional information concerning route reconnaissance.

Conducting the Route Reconnaissance - A Small Unit Perspective

Reconnaissance teams develop the picture of the battlefield for the commander. Reconnaissance provides information that is critical to the process of intelligence preparation of the battlespace (IPB). The commander employs his intelligence, surveillance, and reconnaissance (ISR) assets as either a "recon push" or "recon pull".

Recon push

Recon push means that prior to a battle, the commander identifies one or more named area of interest (NAI) for a recon mission. In this manner, the recon team is pushed toward an objective within the area of operation (AO) and will develop the situation for the commander.

Recon Pull

Recon pull means that during a battle, the commander identifies an enemy or geographical objective for one or more recon teams. In this manner, each recon team has significant latitude of movement throughout the AO in order to gain contact with the enemy objective, develop the situation, and pull the main force, altering the direction of advance in order to engage the enemy.

Steps

1. In the case of route recon, the patrol typically begins from the assembly area (AA) and the start point serves in the same manner as a release point. During a route recon, a highly mobile reserve force may follow a safe distance behind the recon team. In such cases, the PL issues a contingency plan for both the recon team and the reserve force.

2. The patrol moves along a designated direction of advance or LOC. At each phase line and NAI, the recon team conducts a security halt for the entire patrol—including the reserve force. This security halt becomes an ERP. From the ERP, the recon team gathers information using the butterfly technique.

3. The distance each recon team will travel out from the route depends greatly on the visibility of the terrain and/or the size of the obstacle. The recon team gathers all information regarding road conditions, obstacles, bridges, enemy activity, civilian traffic, and natural choke points along parallel terrain where the enemy might impose a threat along the route.

4. Upon returning to the security halt, each recon team disseminates information among the patrol members. A written record is kept to log all information and the PL will report to higher command after each rendezvous back at the ERP.

5. The recon patrol continues on route to the next phase line or NAI and the entire process repeats itself until the patrol has adequately navigated the entire route. At that time, the PL may be required to move the recon patrol back to the FLOT or the patrol may be tasked to another mission. In either case, the PL must make a full report to higher command.

B. The Zone Reconnaissance

Zone reconnaissance is a form of reconnaissance that involves a directed effort to obtain detailed information on all routes, obstacles, terrain, and enemy forces within a zone defined by boundaries. It is appropriate when the enemy situation is vague, existing knowledge of the terrain is limited, or combat operations have altered the terrain. A zone reconnaissance may include several route or area reconnaissance missions assigned to subordinate units.

A zone reconnaissance is normally a deliberate, time-consuming process. It takes more time than any other reconnaissance mission, so the commander must allow adequate time to conduct it. A zone reconnaissance is normally conducted over an extended distance. It requires all ground elements executing the zone reconnaissance to be employed abreast of each other. However, when the reconnaissance objective is the enemy force, a commander may forgo a detailed reconnaissance of the zone and focus his assets on those named areas of interest (NAI) that would reveal enemy dispositions and intentions.

Zone Reconnaissance Tasks

Unless the commander orders otherwise, a unit conducting a zone reconnaissance performs the following tasks within the limits of its capabilities. A commander issue guidance on which tasks the unit must complete or the priority of tasks, which is usually clear from the reconnaissance objective. Tasks include:

- Find and report all enemy forces within the zone
- Clear all enemy forces in the designated AO within the capability of the unit conducting reconnaissance
- Determine the trafficability of all terrain within the zone, including built-up areas
- Locate and determine the extent of all contaminated areas in the zone
- Evaluate and classify all bridges, defiles, overpasses, underpasses, and culverts in the zone
- Locate any fords, crossing sites, or bypasses for existing and reinforcing obstacles (including built-up areas) in the zone
- Locate all obstacles and create lanes as specified in execution orders
- Report the above information to the commander directing the zone reconnaissance, to include providing a sketch map or overlay

Conducting the Zone Reconnaissance - A Small Unit Perspective

A zone recon collects detailed information of the terrain, obstacles, routes, and enemy forces within a specified zone that has been designated by boundaries on a map.

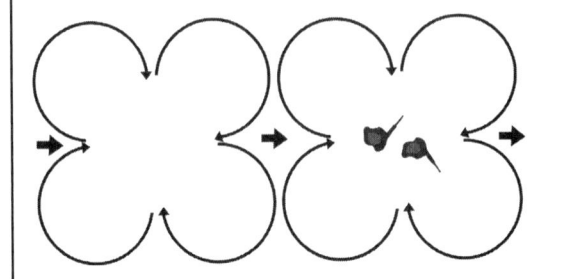

This method of conducting a zone reconnaissance uses a series of rally points to stop and then dispatch the recon teams. The patrol then moves to the next rally point. Zone recon tends to be a time consuming mission. (Ref: FM 7-92, chap 4, fig. 4-6)

1. Upon occupying the ORP, the PL confirms the location of the recon objective. All special equipment is prepared for use and plans are finalized. In this case, the ORP acts more like an assault position in that the patrol will not defend the ORP, nor will they return to the ORP.

The converging route method uses multiple recon teams—and therefore requires larger numbers of troops. Rally points are established and the multiple recon teams move along a designated lane or terrain feature. (Ref: FM 7-92, chap 4, fig. 4-7)

2. **Converging Route Method:** This method uses multiple recon teams that do NOT meet again until they rendezvous at the far side of the zone! This means that a rally point is clearly established on an easily recognized terrain feature or landmark at the far side of the recon objective. Additionally, a near recognition signal is established to ensure that the fireteams do not mistakenly fire upon each other!
3. At the release point, the PL divides the patrol into the assigned recon teams. The PL must travel with one of these teams. The PL issues each recon team a contingency plan.
4. The recon teams depart the release point and move along parallel directions of advance through the specified zone. The recon teams stop at each phase line and NAI to conduct recons using the butterfly technique—which makes use of a series of designated, en route rally points. The recon team pays particular attention to signs of enemy activity:
   - Fresh trash and cigarette butts indicate recent enemy activity
   - Boot prints and bent vegetation indicate the direction of enemy travel
   - The number of pressed vegetation spots at rest stops indicate the size of enemy patrols

Ultimately, the recon teams look for enemy movement, the size of their patrols, and established routes of movement. Also of great importance is the type of activity. For example, is the enemy running re-supply routes or ambush patrols? All enemy resources—outposts, water points, and obstacles such as a minefield—are thoroughly investigated.

5. Each time the recon teams reassemble in a rally point, information is disseminated and reported back to the PL, if possible. The PL in turn makes reports back to higher command.
6. The recon teams rendezvous at the far end of the objective and the PL takes charge of all collected information, disseminating the information to the patrol members and issuing another report to higher command.
7. The patrol moves back to the FLOT via an alternative route. It is generally not a good idea to return using the same route as the reconnaissance. To do so would invite the possibility of an enemy ambush, if in fact the enemy observed your movement. Upon arrival, the PL reports directly to the commander.

(Tactical Enabling Tasks) II. Reconnaissance 5-15

C. The Area Reconnaissance

Area reconnaissance is a form of reconnaissance that focuses on obtaining detailed information about the terrain or enemy activity within a prescribed area. This area may include a town, a ridgeline, woods, an airhead, or any other feature critical to operations. The area may consist of a single point, such as a bridge or an installation. Areas are normally smaller than zones and are not usually contiguous to other friendly areas targeted for reconnaissance.

Area Reconnaissance Tasks
The tasks for an area reconnaissance are the same as for a zone reconnaissance, to include:
- Find and report all enemy forces within the zone
- Clear all enemy forces in the designated AO within the capability of the unit conducting reconnaissance
- Determine the trafficability of all terrain within the zone, including built-up areas
- Locate and determine the extent of all contaminated areas in the zone
- Evaluate and classify all bridges, defiles, overpasses, underpasses, and culverts in the zone
- Locate any fords, crossing sites, or bypasses for existing and reinforcing obstacles (including built-up areas) in the zone
- Locate all obstacles and create lanes as specified in execution orders
- Report the above information to the commander directing the zone reconnaissance, to include providing a sketch map or overlay

Conducting the Area Reconnaissance - A Small Unit Perspective

The area recon has the objective of obtaining detailed information on an identified terrain feature, man-made feature, or enemy force within a specified area. This area is much smaller in terms of space than the objectives for the other forms of recon and thereby allows for a much smaller recon team.

1. Upon occupying the ORP, the PL confirms the location of the recon objective. All special equipment is prepared for use and plans are finalized. The PL issues contingency plans to the recon team(s) and to the security team left to defend the ORP.

2. The PL moves to the release point and dispatches the recon team(s) to approach the objective. Typically, the PL moves back to the ORP or may opt to accompany one of the recon teams if there are limited troops available. This decision is made in the planning phase of the mission.

Single-Team Method
3. Using the clock method, the recon team approaches the recon objective from its closest point. This point now becomes the six o'clock position of the objective. The recon team will observe the enemy force, the enemy activity, and obstacles using stealth.

4. The recon team records all pertinent information:
- A sketch of their vantage point, including terrain and enemy structures
- An exact number of enemy personnel sighted
- A descriptive list of all equipment, uniforms and markings
- The time of guard shift, eating, and sleeping shift rotations
- The direction, time, and size of any patrols coming or going from the area

5. The recon team then withdraws and maneuvers to the approximate nine o'clock position. The recon team repeats their observation and recording activities. The recon team continues on to the twelve o'clock position and then the three o'clock position, obtaining information from all four vantage points—if such a maneuver is possible. At a bare minimum, the recon team must observe the targeted area from two vantage points.

Double-Team Method

3. When the troops are available, the PL may choose to dispatch two recon teams to the objective. This speeds the gathering of information, but also creates a very real danger of fratricide. To coordinate this so that there is little chance of the two recon teams actually making contact, the teams are assigned opposite vantage points.

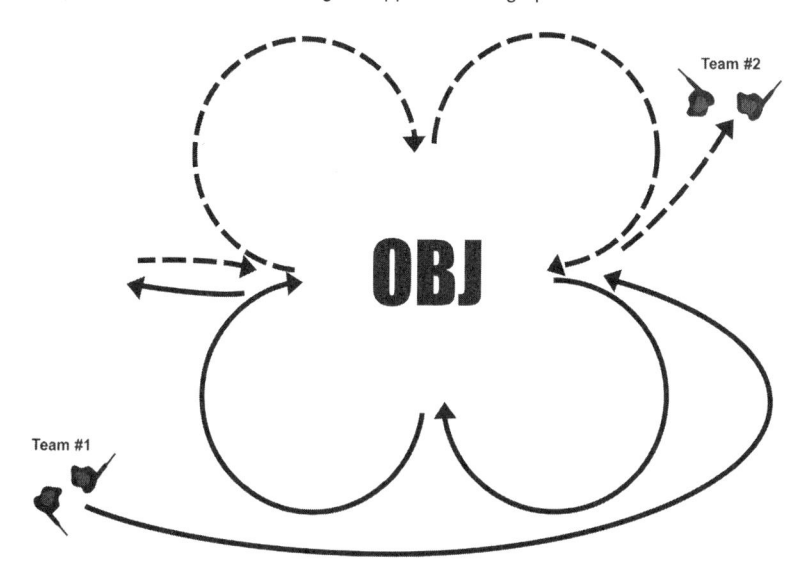

The double team method of the area recon uses multiple, small teams to capitalize on time and place the "smallest footprint" forward. This is important when stealth is a factor. (Ref: FM 7-92, chap 4, fig. 4-5)

4. Both teams are dispatched together and proceed toward the closest point of the targeted area. Upon locating this six o'clock position, a designated recon team would then withdraw slightly and proceed to the approximate twelve o'clock position.

5. Both recon teams record the same pertinent information as list for the single-team method. However, the recon team at the six o'clock position will proceed only to the vantage point of the nine o'clock position and then return to the ORP. Likewise, the recon team at the twelve o'clock position will proceed only to the three o'clock vantage point, and then return to the ORP.

Dissemination of Information

6. Once the recon team(s) returns to the ORP, all information is immediately disseminated among every member of the recon patrol. If there is time, additional copies of the sketches and record lists should be made. This will increase the likelihood of the gathered information making it back to the commander if the patrol is ambushed in route.

7. The recon team returns to the forward line of troops (FLOT) via a pre-designated route. Upon arrival, the PL reports directly to the commander.

D. Reconnaissance in Force (RIF)

A reconnaissance in force is a deliberate combat operation designed to discover or test the enemy's strength, dispositions, and reactions or to obtain other information. Battalion-size task forces or larger organizations usually conduct a reconnaissance in force (RIF) mission. A commander assigns a RIF mission when the enemy is known to be operating within an area and the commander cannot obtain adequate intelligence by any other means. A unit may also conduct a RIF in restrictive-type terrain where the enemy is likely to ambush smaller reconnaissance forces. A RIF is an aggressive reconnaissance, conducted as an offensive operation with clearly stated reconnaissance objectives. The overall goal of a RIF is to determine enemy weaknesses that can be exploited.

Reconnaissance in Force Tasks

A unit conducting a RIF performs the following tasks within the limits of its capabilities. If a unit does not have the time or resources to complete all of these tasks, it must inform the commander assigning the mission. He must then issue further guidance on which tasks the unit must complete or the priority of tasks, which is usually clear from the reconnaissance objective. Tasks include:

- Penetrating the enemy's security area and determining its size and depth
- Determining the location and disposition of enemy main positions
- Attacking enemy main positions and attempting to cause the enemy to react by using local reserves or major counterattack forces, employing fire support assets, adjusting positions, and employing specific weapon systems
- Determining weaknesses in the enemy's dispositions to exploit

E. Special Reconnaissance

Special reconnaissance includes reconnaissance and surveillance actions conducted as a special operation in hostile, denied, or politically sensitive environments to collect or verify information of strategic or operational significance, employing military capabilities not normally found in conventional forces (JP 3-05).

On Point

A commander needs to know what lies ahead of the friendly force before an engagement, during an engagement, and after an engagement. The reconnaissance team acts as the eyes and ears for the commander. This is absolutely essential for situational awareness.

Reconnaissance is the means by which a commander "sees" the battlefield. In spite of all the wonderful tools available for conducting surveillance, the task of recon is by and large a human endeavor. That means boots on the ground.

Contact with the enemy does not necessarily mean to engage the enemy in a firefight. For the sake of reconnaissance, contact means to observe the enemy, develop the situation, and pass back the pertinent information that the commander needs in order to make a sound tactical decision.

The principles of Mission, Enemy, Terrain and weather, Time available, Troops available, Civilians on the battlefield (METT-TC) dictate the reconnaissance effort. Though each principle consideration is important, the consideration of time is critical in establishing how much stealth is required.

- Unlimited time + great detail needed means that considerable stealth is required
- Limited time + little detail needed means that considerable speed is required

Chap 5
Tactical Enabling Tasks
III. Relief in Place

Ref: ADRP 3-90, Offense and Defense (Aug '12), pp. 5-5 to 5-6. and FM 3-21.10 (FM 7-10) The Infantry Rifle Company, pp. 8-8 to 8-12.

A relief in place is an operation in which, by the direction of higher authority, all or part of a unit is replaced in an area by the incoming unit. A commander conducts a relief in place as part of a larger operation, primarily to maintain the combat effectiveness of committed units. A relief in place may also be conducted--

- To reorganize, reconstitute, or re-equip a unit that has sustained heavy losses
- To rest units that have conducted sustained operations
- To establish the security force or the detachment left in contact (DLIC) during a withdrawal operation
- To allow the relieved unit to conduct another operation

A unit may relieve another unit by assuming their same defensive position, or the new unit may opt to develop new defensive positions that afford it a better means to coordinate fires against the enemy within the engagement area. (Photo by Jeong, Hae-jung).

The higher headquarters directs when and where to conduct the relief and establishes the appropriate control measures. Normally, the unit relieved is defending. However, a relief may set the stage for resuming offensive operations. A relief may also serve to free the relieved unit for other tasks, such as decontamination, reconstitution, routine rest, resupply, maintenance, or specialized training. Sometimes, as part of a larger operation, a commander wants the enemy force to discover the relief, because that discovery might cause it to do something in response that is prejudicial to its interest, such as move reserves from an area where the friendly commander wants to conduct a penetration.

Conducting the Relief in Place - A Small Unit Perspective

I. Organization

Both units involved in a relief in place should be of similar type—such as mounted or dismounted—and task organized to help maintain operations security (OPSEC). The relieving unit usually assumes as closely as possible the same task organization as the unit being relieved. It assigns responsibilities and deploys in a configuration similar to the relieved unit.

The relieving unit establishes advance parties to conduct detailed coordination and preparations for the operation, down to the company level and possibly to the platoon level. These advance parties infiltrate forward to avoid detection. They normally include the echelon's tactical command post, which co-locates with the main headquarters of the unit being relieved. The commander may also attach additional liaison personnel to subordinate units to ensure a smooth changeover between subordinate units.

II. Planning & Preparation

Once ordered to conduct a relief in place, the commander of the relieving unit contacts the commander of the unit to be relieved. The co-location of unit command posts also helps achieve the level of coordination required. If the relieved unit's forward elements can defend the AO, the relieving unit executes the relief in place from the rear to the front. This facilitates movement and terrain management.

Hasty or Deliberate

A relief is either deliberate or hasty, depending on the amount of planning and preparations. The major differences are the depth and detail of planning and, potentially, the execution time. Detailed planning generally facilitates shorter execution time by determining exactly what the commander believes he needs to do and the resources needed to accomplish the mission. Deliberate planning allows the commander and his staff to identify, develop, and coordinate solutions to most potential problems before they occur and to ensure the availability of resources when and where they are needed.

In a deliberate relief, units exchange plans and liaison personnel, conduct briefings, perform detailed reconnaissance, and publish orders with detailed instructions. In a hasty relief, the commander abbreviates the planning process and controls the execution using oral and fragmentary orders.

Preparations begin with an exchange of liaison officers and NCOs from both the forward unit and the relieving unit. The effort is to identify maneuver control measures. At a minimum this means identifying the AO boundaries, assembly area (AA), routes, release points, and battle positions. Furthermore, each battle position must identify the engagement area, target reference points (TRP), defensive fires measures, and fire support measures to the incoming friendly forces.

The final preparation focuses on whether there will be new positions created, or old positions simply occupied by the incoming friendly force. Obviously, the activity and noise inherent in the construction of new fighting positions is a dead giveaway that a relief in place is underway. If stealth is an important factor, then simply assuming the older battle positions would be best. However, often enough the relief in place also has the intention of developing a more advantageous battle position. This must also be planned and coordinated.

III. Execution

Techniques: Sequential, Simultaneous or Staggered

There are three techniques for conducting a relief: sequentially, simultaneously, or staggered. A sequential relief occurs when each element within the relieved unit is relieved in succession, from right to left or left to right, depending on how it is deployed. A simultaneous relief occurs when all elements are relieved at the same time. A staggered relief occurs when the commander relieves each element in a sequence determined by the tactical situation, not its geographical orientation. Simultaneous relief takes the least time to execute, but is more easily detected by the enemy. Sequential or staggered reliefs can take place over a significant amount of time.

In practice, small units almost always conduct the relief simultaneously. While it is true at higher echelons, larger units may phase the relief in place sequentially or staggered, small units such as companies and platoons almost always assume their position en mass in a simultaneous manner.

1. Upon receiving the warning order (WARNO), the fresh unit will exchange of liaison officers and/or NCOs with the forward unit. This is done to achieve three effects:
   - To identify the forward unit's battle positions as well as friendly and enemy situations
   - To form the relieving unit into a similar task organization as the forward unit
   - To coordinate link up operations

2. The plan is developed and rehearsed. This plan will balance the need for speed with the need for stealth. Similarly, the plan will dictate whether new battle positions will be created, or the forward unit's battle positions will be assumed instead. There is considerable coordination that must take place at headquarters-level to ensure that the relief takes place smoothly; that security is maintained for force protection; and that any chance of fratricide is mitigated—particularly if the relief will take place in hours of limited visibility.

3. An advanced guard of the fresh unit conducts the relief first. They will move forward of the forward line of troops (FLOT) to conduct screening operations. This activity is intended to deny observation to enemy reconnaissance teams and provide early warning of an enemy attack.

4. The fresh unit moves into the AO and halts in the AA. The liaison coordinates for a guide, and the fresh unit moves along its assigned route to a release point immediately behind the battle position. At the release point, the two leaders meet (from the incoming fresh unit and the forward unit) to conduct final coordination.

5. At a previously designated time, signal, or condition, the subordinate platoons, squad, and fireteams guide into and assume the fighting positions of the forward unit—with the forward unit troops still in place!

6. Once the fresh unit has assumed all of the fighting positions within the sector, the Forward unit commander passes command of the battle position to the new unit commander. The passage of command is communicated by radio, field phone, or a pre-designated signal.

7. On order, the relieved unit now begins to egress back through the release point and follows the same route back to their assigned AA. Typically the last fighting positions to withdraw are the crew-served weapon positions. Once all other relieved troops have fallen back, these crews will follow suit.

8. A variation on the relief in place operation may take place if the intent of the fresh unit is to immediately transition into offensive action. In this case, the forward unit does not relinquish command, and at a minimum, the crew-served weapons are maintained in place to provide cover fire for the fresh unit's advance into the engagement area.

On Point

Units cannot remain on the front line indefinitely. Troops become exhausted and complacent, and the decision-making process is degraded. For this reason, a scheduled relief of the forward unit may take place. There are additional reasons a unit may require relief. In the event the forward unit has come under chemical attack, or experienced unusually high casualties the forward unit may also be relieved.

The relief in place operation is directed by a higher authority for a fresh unit to assume the duties and geographic area of operations (AO) of the forward unit. This may be conducted deliberately or hastily depending on the amount of time available for planning. As a general rule of thumb, the more planning, the quicker and smoother the relief in place goes.

The enemy in front of the unit to be relieved almost always detects the relief in place. This is due to the increased activity and noise, as well as the appearance of new or different equipment. The hastier the relief in place is conducted, the quicker the enemy is to recognize the relief.

The relief in place should be a routine operation. However, anecdotal evidence indicates a disproportional amount of incidents of fratricide occur during relief in place and passage of lines operations. The relief in place leaves the forces massed to the point of having little to no room for maneuver and vulnerable to enemy attack.

To complicate things further, the relief in place is often conducted during conditions of limited visibility, and when relieving the FLOT there is always the possibility of a firefight. That does not present such a significant threat to the forward unit, per se. After all, they have a better grasp of situational awareness—the direction and capability of the enemy, assigned sectors of fire, and an understanding of where the other friendly fighting positions are located. Furthermore, the forward unit troops are relatively secure within their fighting positions for most of the relief in place.

This is not true for the relieving unit coming in fresh. They are not familiar to the enemy positions, directions, or even capabilities. The relieving unit is not fully aware of the location of every forward unit's position. And the relieving unit is moving forward—outside the protection of the fighting positions.

Both units should make every effort to keep the enemy from knowing about the relief. Try to conduct the relief during limited visibility to reduce the risk of discovery by a capable threat.

The dispositions, activities, and radio traffic of the relieved unit must be maintained throughout the relief. Both companies should be on the relieved company's net. The relieved company continues routine traffic, which the relieving company monitors. Once the relief is complete and on a prearranged signal, the relieving company changes to their assigned frequency. Security activities, such as OPs and patrols, must maintain the established schedule. This might require some personnel from the relieving unit being placed under operational control (OPCON) of the relieved unit before the relief.

Chap 5
Tactical Enabling Tasks
IV. Passage of Lines

Ref: ADRP 3-90, Offense and Defense (Aug '12), pp. 5-5 to 5-6; FM 7-85 Ranger Unit Operations, chap 6; and FM 3-21.10, The Infantry Rifle Company, pp. 8-13 to 8-17.

Passage of lines is an operation in which a force moves forward or rearward through another force's combat positions with the intention of moving into or out of contact with the enemy. A passage may be designated as a forward or rearward passage of lines (JP 1-02). A commander conducts a passage of lines to continue an attack or conduct a counterattack, retrograde security or main battle forces, and anytime one unit cannot bypass another unit's position. It involves transferring the responsibility for an area of operations between two commanders. That transfer of authority usually occurs when roughly two-thirds of the passing force has moved through the passage point. If not directed by higher authority, the unit commanders determine—by mutual agreement—the time to pass command.

The patrol forms into a tight 360-degree assembly area just behind the forward line of troops. The patrol leader coordinates with the forward unit commander. (Photo by Jeong, Hae-jung).

A passage of lines occurs under two basic conditions. A forward passage of lines occurs when a unit passes through another unit's positions while moving toward the enemy. A rearward passage of lines occurs when a unit passes through another unit's positions while moving away from the enemy. Reasons for conducting a passage of line include—

- Sustain the tempo of an offensive operation
- Maintain the viability of the defense by transferring responsibility from one unit to another
- Transition from a delay or security operation by one force to a defense
- Free a unit for another mission or task

(Tactical Enabling Tasks) IV. Passage of Lines 5-23

I. Conducting a Passage of Lines - A Small Unit Perspective

Passage of lines consists of essentially two tasks, (1) coordinating the time and place of the *departure*, and (2) coordinating the time, place and signals of the *reentry* through the FLOT. This means we identify a time and a passage point for both phases. The forward unit commander identifies the passage lane, which is the precise route our patrol takes through his defensive position. Use the following for planning a passage of lines:

1. Contact and coordinate with the forward unit commander
2. Move to an assembly area (AA) behind the passage point
3. Link up with the guide to depart the FLOT
4. Conduct a security halt past the forward edge of battle area (FEBA)
5. Complete the mission
6. Return to the passage point
7. Render the far & near recognition signal to the forward unit commander
8. Link up with the guide to reenter the FLOT
9. Move into the AA to debrief the patrol

Let's break it down into two parts, departing the FLOT and reentering the FLOT. The process is more detailed; however the list will make sense as the mission planning is conceptualized.

Departing the Forward Line of Own Troops (FLOT)

1. Communicate and coordinate with the forward unit commander. The PL coordinates the time and place of the patrol's departure and reentry with the forward unit commander. The PL chooses an appropriate time. The forward unit commander chooses the appropriate passage lane through his defense. The forward unit commander also assigns a guide to lead the patrol through the wire and mine obstacles forward of the FLOT.

2. Move to the AA behind the passage point. Here the final planning, rehearsals, and coordination with the forward unit commander takes place. For this coordination, the forward unit commander supplies the following information to the patrol:

- An orientation on terrain
- Known or suspected enemy positions
- Recent enemy activity
- The location of friendly OP/LP and obstacles—wire and minefields
- Available combat support—guides, fire support, medevac and reaction forces

The PL provides the following information to the forward unit commander:

- The patrol's unit designation
- The size of the patrol
- The departure and reentry times
- All coordinating signals—near and far recognition

3. Link up with the guide to depart the FLOT through the passage lane. Once the patrol has occupied the AA, the forward unit commander links the PL with a guide. The PL restates the patrol's departure and reentry times, and near and far recognition signals for the guide. The guide is then introduced to the patrol's pointman and dragman. This allows the guide to recognize the beginning and ending of the patrol as he counts each member through the passage lane.

The PL also establishes a time limit for which the guide will wait on the far side of the passage lane. This measure allows the guide to lead the patrol back through the passage lane if the patrol is attacked in the FEBA and needs to reenter the FLOT quickly!

4. Conduct a security halt past the FEBA. As soon as the patrol passes through the passage lane of the FLOT, the patrol enters the FEBA. The patrol moves to the far side of the FEBA, seeks adequate concealment, and conducts a security halt. This first security halt is called the "listening halt." Every member of the patrol sits comfortably and then removes their headgear. Making no noise, the patrol must listen to the indigenous sounds for about five minutes. This lets the patrol member's eyes and ears adjust to the new environment.

5. Complete the Mission. When the PL is comfortable that the patrol is safe and has adjusted to the noise, sights, and smells, the patrol continues towards the objective. When the mission is complete, the patrol will return to the FLOT. Under normal circumstances the patrol returns using a different route than the patrol's advance. This reduces the opportunity for the enemy to ambush the patrol.

It might be worth mentioning here that depending on the nature of the mission, the patrol might not return to the FLOT…in which case such coordination won't be necessary. Recognizing that exception, the norm is to always plan and coordinate to return.

Reentering the Forward Line of Own Troops (FLOT)

1. Return to the passage point. The patrol returns to a secure position on the far side of the FEBA using a pre-determined route. From this position the PL contacts the forward unit commander and coordinates to reenter the FLOT.

2. Render the far & near recognition signal to the forward unit commander. Once in position, the patrol makes contact with the forward unit commander/guide using the designated far recognition signal. It is the patrol's intention to make contact with the far recognition signal first. Subsequently, the patrol makes contact using the near recognition signal. The far recognition signal uses distance signaling, such as a radio broadcast, or smoke signal, or flare/light signal. The near recognition signal uses verbal contact, such as a password, or visual recognition of each other.

If for any reason contact cannot be made with the far recognition signal, the patrol dispatches a small element forward to make contact with the FLOT using the near recognition signal to coordinate our reentry.

During nighttime operations, if the patrol cannot make contact with the forward unit commander/guide using the far recognition signal, it is customary that the patrol waits in security position until daylight. Only during daylight will the patrol render a near recognition signal without first making contact with the far recognition signal.

3. Link up with the guide to reenter the FLOT through the passage lane. After the patrol makes contact with the FLOT, a security team crosses the FEBA to render the near recognition signal with the guide. A member of the security team then returns to the patrol and leads the remainder of the patrol across the FEBA to the guide waiting at the passage lane. The PL links up with the guide and counts the patrol members in *by name*. This lets the guide know that no enemy has slipped into the patrol and infiltrated their defensive line.

4. Move into the AA to debrief the patrol. The patrol moves back into the AA behind the FLOT. The patrol is debriefed either by the forward unit commander or representative of higher command. The patrol discusses the nature and findings of the mission on a "need to know basis."

II. Organization

A unit may participate in a passage of lines as either the passing or stationary force. Except for co-locating command posts and providing for guides by the stationary force, conducting a passage of lines does not require a special task organization.

The organization of the patrol depends more on the primary mission than the task of passage of lines. A forward passing unit's order of march is generally reconnaissance and security elements first.

The forward friendly unit provides a guide through friendly obstacles (mines and wire). The guide waits at this position for a pre-determined length of time in case the patrol needs to fall back through the passage lane. (Photo by Jeong, Hae-jung).

On Point

A passage of lines is a complex process at higher levels of command in which the commander must move hundreds or even thousands of troops through the forward lines of another friendly unit. When two friendly units converge upon the same space on a battlefield, there is always the potential to mistake each other for enemy. Fratricide is a very real danger in these cases, particularly at night.

For small tactical units, such as the fire team, squad, platoon, and company the process of departing and reentering the forward line is much simpler—though the danger of fratricide is the same. For this reason, the patrol leader coordinates carefully with the forward unit commander and supporting unit commander.

The process of a passage of lines serves as coordination between two units when one unit either moves forward or rearward through the defensive line of a stationary unit. More specifically, a unit may depart forward of the forward line of troops (FLOT) in order to maneuver against an enemy force. Conversely, a unit may reenter rearward of the FLOT in order to maneuver away from an enemy force.

Chap 5

Tactical Enabling Tasks

V. Encirclement Opns

Ref: ADRP 3-90, Offense and Defense (Aug '12), chap 5, pp. 5-6 to 5-7.

Encirclement operations are operations where one force loses its freedom of maneuver because an opposing force is able to isolate it by controlling all ground lines of communications and reinforcement. A unit can conduct offensive encirclement operations designed to isolate an enemy force or conduct defensive encirclement operations as a result of the unit's isolation by the actions of an enemy force. Encirclement operations occur because combat operations involving modernized forces are likely to be chaotic, intense, and highly destructive, extending across large areas containing relatively few units as each side maneuvers against the other to obtain positional advantage.

I. Offensive Encirclement Operations

The commander conducts offensive encirclements to isolate an enemy force. Typically, encirclements result from penetrations and envelopments, or are extensions of exploitation and pursuit operations. As such, they are not a separate form of offensive operations but an extension of an ongoing operation. They may be planned sequels or result from exploiting an unforeseen opportunity. They usually result from the linkup of two encircling arms conducting a double envelopment. However, they can occur in situations where the attacking commander uses a major obstacle, such as a shoreline, as a second encircling force. Although a commander may designate terrain objectives in an encirclement, isolating and defeating enemy forces are the primary goals. Ideally, an encirclement results in the surrender of the encircled force. This minimizes friendly force losses and resource expenditures.

II. Defending Encircled

An encircled force can continue to defend encircled, conduct a breakout, exfiltrate toward other friendly forces, or attack deeper into enemy-controlled territory. The commander's form of maneuver once becoming encircled depends on the senior commander's intent and the mission variables of mission, enemy, terrain and weather, troops and support available, time available, and civil considerations (METT-TC), including the—

- Availability of defensible terrain
- Relative combat power of friendly and enemy forces
- Sustainment status of the encircled force and its ability to be resupplied, including the ability to treat and evacuate wounded Soldiers
- Morale and fighting capacity of the Soldiers

Encirclement of a friendly force is likely to occur during highly mobile and fluid operations, or when operating in restrictive terrain. A unit may find itself encircled as a result of its offensive actions, as a detachment left in contact, when defending a strong point, when occupying a combat outpost, or when defending an isolated defensive position. The commander anticipates becoming encircled when assigned a stay-behind force mission, or when occupying either a strong point or a combat outpost. The commander then makes the necessary preparations.

The senior commander within an encirclement assumes command over all encircled forces and takes immediate action to protect them. In the confusion leading to an encirclement, it may be difficult to even determine what units are being encircled, let alone identify the senior commander. However, the senior commander must be determined as quickly as possible. When that commander determines the unit is about to be encircled, the commander must decide quickly what assets stay and what assets leave. The commander immediately informs higher headquarters of the situation. Simultaneously, the commander directs the accomplishment of the following tasks—

- Establish security
- Reestablish a chain of command
- Establish a viable defense
- Maintain morale

The commander positions security elements as far forward as possible to reestablish contact with the enemy and provide early warning. Vigorous patrolling begins immediately. Each unit clears its position to ensure that there are no enemy forces within the perimeter. Technical assets, such as Joint Surveillance Target Attack Radar System (JSTARS) and electronic warfare systems, augment local security and locate those areas along the perimeter where the enemy is deploying additional forces.

The commander reestablishes unity of command. The commander reorganizes any fragmented units and places Soldiers separated from their parent units under the control of other units. The commander establishes a clear chain of command throughout the encircled force, reestablishes communications with units outside the encirclement, and adjusts support relationships to reflect the new organization.

A. Breakout from an Encirclement

A breakout is an operation conducted by an encircled force to regain freedom of movement or contact with friendly units. It differs from other attacks only in that a simultaneous defense in other areas of the perimeter must be maintained. A breakout is both an offensive and a defensive operation. An encircled force normally attempts to conduct breakout operations when one of the following four conditions exist:

- The commander directs the breakout or the breakout falls within the intent of a higher commander
- The encircled force does not have sufficient relative combat power to defend itself against enemy forces attempting to reduce the encirclement
- The encircled force does not have adequate terrain available to conduct its defense
- The encircled force cannot sustain itself long enough to be relieved by forces outside the encirclement

B. Exfiltration

If the success of a breakout attack appears questionable, or if it fails and a relief operation is not planned, one way to preserve a portion of the force is through organized exfiltration.

Friendly forces exfiltrate when they have been encircled by enemy forces and cannot conduct a breakout or be relieved by other friendly forces. Forces returning from a raid, an infiltration, or a patrol behind enemy lines can also conduct an exfiltration. The commander exfiltrates an encircled force to preserve a portion of the force; it is preferable to the capture of the entire force. A force exfiltrates only after destroying or incapacitating all equipment (less medical) it must leave behind. Only as a last resort, when the alternative is the capture of the entire force, does a force conducting an exfiltration leaves its casualties in place with supplies, chaplain support, and medical attendants.

FM 3-90, app B, describes exfiltration in more detail.

Chap 5

Tactical Enabling Tasks
VI. Troop Movement

Ref: ADRP 3-90, Offense and Defense (Aug '12), chap 5, pp. 5-4 to 5-5.

Troop movement is the movement of troops from one place to another by any available means. The ability of a commander to posture friendly forces for a decisive or shaping operation depends on the commander's ability to move that force. The essence of battlefield agility is the capability to conduct rapid and orderly movement to concentrate combat power at decisive points and times. Successful movement places troops and equipment at their destination at the proper time, ready for combat. The three types of troop movement are administrative movement, tactical road march, and approach march.

I. Methods of Troop Movement

Troop movements are made by dismounted and mounted marches using organic combat and tactical vehicles and motor transport air, rail, and water means in various combinations. The method employed depends on the situation, the size and composition of the moving unit, the distance the unit must cover, the urgency of execution, and the condition of the troops. It also depends on the availability, suitability, and capacity of the different means of transportation. Troop movements over extended distances have extensive sustainment considerations. When necessary, dismounted and mounted marches can be hurried by conducting a forced march.

A. Administrative Movement

A movement in which troops and vehicles are arranged to expedite their movement and conserve time and energy when no enemy interference, except by air, is anticipated (JP 1-02).

B. Tactical Road March

A tactical road march is a rapid movement used to relocate units within an area of operations to prepare for combat operations. Units maintain security against enemy air attack and prepare to take immediate action against an enemy ambush, although they do not expect contact with enemy ground forces. (If the moving unit anticipates making contact with significant enemy ground forces, it will use a mix of combat formations and movement techniques.)

The primary consideration of the tactical road march is rapid movement. However, the moving force employs security measures, even when contact with enemy ground forces is not expected. Units conducting road marches may or may not be organized into a combined arms formation. During a tactical road march, the commander is always prepared to take immediate action if the enemy attacks.

C. Approach March

An approach march is the advance of a combat unit when direct contact with the enemy is intended. However, it emphasizes speed over tactical deployment. Armored, Stryker, and infantry forces conduct tactical road marches and approach marches.

II. Movement Techniques
Ref: ADP 3-90, Offense and Defense (Aug '12), p. 5-5.

The commander uses the combat formations described in FM 3-90 in conjunction with three movement techniques: traveling, traveling overwatch, and bounding overwatch. The following figure illustrates when a unit is most likely to use each technique.

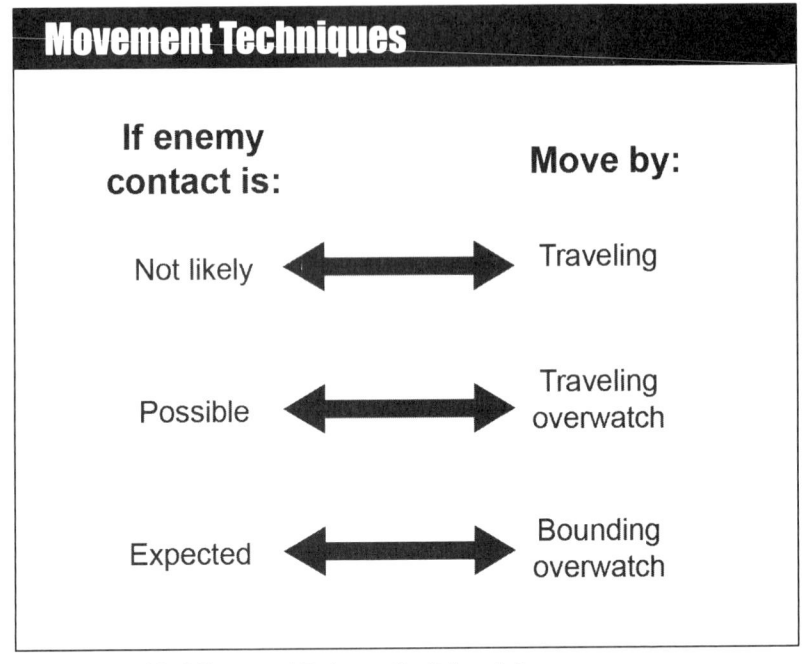

Movement Techniques

If enemy contact is:	Move by:
Not likely	Traveling
Possible	Traveling overwatch
Expected	Bounding overwatch

Ref: ADRP 3-90, Offense and Defense, fig. 5-1, p. 5-5.

See also pp. 8-7 to 8-10 for discussion of the traveling techniques as applied in patrols and patrolling (dismounted).

Chap 6: Special Purpose Attacks

Ref: FM 3-90 Tactics, pp. 5-29 to 5-40.

An attack is an offensive operation that destroys or defeats enemy forces, seizes and secures terrain, or both. Movement, supported by fires, characterizes the conduct of an attack. However, based on his analysis of the factors of METT-TC, the commander may decide to conduct an attack using only fires. An attack differs from a MTC because enemy main body dispositions are at least partially known, which allows the commander to achieve greater synchronization. This enables him to mass the effects of the attacking force's combat power more effectively in an attack than in a MTC.

Special purpose attacks are ambush, spoiling attack, counterattack, raid, feint, and demonstration. The commander's intent and the factors of METT-TC determine which of these forms of attack are employed. He can conduct each of these forms of attack, except for a raid, as either a hasty or a deliberate operation.

This chapter specifically discusses "special purpose attacks." Chap. 2 discusses the attack (pp. 2-13 to 2-18) and other forms of the offense.

Special Purpose Attacks

- **I.** Ambush
- **II.** Raid
- **III.** Spoiling Attack
- **IV.** Counterattack
- **V.** Demonstration
- **VI.** Feint

Ref: FM 3-90 Tactics, pp. 5-29 to 5-40.

I. Ambush

An ambush is a form of attack by fire or other destructive means from concealed positions on a moving or temporarily halted enemy. It may include an assault to close with and destroy the engaged enemy force. In an ambush, ground objectives do not have to be seized and held.

Note: See pp. 6-3 to 6-16 for further discussion on the ambush.

Special Purpose Attacks 6-1

II. Raid

A raid is a form of attack, usually small scale, involving a swift entry into hostile territory to secure information, confuse the enemy, or destroy installations. It ends with a planned withdrawal from the objective area on mission completion. A raid can also be used to support operations designed to rescue and recover individuals and equipment in danger of capture.

Note: See pp. 6-17 to 6-22 for further discussion on the raid.

III. Spoiling Attack

A spoiling attack is a form of attack that preempts or seriously impairs an enemy attack while the enemy is in the process of planning or preparing to attack. The objective of a spoiling attack is to disrupt the enemy's offensive capabilities and timelines while destroying his personnel and equipment, not to secure terrain and other physical objectives. A commander conducts a spoiling attack whenever possible during friendly defensive operations to strike the enemy while he is in assembly areas or attack positions preparing for his own offensive operation or is temporarily stopped. It usually employs heavy, attack helicopter, or fire support elements to attack enemy assembly positions in front of the friendly commander's main line of resistance or battle positions.

IV. Counterattack

A counterattack is a form of attack by part or all of a defending force against an enemy attacking force, with the general objective of denying the enemy his goal in attacking. The commander directs a counterattack—normally conducted from a defensive posture—to defeat or destroy enemy forces, exploit an enemy weakness, such as an exposed flank, or to regain control of terrain and facilities after an enemy success. A unit conducts a counterattack to seize the initiative from the enemy through offensive action. A counterattacking force maneuvers to isolate and destroy a designated enemy force. It can attack by fire into an engagement area to defeat or destroy an enemy force, restore the original position, or block an enemy penetration. Once launched, the counterattack normally becomes a decisive operation for the commander conducting the counterattack.

V. Demonstration

A demonstration is a form of attack designed to deceive the enemy as to the location or time of the decisive operation by a display of force. Forces conducting a demonstration do not seek contact with the enemy.

VI. Feint

A feint is a form of attack used to deceive the enemy as to the location or time of the actual decisive operation. Forces conducting a feint seek direct fire contact with the enemy but avoid decisive engagement. A commander uses them in conjunction with other military deception activities. They generally attempt to deceive the enemy and induce him to move reserves and shift his fire support to locations where they cannot immediately impact the friendly decisive operation or take other actions not conducive to the enemy's best interests during the defense.

The principal difference between these forms of attack is that in a feint the commander assigns the force an objective limited in size, scope, or some other measure. Forces conducting a feint make direct fire contact with the enemy but avoid decisive engagement. Forces conducting a demonstration do not seek contact with the enemy. The planning, preparing, and executing considerations for demonstrations and feints are the same as for the other forms of attack.

Special Purpose Attacks
I. Ambush

Ref: FM 3-90 Tactics, pp. 5-29 to 5-34; FM 7-85 Ranger Unit Operations, chap 6; and FM 3-21.8 (FM 7-8) The Infantry Rifle Platoon and Squad, pp. 7-26 to 7-29.

An ambush is a form of attack by fire or other destructive means from concealed positions on a moving or temporarily halted enemy. It may take the form of an assault to close with and destroy the enemy, or be an attack by fire only. An ambush does not require ground to be seized or held. Ambushes are generally executed to reduce the enemy force's overall combat effectiveness. Destruction is the primary reason for conducting an ambush. Other reasons to conduct ambushes are to harass the enemy, capture the enemy, destroy or capture enemy equipment, and gain information about the enemy. Ambushes are classified by category (deliberate or hasty), formation (linear or L-shaped), and type (point, area, or antiarmor).

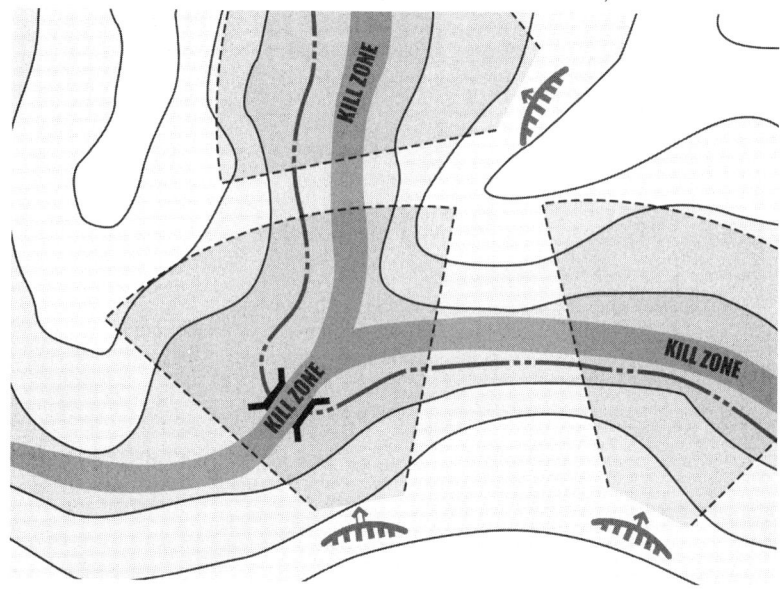

Area ambushes trap the enemy in a network of attacks from multiple concealed positions. These ambushes are carefully oriented to avoid "friendly fire" and yet maximize combat power against the enemy. (Ref: FM 7-8, chap 3, fig. 3-15).

The execution of an ambush is offensive in nature. However, a unit may be directed to conduct an ambush during offensive or defensive operations. An ambush normally consists of the following actions:
- Tactical movement to the objective rally point (ORP)
- Reconnaissance of the ambush site
- Establishment of the ambush security site
- Preparation of the ambush site
- Execution of the ambush
- Withdrawal

The intent of any ambush is to kill enemy troops and destroy enemy equipment. From a small unit perspective, how that is achieved and to what extent determines the difference in employing either a near ambush or a far ambush.

Near Ambush

The near ambush has the expressed purpose of destroying the target. This often requires an assaulting force to literally overrun the target after the initial volley of fire has inflicted tremendous damage. Again, the intent is to destroy everything.

Since the patrol will overwhelm the target, the patrol will get as close as possible to the enemy. This close proximity also means that friendly forces MUST outnumber the enemy target.

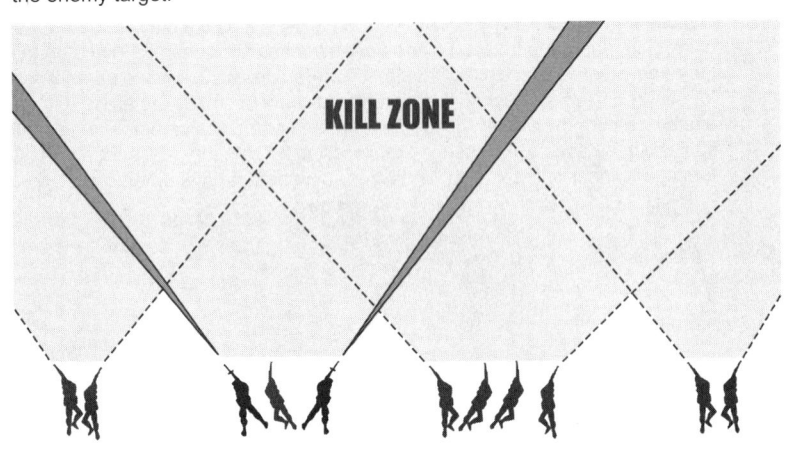

The near ambush has the express intent of overwhelming and destroying the enemy force. The near ambush masses close to the kill zone and requires careful fire coordination. The linear method offers the greatest simplicity. (Ref: FM 7-85, chap 6, fig. 6-1).

Far Ambush

The far ambush has only the purpose of injuring and/or delaying the target. This rarely ever calls for an assaulting force—since the patrol doesn't seek the complete destruction of the enemy force there is no need to risk the loss of friendly troops. The far ambush simply intends to *harass*.

A far ambush team can engage an enemy patrol of any size or type. It does not matter if the enemy force is larger than the patrol because significant distances are used, as well as natural obstacles of the terrain, and established routes of withdraw that allows the ambush patrol to escape before the enemy has time to organize an effective counterattack.

I. Organization

An ambush patrol will be broken into multiple teams, each with a very specific set of responsibilities. The Infantry platoon is normally task-organized into assault, support, and security elements for execution of the ambush.

Each team must be assigned a leader. The ambush will require special equipment for each assigned team. This equipment should be made available for the rehearsal as well, to ensure everything functions according to the execution plan. Each team has a specific set of duties.

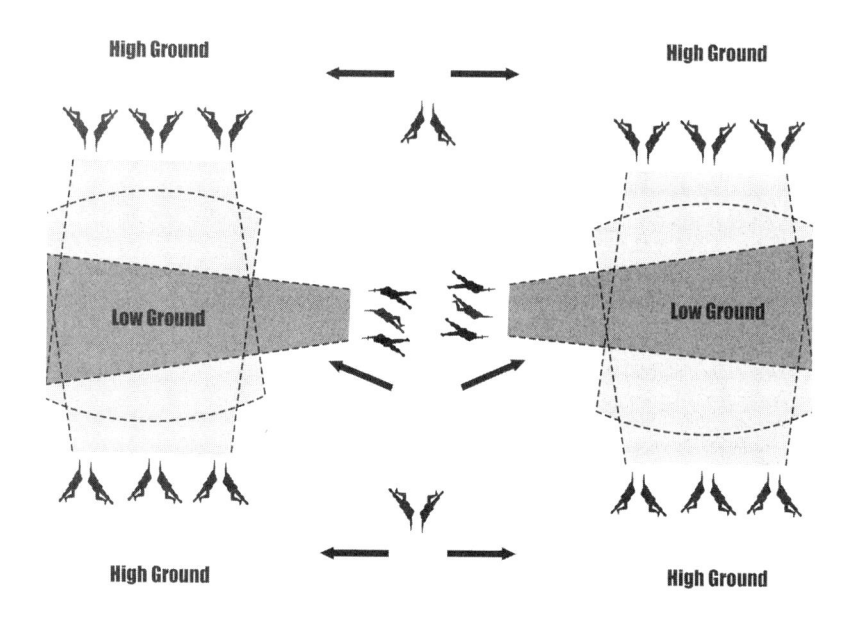

The parallel ambush method is a favorite among OPFOR militaries. It uses the high ground and plunging fires to prevent fratricide. Support teams may be used in the middle to stop the enemy advance.

The small unit leader considers the factors of METT-TC to determine the required formation:

Linear
In an ambush using a linear formation, the assault and support elements deploy parallel to the enemy's route. This position forces the enemy on the long axis of the kill zone, and subjects the enemy to flanking fire. The linear formation can be used in close terrain that restricts the enemy's ability to maneuver against the platoon, or in open terrain (provided a means of keeping the enemy in the kill zone can be effected).

L-Shaped
In an L-shaped ambush the assault element forms the long leg parallel to the enemy's direction of movement along the kill zone. The support element forms the short leg at one end of and at a right angle to the assault element. This provides both flanking (long leg) and enfilading (short leg) fires against the enemy. The L-shaped ambush can be used at a sharp bend in a road, trail, or stream. It should not be used where the short leg would have to cross a straight road or trail. The platoon leader must consider the other factors of METT-TC before opting for the L-shaped formation. Special attention must be placed on sectors of fire and surface danger zone (SDZ) of weapons because of the risk of fratricide.

V-Shaped Ambush
The V-shaped ambush assault elements are placed along both sides of the enemy route so they form a V. Take extreme care to ensure neither group fires into the other. This formation subjects the enemy to both enfilading and interlocking fire.

Note: See pp. 6-6 to 6-7 for discussion of the organization of the near ambush; pp. 6-8 to 6-9 for organization of the far ambush. See p. 6-16 for a discussion of the categories of ambushes.

A. Organization - The Near Ambush

An ambush patrol will be broken into multiple teams, each with a very specific set of responsibilities. The team must be assigned a leader. The team must also be allowed to work together at least during the rehearsals prior to the mission. This practice lets members of the patrol understand their role within that specific team, and shows how the multiple teams fit into the bigger picture of the mission.

The ambush will require special equipment for each assigned team. This equipment should be made available for the rehearsal as well, to ensure everything functions according to the execution plan.

The near ambush breaks into three teams—the security team, the support team, and the assault team. Each team has a specific set of duties, and the security and assault teams break down into further specialty teams.

The near ambush is broken into three teams, security, support, and assault. The security team is typically broken down further into a left, right, and sometimes a rear security team. (Photo by Jeong, Hae-jung).

1. The Security Team
The **security team** is responsible for the ambush's left, right, and sometimes rear security before the entire ambush is positioned. As is often the case, the security team is also assigned to specific security details while the ambush patrol maneuvers to and from the intended ambush site. Additionally, a security team will be placed in an overwatch position after the leader's recon confirms the exact site of the ambush. This security team will remain at the overwatch position while the patrol leader returns to the objective rally point (ORP) and brings the remainder of the ambush patrol forward.

The security team breaks into the "left security team" and the "right security team" at the ambush site. The security team leader will take a position with one security team or the other. In the event that a "rear security team" is required to remain at the release point, the security team leader will stay at that position.

2. The Support Team

The **support team** is responsible for delivering effective, heavy weapon fires against the enemy. The support team rarely breaks into separate teams. They are positioned in the ambush formation so that they may deliver accurate fire to the entire kill zone. The support team is typically comprised of machine gunners, grenadiers, and marksmen.

3. The Assault Team

The **assault team** is responsible for augmenting the fires of the support team against the enemy during the initial volley. This act significantly impacts the target in a physical sense, and also adds to the psychological shock of the ambush's violence of action. If a signal is given to sweep the kill zone after the initial volley of fire, the assault team must move forward to attack any surviving enemy in or around the kill zone.

Specialty Teams

The assault team also breaks into specialty teams. These teams almost always include an enemy prisoners of war (EPW) search team to look for priority intelligence requirements (PIR) such as enemy maps, orders, radio frequencies, and the like—and an aid and litter team to evacuate any friendly casualties. Additional specialty teams can include a demo team to destroy or booby-trap enemy weapons and equipment with explosives, or a grab team to take EPW.

As a general rule, the ambush patrol leader will take a position with the support team in order to better initiate and coordinate the fires of the ambush. This is just a guideline, however, and the patrol leader may determine where they are best in control.

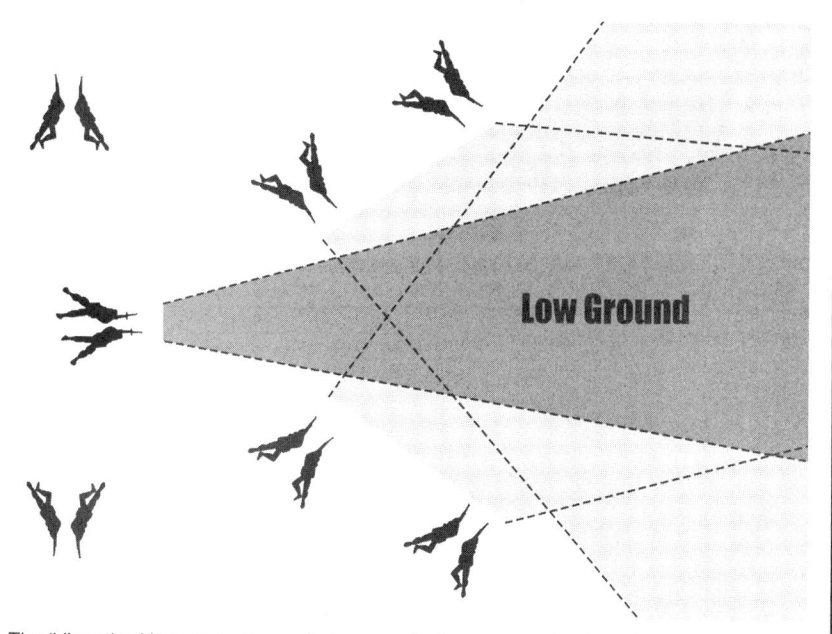

The 'V' method is commonly used when conducting near ambushes along a single draw of a ridge. By occupying the high ground of the spurs and directing fire downward, this method is extremely effective and safe. (Ref: FM 7-85, chap 6, fig. 6-3).

B. Organization - The Far Ambush

The far ambush breaks into only two teams—the security team and the support team. However, these teams may be spread out over large distances and assume very specific responsibilities in regards to a particular target.

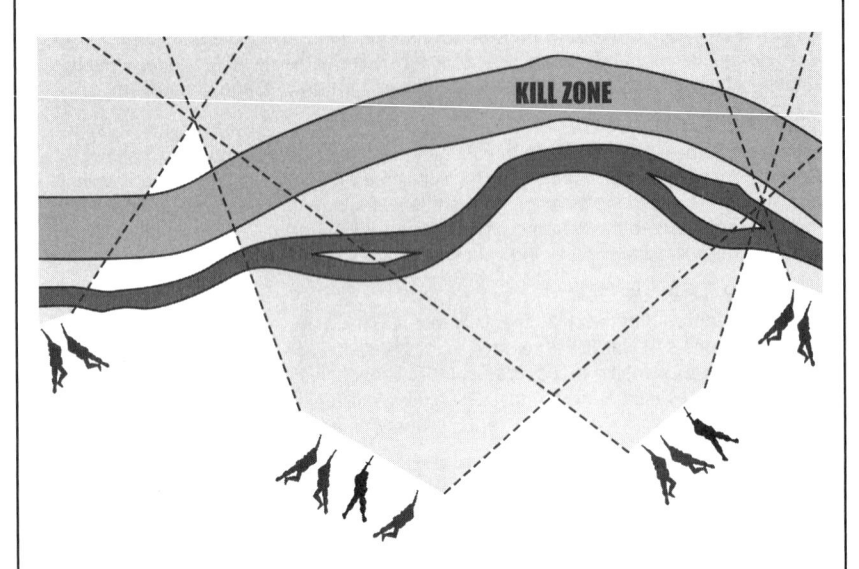

The far ambush keeps considerable distance between the ambush patrol and the enemy in the kill zone. Note that the creek offers protection from counterattack. The 'L' method is common for the far ambush. (Ref: FM 7-85, chap 6, fig. 6-2).

1. The Security Team

The **security team**, in a very similar manner to the near ambush, assumes responsibility for the far left and right sides of the far ambush formation. There is also a greater likelihood for the need of a rear security team on the far ambush to protect the multiple teams as they maneuver into their escape routes.

Also, the security team is responsible for security details as the ambush patrol moves to and from the ambush site. Again, the security team responsibilities are almost identical, regardless of whether we employ a near or far ambush. The major difference for the security team is that the far ambush requires the security teams to be very far apart from each other—often operating without visibility of each other. This takes considerable coordination in that the security teams must protect the ambush force, and ensure no friendly fire incidents occur.

2. The Support Team

The **support team** is, again, responsible for delivering effective fires against the enemy. However, in the far ambush, the support team is often broken into separate teams in order to deliver accurate fire to the entire kill zone. The support team is typically comprised of machine gunners, grenadiers, and marksmen. Depending on the type of target, the support team for a far ambush may also include missile launchers, combat engineers, and even mortar crews.

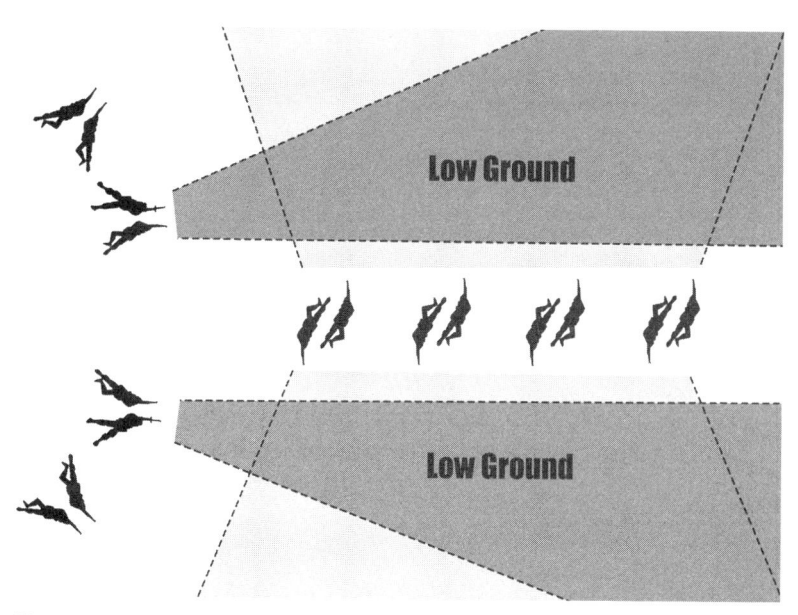

The 'T' method is a variation of the 'Z' method, but places the support teams at the same end of the ambush line. It is used for near ambushes along a spur when we are uncertain which side of the spur the enemy is traveling. (Ref: FM 7-85, chap 6, fig. 6-3).

III. Planning & Preparation

Surprise, coordinated fires, and control are the keys to a successful ambush. Surprise allows the ambush force to seize control of the situation. If total surprise is not possible, it must be so nearly complete that the target does not expect the ambush until it is too late to react effectively. Thorough planning, preparation, and execution help achieve surprise.

The commander conducts a leader's reconnaissance with key personnel to confirm or modify his plan. This reconnaissance should be undetected by the enemy to preclude alerting him. If necessary, the commander modifies the ambush plan and immediately disseminates those changes to subordinate leaders and other affected organizations. The leader's key planning considerations for any ambush include:

- Cover the entire kill zone (engagement area) by fire
- Use existing terrain features (rocks or fallen trees, for example) or reinforcing obstacles (Claymores or other mines) orienting into the kill zone to keep the enemy in the kill zone
- Determine how to emplace reinforcing obstacles on the far side of the kill zone
- Protect the assault and support elements with mines, Claymores, or explosives
- Use the security element to isolate the kill zone
- Establish rear security behind the assault element
- Assault into the kill zone to search dead and wounded, to assemble prisoners, and to collect equipment. The assault element must be able to move quickly on its own through the ambush site protective obstacles.
- Time the actions of all elements of the platoon to prevent the loss of surprise

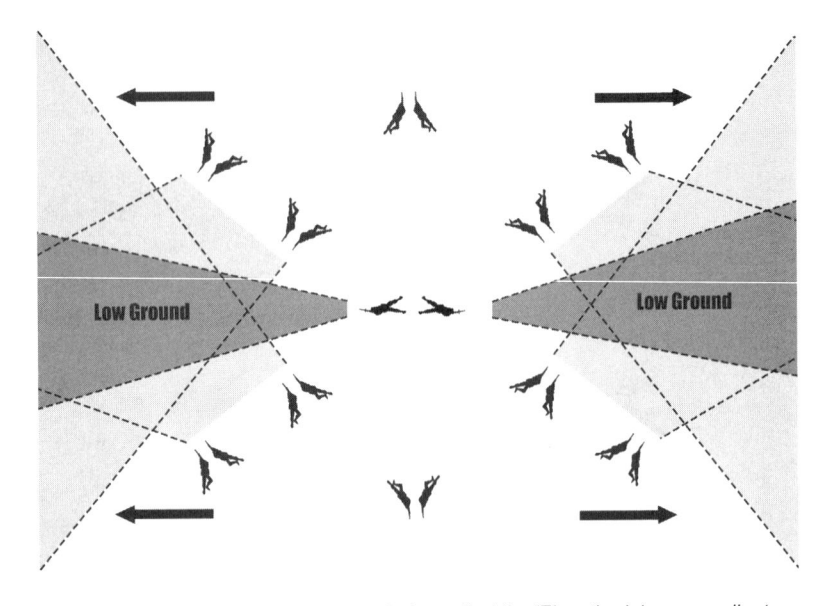

The 'X' method, as well as a close variation called the 'Z' method, is an excellent choice when the enemy may approach from multiple directions such as along converging draws or intersections. This allows the ambush team to cover multiple avenues of approach. (Ref: FM 7-85, chap 6, fig. 6-3).

A. Near Ambush

Physically, a near ambush kill zone is *close enough for the assault team to rush across*. In most types of terrain and vegetation, that means hand grenade range—35 meters. As the density of vegetation increases, the ability to rush forward decreases. In thick vegetation such as a jungle, a near ambush may need to be employed as close as 10 meters from the kill zone. Conversely, in wide-open terrain, a near ambush could be conducted =40 or even 50 meters from the kill zone. It's relative to the troop's ability to rush, taking into consideration terrain and vegetation.

In preparing a near ambush, employ obstacles on the far side of the kill zone. That is because the intent is to destroy the enemy target and not allow them to escape. These obstacles can be naturally occurring—such as a steep hill or cliff, a large body of water, or a wide open field that exposes the enemy troops as they attempt to run away. These obstacles can also be man-made—such as antipersonnel mines or well-camouflaged wire obstacles. The best technique is to employ multiple obstacles, including man-made and naturally occurring types.

B. Far Ambush

Physically, a far ambush kill zone is determined to be too far a distance to rush across. In almost all terrain, that would include distances significantly greater than 35 meters. It is true that within the most densely vegetated environment, an ambush at 50 or even 40 meters might be considered a far ambush. But typically, far ambushes make use of distances of 70 meters and up to 700 meters from the kill zone.

The far ambush is used to harass, injure, or delay the target. Due to the significant distances, there is no need to outnumber the enemy target. However, in order to effectively inflict damage, it will be beneficial to have some identification of the target. For example, is the target a supply convoy…or is it a combat patrol? Does the enemy employ armored vehicles, thin-skinned vehicles, or only foot patrols?

In preparing a far ambush, employ obstacles on the near side of the kill zone. Carefully plan and identify escape routes. The effort here is to allow the ambush patrol to escape before the enemy target recovers from the initial volley of fire and counterattacks. The obstacles in between the ambush patrol and the enemy—whether they are man-made or naturally occurring obstacles—serve to delay the enemy counterattack. The identified escape routes speed the escape and make coordination much more simplified.

III. Conducting the Ambush - A Small Unit Perspective

Fire discipline is a key part of any ambush. Fire must be withheld until the ambush commander gives the signal to initiate the ambush. That signal should be fire from the most deadly weapon in the ambush. Once initiated, the ambush unit delivers its fires at the maximum rate possible given the need for accuracy. Otherwise, the assault could be delayed, giving the target time to react and increasing the possibility of fratricide. Accurate fires help achieve surprise as well as destroy the target. When it is necessary to assault the target, the lifting or shifting of fires must be precise. The assault element does not conduct its assault until enemy fires or resistance has been negated or eliminated.

If the ambush fails and the enemy pursues the ambush force, it may have to withdraw by bounds. The ambush force should use smoke to help conceal its withdrawal. Activating limited-duration minefields along the withdrawal routes after the passage of the withdrawing ambush force can help stop or delay enemy pursuit. The commander positions the support element to assist in the withdrawal of the assault element.

Note: See pp. 6-12 to 6-13 for discussion of conducting the near ambush; pp. 6-14 to 6-15 for conducting the far ambush.

After the initial volley of fire, the commander may give the signal for the assault team to attack across the kill zone. The support team and security teams must first lift or shift their fires to prevent fratricide. (Photo by Jeong, Hae-jung).

(Special Purpose Attacks) I. Ambush 6-11

A. Conducting the Near Ambush

The near ambush is used to destroy the enemy target. Necessarily, it is preferred that the ambush patrol outnumber the enemy by a 2:1 ratio. If they are more evenly matched, that is acceptable. To achieve numerical superiority, the PL needs to know the approximate size of the enemy patrols.

1. The patrol occupies an ORP either one terrain feature or approximately 300 meters away from the intended ambush site. The PL assembles the leader's recon team and issues a five-point contingency plan to the assistant patrol leader (APL) prior to leaving on the recon. At a minimum, the leader's recon will include the PL, a two-man security team, and either the leader of the assault team or the support team—typically the team the PL will *not* accompany.

2. At the designated ambush site, the PL ensures that it is an appropriate terrain using the considerations of Observation, Cover & concealment, Obstacles, Key terrain, and Avenues of approach (OCOKA). The PL does this without contaminating the kill zone—meaning he shouldn't actually walk through or onto the kill zone, but should move around to view it from the far right, far left, and from the middle of the ambush formation. If the terrain is not suitable for a near ambush, the PL chooses an appropriate site nearby.

3. The PL will then post the two-man security team far back, where they are easily concealed but can still view the kill zone. This post will later become the release point. The security team sits back-to-back, with one man facing the kill zone and the other facing back in the direction of the ORP. The PL will leave these men with a five-point contingency plan and return to the ORP. The security team must monitor all enemy activity and report this to the PL upon his return. It will be critical to know if the enemy has stopped on or near the ambush site.

4. After returning to the ORP, the PL coordinates any changes to the original plan with every member of the ambush patrol. Final preparations are conducted in the ORP and the PL pulls together the patrol. The order of march will be the PL, security team, support team, and lastly the assault team. This is the exact order because the ambush must be placed into position using this sequence.

5. The PL leads the patrol to the security team at the release point. He links up the two-man security team with their security team leader. If the intended location of the left and right side security areas can be seen from the release point, the PL will have the security team leader place his teams into position. If the locations cannot be seen, the PL positions the left and right security teams, taking the Security team leader with him.

6. Once the left and right security teams are in position, the PL returns to the release point and picks up the support team. He positions them into the formation, typically in front of the release point. The PL returns back to the release point and picks up the assault team. They, too, are placed in formation and assume the opposite (left or right) side of the support team.

7. With everyone in place, the PL will take his place as the leader of either the support or the assault team, as determined in the operation order (OPORD). The PL will conduct a communication system check, and then the team awaits the enemy. Security is kept at 100 percent.

8. The ambush patrol continues to wait in position until:

- The PL gives the "end time" signal that indicates the patrol must return
- The PL gives the "no fire" signal and allows a larger enemy force to pass
- OR...the ambush is initiated

9. The ambush patrol fires upon the kill zone only when:

- The PL initiates fire against the enemy in the kill zone
- OR—the enemy discovers the ambush patrol

6-12 (Special Purpose Attacks) I. Ambush

After the initial volley of fire, the commander may give the signal for the assault team to attack across the kill zone. The support team and security teams must first lift or shift their fires to prevent fratricide. (Photo by Jeong, Hae-jung).

Actions on the Objective

1. The actions on the objective will not take long. All members of the ambush patrol fire into the kill zone—regardless of whether or not they see a specific target. The only exceptions may be the security teams who may not have a clear view of the kill zone.

2. The PL gives the signal to cease fire, to shift fires, or to lift fires. At this moment, the PL decides if it is reasonably safe for the assault team to move across the kill zone.

3. If the assault team is sent across the kill zone, it conducts the following actions in this order:

 • Sweep the kill zone on line, being certain to double-tap all enemy
 • Secure the far side of the kill zone
 • Send the necessary specialty teams back into the kill zone to search for PIR, aid friendly casualties, or to destroy enemy equipment

4. The assault team leader then gives the thumbs-up signal to the PL indicating that the far side of the kill zone is secure and that the specialty teams have finished their tasks. The assault team secures the far side of the kill zone until the PL signals to fall back.

5. The PL gives the signal for the assault team to fall back through the release point to the ORP. The assault team does this without hesitation and doesn't wait for other teams.

6. The PL gives the signal for the support team to fall back through the release point to the ORP. The support team does this immediately.

7. The PL will give the signal for the security team to fall back through the release point and move to the ORP. The PL will wait for the security team at the release point and move back to the ORP with this element. In this manner, the ambush has displaced in the exact reverse order that it was emplaced.

Specialty teams, such as the PIR search team, come from within the assault team. This is partly because the assault team tends to be the largest team on the ambush, but also because it is the only team that enters the kill zone. (Photo by Jeong, Hae-jung).

Reconsolidate & Reorganize

1. The first element to get back to the ORP is the assault team. The assault team leader forms the ORP into a 360° security area and continues to shape the ORP until a senior leader replaces him. Redistribute ammunition and water; be sure key weapons are manned.

2. All troops must be accounted for and casualties must receive medical aid. Following accountability, the PL will ask for a sensitive equipment check.

3. The PL will facilitate the dissemination of PIR to all members of the patrol. This is necessary because—should the patrol later become engaged and take casualties—the PIR must be relayed to higher command.

4. The designated route of return and plans to evacuate casualties and/or EPW will be followed in accordance with the OPORD. Continue the mission.

(Special Purpose Attacks) I. Ambush 6-13

B. Conducting the Far Ambush

The goal in the far ambush is to *inflict damage* upon an enemy patrol. There is no requirement to overrun the target—in fact, the far ambush doesn't even have an assault team. It may help to think of the far ambush as a grandiose sniper mission.

1. The PL assembles a team to conduct the leader's recon at an ORP that has been established at an appropriate distance away from the intended ambush site. This team will include the PL and the entire security team. Prior to heading out on the recon, the PL will leave the support team leader with a five-point contingency plan.

2. The PL moves the recon team to the intended ambush site and posts the security team in a position that affords a complete view of the ambush site. Then the PL and the security team leader move to the far left, far right, and middle to determine if the terrain is appropriate, to determine where to place the security and support teams, and to identify routes of escape. Furthermore, the kill zone must have some type of natural obstacle between the enemy and the ambush patrol to slow down the enemy's counterattack. If no natural obstacle exists, then a man-made obstacle will have to be employed and camouflaged. If the terrain is inappropriate for this mission, another location must be chosen nearby.

3. The PL positions the entire security team toward the middle of the ambush line so that they can view the kill zone. This becomes the release point. The PL leaves a five-point contingency plan with the security team leader and returns to the ORP.

4. The PL disseminates any changes to the original plan to all patrol members back at the ORP. Complete final preparations and form the support team. The support team may break into sub-teams to form the order of march according to the OPORD. The PL will link up the support teams with the security team.

5. At the release point, the PL releases each element to the team leader. The security team leader places his teams as directed. The support team leader places the team(s), and the PL moves with whichever support team he decides will give him the best view of the kill zone and command of the elements.

The far ambush places high value on marksmanship skills. The support team is often broken into a series of sniper, machinegun, and rocket teams that fire into the kill zone from multiple, coordinated directions. (Photo by Jeong, Hae-jung).

6. With all elements in place, the PL conducts a communication systems check. The entire patrol will wait in place for the ambush to initiate. In the far ambush, due to the distance between elements, the left or right side security will often be the first to see the enemy element. They inform the PL on enemy movement and estimated time of arrival.

7. It is unlikely that the enemy will detect the ambush patrol in a far ambush mission. Still, the same rules apply to the far ambush that applies to the near ambush:

- The PL gives the "time" signal that indicates the patrol must return
- The PL gives the "no fire" signal and allows a large enemy team to pass
- OR...the ambush is initiated

8. The ambush patrol fires upon the kill zone only when:

- The PL initiates fire against the enemy in the kill zone, OR
- if the enemy discovers the ambush patrol

Actions on the Objective
1. The actions on the objective depend upon the enemy's ability to react. The far ambush continues until the enemy escapes from the kill zone, the enemy begins a counterattack, or the PL is satisfied with the effect or the ambush. Do not fire all ammunition. The patrol still has to escape. Combat leaders must keep in mind that the longer the patrol stays, the more vulnerable the patrol becomes to enemy counterattack.

2. Upon the signal to cease fire, the support teams automatically withdraw to the ORP using the designated escape route. There is no need to pass through the release point. This will waste time due to the distance between elements.

3. The PL travels with one of the support teams. When he calculates that he has passed the release point, he signals the security team to withdraw. The security team leader rallies his team in a concealed area behind the release point to make certain to account for the entire team. Then the security team proceeds to the ORP.

Escape routes are an important consideration for any ambush, but they are absolutely vital to the far ambush. Once the far ambush has achieved its effect, defilades, ravines, and draws are used as escape routes. (Photo by Jeong, Hae-jung).

Reconsolidate & Reorganize
1. The PL arrives back at the ORP and forms into a 360° security area until all patrol members are assembled. Ammunition and water are distributed as needed. Key weapons are manned. Casualties receive medical aid.

2. Once every member is accounted for, the PL asks for a sensitive equipment check. Patrol members count their assigned sensitive equipment (weapons, radios, night vision devices, etc.) by physically touching the item.

3. Far ambushes also attempt to gain PIR and all observations are quickly disseminated. The designated route of return and plans to evacuate casualties are conducted according to the OPORD. Continue the mission.

IV. Ambush Categories

Ambushes are classified by category (deliberate or hasty), formation (linear or L-shaped), and type (point, area, or antiarmor). The leader determines the category of ambush through an analysis of the factors of METT-TC. Typically, the two most important factors are time and enemy. The leader's key planning considerations for any ambush include the following:

- Cover the entire kill zone (engagement area) by fire
- Use existing terrain features (rocks or fallen trees, for example) or reinforcing obstacles (Claymores or other mines) orienting into the kill zone to keep the enemy in the kill zone
- Determine how to emplace reinforcing obstacles on the far side of the kill zone
- Protect the assault and support elements with mines, Claymores, or explosives
- Use the security element to isolate the kill zone
- Establish rear security behind the assault element
- Assault into the kill zone to search dead and wounded, to assemble prisoners, and to collect equipment. The assault element must be able to move quickly on its own through the ambush site protective obstacles.
- Time the actions of all elements of the platoon to prevent the loss of surprise

Deliberate

A deliberate ambush is a planned offensive action conducted against a specific target for a specific purpose at a predetermined location. When planning a deliberate ambush, the leader requires detailed information on the—

- Size and composition of the targeted enemy unit
- Weapons and equipment available to the enemy
- Enemy's route and direction of movement
- Times that the targeted enemy unit will reach or pass specified points along the route

Hasty

A hasty ambush is conducted when a unit makes visual contact with an enemy force and has time to establish an ambush without being detected. The conduct of the hasty ambush should represent the execution of disciplined initiative within the parameters of the commander's intent. The actions for a hasty ambush should be established in a unit SOP and rehearsed so Soldiers know what to do on the leader's signal.

Area Ambush

An area ambush (more than one point ambush) is not conducted by a unit smaller than a platoon. This ambush works best where enemy movement is restricted. Once the platoon is prepared, the area ambush is conducted the same as a point ambush. The dominating feature of an area ambush is the amount of synchronization between the separate point ambushes.

Area ambushes require more planning and control to execute successfully. Surprise is more difficult to achieve due to the unit's dispersion in the AO. Having more than one ambush site increases the likelihood of being detected by the enemy or civilians. This major disadvantage is offset by the increased flexibility and sophistication available to the leader.

Point Ambushes

Point ambushes are set at the most ideal location to inflict damage on the enemy. Such ambushes must be able to handle being hit by the enemy force from more than one direction. The ambush site should enable the unit to execute an ambush in two or three main directions. The other directions must be covered by security that gives early warning of enemy attack.

Special Purpose Attacks
II. Raid

Ref: FM 3-90 Tactics, pp. 5-38 to 5-39 and FM 7-85 Ranger Unit Operations, chap 5.

The raid is a special purpose attack that falls under the category of strike operations. Raiding patrols infiltrate well into enemy territory. This means the raiding force can expect to be outnumbered, outgunned, and far from help. A successful raid is the hallmark tactic of special operational forces and a crowning achievement for any combat unit. No tactical maneuver requires a more advanced set of skills—intelligence gathering through reconnaissance, brilliant planning, accurate rehearsals, and individual and team skill craft that is the envy of others.

A raid is a limited-objective form of attack entailing swift entry into hostile terrain. A raid operation always ends with a planned withdrawal to a friendly location upon the completion of the assigned mission. It is not intended to hold terrain.

There are many reasons to conduct raids. A patrol might be tasked to destroy key enemy equipment or facilities, temporarily seize key terrain, gather intelligence items, or liberate personnel. While each of these missions differs from the next, they each entail a set of basic considerations.

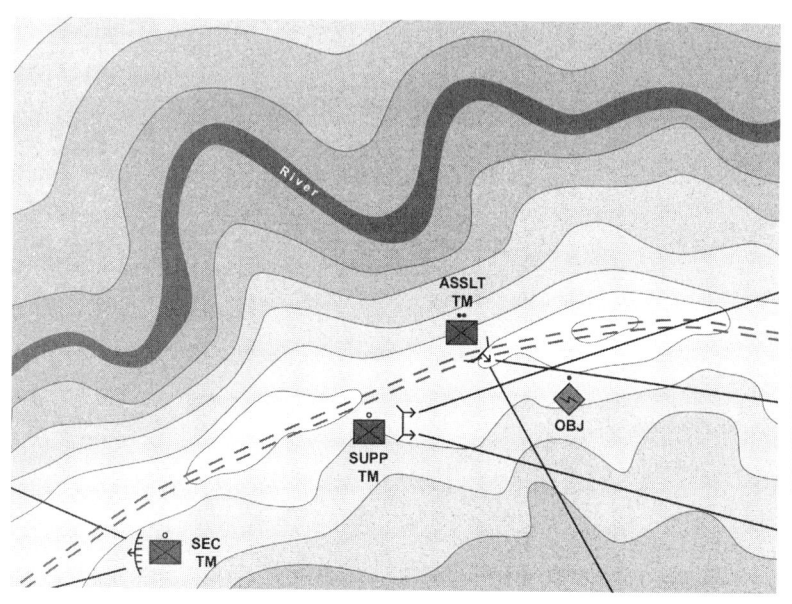

The raid has the same basic teams as the ambush—security, support, and assault. The principle difference is that the raid is conducted on established, stationary targets, whereas the ambush is used on moving or temporarily halted targets. (Ref: FM 7-85, chap 5, fig. 5-1).

I. Organization

In a very similar manner as the ambush force, the raiding force breaks down into three main elements: the security team, the support team, and the assault team.

1. Security Team
The security team is most commonly deployed to the left, right and sometimes rear of the raiding formation as it is deployed around the objective. They carry rifles, light machineguns, anti-personnel mines and possibly some anti-armor capabilities. Their main purpose is to isolate the target, prevent any enemy reinforcements, and to seal the escape of any enemy running from the objective.

2. Support Team
The **support team** is commonly deployed center of the raiding formation in such a manner that they have a clear view of the objective. They carry heavier mass-casualty producing weapons, such as machineguns, grenade launchers, or missiles. This team is primarily responsible for the shock effect, as well as inflicting as many casualties upon the enemy as possible to ensure the success of the assaulting team.

3. Assault Team
The assault team is deployed as closely to the objective as stealth and coordinated fire support allow. They are lightly armed with rifles and carbines but may have special equipment. This team is responsible for the destruction, capture, or liberation of the target. Upon assaulting across the objective, they are also the most exposed—and least armed element of the raid.

After the leader's recon, the patrol leader leaves a 2-man security team overlooking the objective from a concealed position. This position later becomes the release point for the raid. The PL returns to the ORP to finalize plans. (Photo by Jeong, Hae-jung).

Raids also have sub-groups within the assault team. Examples of these specialty teams would include grab teams for abduction, prisoner search teams for intelligence requirements, aid & litter teams for wounded, demolition or explosive ordinance disposal (EOD) teams, and possibly even chemical biological radiological and nuclear (CBRN) recovery teams.

The assault team moves into position last. The assault team is lightly armed and may require special equipment for breaching obstacles. After the support team shifts fire, the assault team sweeps the objective and destroys the enemy. (Photo by Jeong, Hae-jung).

II. Planning & Preparation

The main differences between a raid and other attack forms are the limited objectives of the raid and the associated withdrawal following completion. Raids might be conducted in daylight or darkness, within or beyond supporting distance of the parent unit. When the area to be raided is beyond supporting distance of friendly lines, the raiding party operates as a separate force. A specific objective is normally assigned to orient the raiding unit. During the withdrawal, the attacking force should use a route or axis different from that used to conduct the raid itself.

The raid requires the most up-to-date information, which aids in the development of the intelligence preparation of the battlespace (IPB). This translates into effective maneuver control measures, fire control measures, and nothing less than a brilliant plan that makes ample use of stealth and timing. Rehearsals are an essential part of the raid preparations.

The enemy will likely have good communications with reinforcements in the immediate vicinity, making the raid patrol all the more vulnerable to a counterattack. This fact dictates that severe time constraints exist from the moment the raid begins, until the moment the patrol withdraws from the objective.

A raid will be much more successful if the patrol has the element of surprise. To maximize this surprise, the raid is conducted at a time when the enemy is least likely to expect an attack. Attack when visibility is limited and, if possible, attack from an unexpected avenue of approach—such as from seemingly impassable terrain. This requires in depth knowledge of the terrain surrounding the objective as well.

(Special Purpose Attacks) II. Raid 6-19

III. Conducting the Raid - A Small Unit Perspective

Surprise, firepower, and a tenacious attack stun and disorient the enemy. The psychological effect of violence should not be underestimated. If the enemy on the objective believe that the raid patrol is actually much larger due to the use of massed firepower and violence, the enemy is less likely to stand and defend against the assaulting team. Furthermore, after the patrol has withdrawn from the objective, the enemy will pursue the raid patrol much less aggressively if they believe the patrol is very large.

The ORP must be secured. The patrol leader leaves the assistant patrol leader in the ORP with a security force. The rest of the patrol will drop any unnecessary gear at the ORP, and return to the ORP after the raid is complete. (Photo by Jeong, Hae-jung).

A. Infiltrate to the Objective

1. After infiltrating into enemy territory and the objective rally point (ORP) has been properly occupied and secured, the patrol leader (PL) conducts a leader's recon of the objective. At a minimum, the PL takes a two-man security team, and the leader of the support or assault team—whichever one the PL will *not* be positioned with during the raid. It generally is considered a good idea to bring all three element leaders, if possible.

2. The PL leaves a contingency plan with the assistant patrol leader (APL) in the ORP before departing. The leader's recon conducts a physical inspection of the objective, making certain that everything is as planned. If not, the PL improvises any change to the plan...and since there will be no time or space for rehearsals, the option to change the plan should only be used in extreme circumstances.

3. The PL leaves a two-man security team with communication and a contingency plan. The security team is positioned in such a manner that they maintain constant observation of the objective (this spot will later become the release point) to ensure that enemy reinforcements do not arrive and that the target doesn't leave. The PL and element leader(s) move back to the ORP.

4. The PL issues any changes to the plan in the ORP and finalizes all preparations. Picking up the rest of the patrol, the PL leads them out of the ORP and towards the release point in the following order:

- PL leads at point
- Security team
- Support team
- Assault team pulls up the drag

The patrol leader moves the patrol forward to link up with the release point. From the release point the patrol waits while the security teams carefully and quietly move to their assigned locations. (Photo by Jeong, Hae-jung).

6-20 (Special Purpose Attacks) II. Raid

5. Upon reaching the release point, the PL checks with the on-site security team to make certain everything is okay. The PL then links the entire security team up with the leader and gives them time to move into their designated position. Unlike the ambush, the PL does NOT have the option of positioning each element. On a raid, the elements have all rehearsed exhaustively on where to go. Time must be allowed for each team to stealthily position themselves.

6. The PL releases the support team next and may travel with that element if he has not assigned himself to the assault team. The assault team takes position last and due to the close proximity of their position to the objective, they must be given ample time to move.

7. All elements wait for the signal to commence fire. This may be designated by:
- A fixed time, OR
- A designated signal, OR
- The PL may issue the "No Fire" signal (in which case the patrol withdraws)

B. Actions on the Objective

Just as with the ambush, the initiating volley of fire must physically and psychologically overwhelm the enemy force. Upon the initiating shot, every member of the patrol immediately opens fire on the objective. Failure to immediately suppress the enemy means failure for the raid patrol. If the enemy gains the initiative, the patrol will likely be destroyed.

The support team takes position after the security team sets into place. Because the support team carries crew-served weapons that are heavy, more time must be allotted for their movement. (Photo by Jeong, Hae-jung).

1. After effectively devastating the objective with a heavy volume of fire, the PL gives a designated signal to lift or shift fires. Now, a common misconception is that the term "lift or shift" fires actually means "cease-fire."

Shifting fires means that the support team's direction of fire will shift either left or right in order to suppress fleeing or reinforcing enemy. If shifting of fires cannot be done safely, then **lifting** fires means the support team will continue to fire harmlessly over everyone's heads.

2. The assault team begins their choreographed attack across the objective. The assault team crosses the objective in pre-arranged buddy teams and...
- Double-taps all enemy combatants
- Secures the far side of the objective, AND
- Conducts the sub-tasks of the specialty teams

3. Once the specialty teams are finished with the assigned tasks, the assault team leader gives the PL the signal that they have accomplished the task and are ready to move.

4. The PL then gives three designated signals. The first notifies the assault team to fall back through the release point to the ORP. The second notifies the support team to do the same. The third signal notifies the security team to fall back via their designated route to the ORP.

5. In the ORP, subordinate leaders reconsolidate and reorganize the patrol. The APL accounts for all members and equipment. All crew served weapons and priority equipment is reassigned if there have been casualties. Friendly casualties are cared for in accordance with the operations order (OPORD). Water and ammunition is redistributed.

6. The patrol then falls back to a pre-designated position, usually one terrain feature back from the ORP. The patrol stops to disseminate all information and PIR regarding the raid amongst every patrol member. This is done prior to returning to the FEBA.

(Special Purpose Attacks) II. Raid 6-21

Once the raid has been successfully executed, the patrol moves to an extraction point. The key to achieving elusiveness for a raiding patrol is a stealthy infiltration, and a highly mobile extraction. (Dept. of Army photo by Michael Guillory).

On Point

A raid is a surprise attack to temporarily overwhelm stationary enemy targets, usually deep within enemy territory. Raids are conducted for a multitude of reasons—most commonly to seize or destroy enemy assets, collect valuable information, or rescue personnel aligned with our cause.

A simplified chain of command is an essential organizational requirement. A raid usually requires a force carefully tailored to neutralize specific enemy forces operating in the vicinity of the objective and to perform whatever additional functions are required to accomplish the objective of the raid. These additional functions can consist of the demolition of bridges over major water obstacles or the recovery of an attack helicopter pilot shot down forward of the forward line of own troops (FLOT). The commander incorporates any necessary support specialists during the initial planning stage of the operation.

When a commander and his staff plan a raid, they develop COAs that meet ethical, legal, political, and technical feasibility criteria. Planners require precise, time-sensitive, all-source intelligence. The planning process determines how C2, sustainment, target acquisition and target servicing will occur during the raid. Techniques and procedures for conducting operations across the FLOT, given the specific factors of METT-TC expected to exist during the conduct of the raid, are also developed. The commander and his staff develop as many alternative COAs as time and the situation permit. They carefully weigh each alternative. In addition to those planning considerations associated with other offensive operations, they must determine the risks associated with conducting the mission and possible repercussions.

Time permitting, all elements involved in a raid should be fully rehearsed in their functions. The key elements in determining the level of detail and the opportunities for rehearsal prior to mission execution are time, OPSEC, and deception requirements.

Chap 7
Urban & Regional Environments

Army doctrine addresses five regional environments: desert, cold, temperate, mountain, and jungle. Another area of special consideration involves urban areas*.

Relative Units of Control, Action & Maneuver

	Unit of Control	Unit of Action	Unit of Maneuver	Relative Command & Control
Desert Region	JTF	BDE	BN/CO	Decreasing Unit Size
Cold Region	BDE	BN	CO/PLT	
Temperate Region	BDE	BN	CO/PLT	
Urban Area	BN	CO	PLT/SQD	
Mountain Region	BN	CO	PLT/SQD	Increasing Unit Autonomy
Jungle Region	CO	PLT	SQD/TM	

*Urban Operations

The continued trend worldwide of urban growth and the shift of populations from rural to urban areas continues to affect Army operations. The urban environment, consisting of complex terrain, dense populations, and integrated infrastructures, is the predominant operational environment in which Army forces currently operate. ATTP 3-06.11, Combined Arms Operations in Urban Terrain (Jun '11), establishes doctrine for combined arms operations in urban terrain for the brigade combat team (BCT) and battalion/squadron commanders and staffs, company/troop commanders, small-unit leaders, and individual Soldiers.

See pp. 7-3 to 7-16. See also pp. 7-17 to 7-22 for related discussion of fortified areas.

Desert Operations

Arid regions make up about one-third of the earth's land surface, a higher percentage than that of any other climate. Desert operations demand adaptation to the environment and to the limitations imposed by terrain and climate. Success depends on the an appreciation of the effects of the arid conditions on Soldiers, on equipment and facilities, and on combat and support operations. FM 90-3/FMFM 7-27, Desert Operations (Aug '93), is the Army and Marine Corps' manual for desert operations. It is the key reference for commanders and staff regarding how desert affects personnel, equipment, and operations. It will assist them in planning and conducting combat operations in desert environments.

See pp. 7-23 to 7-30.

Cold Region Operations

When conducting military operations in cold regions, leaders, Soldiers, and Marines must plan to fight two enemies: the cold and the opposing force. Despite the difficulties that cold regions pose, there are armies that have prepared for and can conduct large-scale, sustained operations in cold environments. In contrast, few U.S. Army units or personnel have trained extensively in cold region operations. ATTP 3-97.11/

(Urban & Regional Environments) Overview 7-1

MCRP 3-35D, Cold Region Operations (Jan '11), is the Army's doctrinal publication for operations in the cold region environment. This manual will enable leaders, Soldiers, and Marines to accurately describe cold region environments, their effects on military equipment, impacts these environments have on personnel, and most importantly, how to employ the elements of combat power in cold region environments. It provides the conceptual framework for conventional forces to conduct cold region operations at operational and tactical levels.

See pp. 7-31 to 7-38.

Mountain Operations

With approximately 38 percent of the world's landmass classified as mountains, the Army must be prepared to deter conflict, resist coercion, and defeat aggression in mountains as in other areas. Throughout the course of history, armies have been significantly affected by the requirement to fight in mountains. FM 3-97.6 (90-6), Mountain Operations (Nov '00), describes the tactics, techniques, and procedures that the U.S. Army uses to fight in mountainous regions. It provides key information and considerations for commanders and staffs regarding how mountains affect personnel, equipment, and operations. It also assists them in planning, preparing, and executing operations, battles, and engagements in a mountainous environment. Army units do not routinely train for operations in a mountainous environment. The jungle environment includes densely forested areas, grasslands, cultivated areas, and swamps. Jungles are classified as primary or secondary jungles based on the terrain and vegetation.

See pp. 7-39 to 7-46.

Jungle Operations

Jungles, in their various forms, are common in tropical areas of the world—mainly Southeast Asia, Africa, and Latin America. The climate in jungles varies with location. Close to the equator, all seasons are nearly alike, with rains throughout the year; farther from the equator, especially in India and Southeast Asia, jungles have distinct wet (monsoon) and dry seasons. Both zones have high temperatures (averaging 78 to 95+ degrees Fahrenheit), heavy rainfall (as much as 1,000 centimeters [400+ inches] annually), and high humidity (90 percent) throughout the year. Severe weather also has an impact on tactical operations in the jungle. FM 90-5, Jungle Operations (Aug '93), is the Army's field manual on jungle operations.

See pp. 7-47 to 7-54.

Mission Command Considerations

Commanders of tactical forces will recognize a general tendency of command and control to vary from centralized to decentralized operations that is specific to any given regional or area environment. Such trends represent an historic norm, however the trends are not etched in stone as formalized doctrine. Still, it may help to consider battle command as it fluctuates from unit reliance on highly centralized control of desert operations, to unit autonomy in highly decentralized control of jungle operations.

In the associated graph (previous page), the "unit of control" is the higher command element tasked to an operational objective. The "unit of action" refers to the unit tasked to decisive engagement to achieve the operational objective. And the "unit of maneuver" includes the units responsible for shaping operational success.

Chap 7
I. Urban Operations

Ref: ADRP 3-90, Offense and Defense (Aug '12), chap 5, pp. 5-8 to 5-9. FM 3-21.8 (FM 7-8) The Infantry Rifle Platoon and Squad, pp. 7-36 to 7-47. For further discussion refer to ATTP 3-06.11, Combined Arms Operations in Urban Terrain (Jun '11).

Infantry platoons conduct operations in urban areas using the same principles applicable to other offensive operations. This section explains the general tactics, techniques, and procedures used for a limited attack in an urban area.

Depending on the scale of the operation, Infantry platoons or squads may be required to conduct any or all of the find, fix, fight, and follow-through functions. Leaders should expect trouble in the process of determining the exact location of the enemy and should anticipate enemy knowledge of their movements prior to arriving in the objective area. (Dept. of Army photo by Richard Rzepka, 101st Airborne).

I. Find

The compartmentalized nature of urban terrain, limited observation and fields of fire, and the vast amounts of potential cover and concealment mean that defenders can disperse and remain undetected. The origin of enemy gunfire can be difficult to detect, because distance and direction become distorted by structures. The nature of urban conflicts makes it more difficult for leaders to exercise command and control verbally, and for Soldiers to pass and receive information. Situational understanding is normally limited to the platoon's immediate area.

II. Isolate the Building

The fix function has two aspects: isolating the objective to prevent interference from the outside (while preventing enemy from exiting), and separating forces on the objective from each other (denying mutual support and repositioning). This is accomplished by achieving fire superiority and seizing positions of advantage.

(Urban & Regional Environments) I. Urban Operations 7-3

Urban Operations (UO)
Ref: ADRP 3-90, Offense and Defense (Aug '12), pp. 5-8 to 5-9 and ATTP 3-06.11, Combined Arms Operations in Urban Terrain (Jun '11), introduction.

Commanders conducting major urban operations use their ability to visualize how doctrine and military capabilities are applied within the context of the urban environment. An operational framework is the basic foundation for this visualization. In turn, this visualization forms the basis of operational design and decisionmaking. To accurately visualize, describe, and direct the conduct of operations in an urban environment, commanders and their staffs must understand the basic fundamentals applicable to most urban operations.

Fundamentals of Urban Operations

The impact of the urban operations environment often differs from one operation to the next. However, some fundamentals apply to urban operations regardless of the mission, geographical location, or level of command. Some of these fundamentals are not exclusive to urban environments. Yet, they are particularly relevant to an environment dominated by man-made structures and a dense noncombatant population. Vitally, these fundamentals help to ensure that every action taken by a commander operating in an urban environment contributes to the desired end-state of the major operation.

- Maintain close combat capability
- Avoid the attrition approach
- Control the essential
- Minimize collateral damage
- Preserve critical infrastructure
- Separate noncombatants from combatants
- Restore essential services
- Understand the human dimension
- Create a collaborative information environment
- Transition control

Urban Operational Construct

The five essential components of the urban operational construct are described below.

Understand

Understanding requires the continuous assessment of the current situation and operational progress. Commanders use visualization, staffs use running estimates, and both use the IPB process to assess and understand the urban environment. Commanders and staffs observe and continually learn about the urban environment (terrain, society, and infrastructure) and other mission variables. They use reconnaissance and security forces; information systems; and reports from other headquarters, services, organizations, and agencies. They orient themselves and achieve situational understanding based on a common operational picture and continuously updated CCIR. The commander's ability to rapidly and accurately achieve an understanding of the urban environment contributes to seizing, retaining, and exploiting the initiative during UO.

Shape

Reconnaissance, security, and inform and influence activities are essential to successful UO. These shaping operations set the conditions for decisive operations at the tactical level in the urban area. Isolation, decisive action, minimum friendly casualties, and acceptable collateral damage distinguish success when the AO is properly shaped. Failure to adequately shape the urban AO creates unacceptable risk. Urban shaping operations may include actions taken to achieve or prevent isolation, understand the environment, maintain freedom of action, protect the force, and develop cooperative relationships with the urban population. Some shaping operations may take months to successfully shape the AO.

Engage
In UO, the BCT engages by appropriately applying the full range of capabilities against decisive points leading to centers of gravity. Successful engagements take advantage of the BCT's training; leadership; and, within the constraints of the environment, equipment and technology. Engagement can be active or passive and has many components, but it is characterized by maintaining contact with the threat and population to develop the situation. Successful engagements also require the establishment of necessary levels of control and influence over all or portions of the AO until responsibilities can be transferred to other legitimate military or civilian control. Engagements may range from the overwhelming and precise application of combat power in order to defeat an enemy to large-scale humanitarian operations to HN security force assistance characterized by information and influencing activities.

Consolidate
Forces consolidate to protect and strengthen initial gains and ensure retention of the initiative. Consolidation includes actions taken to eliminate or neutralize isolated or bypassed enemy forces (including the processing of prisoners and civilian detainees) to increase security and protect lines of communications. It includes the sustainment operations, rapid repositioning, and reorganization of maneuver forces and reconnaissance and security forces. Consolidation may also include activities in support of the civilian population, such as the relocation of displaced civilians, reestablishment of law and order, humanitarian assistance and relief operations, and restoration of key urban infrastructure.

Transition
When planning UO, commanders ensure that they plan, prepare for, and manage transitions. Transitions are movements from one phase of an operation to another and may involve changes in the type of operation, concept of the operation, mission, situation, task organization, forces, resource allocation, support arrangements, or mission command. Transitions occur in all operations. However, in UO, they occur with greater frequency and intensity, are more complex, and often involve agencies other than U.S. military organizations. All operations often include a transition of responsibility for some aspect of the urban environment to (or back to) a legitimate civilian authority. Unless planned and executed effectively, transitions can reduce the tempo of UO, slow its momentum, and cede the initiative to the enemy.

Key Tactical Considerations
Commanders and planners of major operations must thoroughly understand the tactical urban battle as well as the effects of that environment on men, equipment, and systems. The complexity of urban environment changes and often compresses many tactical factors typically considered in the planning process. These compressed tactical factors include—
- Time
- Distances
- Density
- Combat power
- Levels of war
- Decision making

Commanders and their staffs should carefully review ATTP 3-06.11 for techniques that support tactical urban operations.

Understanding the Urban Environment

Ref: ATTP 3-06.11, Combined Arms Operations in Urban Terrain (Jun '11), pp. xii to xviii.

Urban operations are among the most difficult and challenging missions a BCT can undertake. Most UO are planned and controlled at division or corps level but executed by BCTs. The unified action environment of UO enables and enhances the capabilities of the BCT to plan, prepare, and execute offensive, defensive, and stability operations. Urban operations are Infantry-centric combined arms operations that capitalize on the adaptive and innovative leaders at the squad, platoon, and company level.

The special considerations in any UO go well beyond the uniqueness of the urban terrain. JP 3-06 identifies three distinguishing characteristics of the urban environment—physical terrain, population, and infrastructure. FM 3-06 identifies three key overlapping and interdependent components of the urban environment: terrain (natural and man-made), society, and the supporting infrastructure.

Terrain

Urban terrain, both natural and man-made, is the foundation upon which the population and infrastructure of the urban area are superimposed. The physical environment includes the geography and man-made structures in the area of operations (AO). A city may consist of a core surrounded by various commercial ribbons, industrial areas, outlying high-rise areas, residential areas, shantytowns, military areas, extensive parklands or other open areas, waterways, and transportation infrastructure. City patterns may consist of a central hub surrounded by satellite areas, or they may be linear, networked, or segmented. They may contain street patterns that are rectangular, radial, concentric, irregular, or a combination of patterns. They may be closely packed where land space is at a premium or dispersed over several square miles. The infinite ways in which these features may be combined make it necessary to approach each urban area as a unique problem.

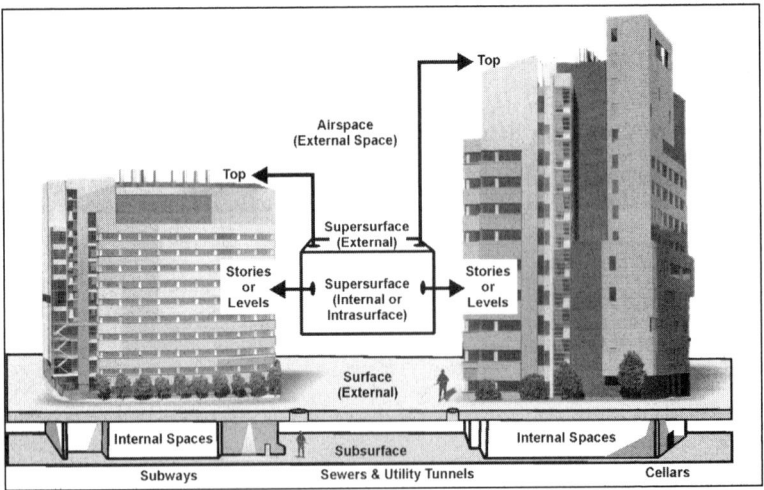

Understanding the physical characteristics of urban terrain requires a multidimensional approach. Commanders operating in unrestricted terrain normally address their AO in terms of air and ground. However, operations within the urban environment provide numerous man-made structures and variables not found in unrestricted terrain. Commanders conducting UO must broaden the scope of their thinking. The total size of the surfaces and spaces of an urban area is usually many times that of a similarly size piece of natural terrain because of the complex blend of horizontal, vertical, interior, exterior, and subterranean forms superimposed on the natural landscape.

Society

Urban operations often require forces to operate in close proximity to a high density of civilians. Even evacuated areas can have a large stay-behind population. The population's presence, attitudes, actions, communications with the media, and needs may affect the conduct of the operation. To effectively operate among an urban population and maintain its goodwill, it is important to develop a thorough understanding of the population and its culture, to include values, needs, history, religion, customs, and social structure.

The demographics of the HN can complicate urban operations. The Army is likely to conduct UO in countries with existing or emerging cultural, ethnic, or religious conflicts. When these conditions exist, the local population may be sympathetic to enemy causes. Refugees and displaced persons are likely to be present. For these and other reasons, cultural awareness is imperative to mission success.

Accommodating the social norms of a population is potentially the most influential factor in conducting UO. Soldiers function well by acting in accordance with American values but may encounter difficulties when applying American culture, values, and thought processes to the populace or individuals the unit and leadership is trying to understand. Defining the structure of the social hierarchy is often critical to understanding the population.

Other considerations include:

- Many governments of developing countries are characterized by nepotism, favor trading, sectarianism, and indifference
- Regardless of causes or political affiliations, civilian casualties are often the focal point of press coverage to the point of ignoring or demeaning any previous accomplishments.
- Religious beliefs and practices are among the most important yet least understood aspects of the cultures of other peoples. In many parts of the world, religious norms are a matter of life and death.
- Another significant problem is the presence of displaced persons within an urban area. Noncombatants without hostile intent can inadvertently complicate UO.

Infrastructure

A city's infrastructure is its foundation. Restoration or repair of urban infrastructure is often decisive to mission accomplishment. During full spectrum operations, destroying, controlling, or protecting vital parts of the urban infrastructure may be a necessary shaping operation to isolate an enemy from potential sources of support. An enemy force may rely on the area's water, electricity, and sources of bulk fuel to support his forces. To transport supplies, the enemy may rely on roads, airfields, sea or river lanes, and rail lines.

Controlling these critical infrastructure systems may prevent the enemy from resupplying his forces. The infrastructure of an urban environment consists of the basic resources, support systems, communications, and industries upon which the population depends. The key elements that allow an urban area to function are significant to full spectrum operations. The force that can control and secure the water, telecommunications, energy production and distribution, food production and distribution, and medical city's boundaries.

All systems fit into six broad categories. Commanders should analyze key facilities in each category and determine their role and importance throughout all phases of UO. (Refer to FM 3-06 for details.) The six categories of infrastructure are—

- Communications and Information.
- Transportation and Distribution
- Energy
- Economics and Commerce
- Administration and Human Services
- Cultural

Threat

Ref: ATTP 3-06.11, Combined Arms Operations in Urban Terrain (Jun '11), pp. xvii to xix.

During UO, units should be prepared to face and defeat traditional, irregular, and hybrid threats.

Traditional
Traditional threats compose regular armed forces employing recognized military capabilities with large formations conducting offensive or defensive operations that specifically confront the BCT's combat power and capabilities.

Irregular
Irregular threats are forces composed of armed individuals or groups who are not members of the regular armed forces, police, or other internal security forces. They engage in insurgency, guerrilla activities, and unconventional warfare as principle activities.

Hybrid
Hybrid threats are likely to simultaneously employ dynamic combinations of traditional and irregular forces, including terrorist and criminal elements to achieve their objectives. They will use an ever-changing variety of conventional and unconventional tactics within the urban AO to create multiple dilemmas for UO forces. Commanders at all levels should organize and equip their forces so they do not rely on a single solution or approach to problem sets. Furthermore, commanders should be prepared to alter plans and operations accordingly when approaches to problems do not work as anticipated. Hybrid threats attempt to avoid confrontation with the UO's combat power and capabilities and may use the civilian population and infrastructure to shield their capabilities from BCT fires. They are most likely based in and target urban areas to take advantage of the density of civilian population and infrastructure.

Potential enemies (traditional, irregular, and hybrid) in UO share some common characteristics. The broken and compartmented terrain is best suited for small-unit operations. Typical urban fighters are organized in squad-size elements and employ small-unit tactics that can be described as guerrilla tactics, terrorist tactics, or a combination of the two. They normally choose to attack (often using ambushes) on terrain that allows them to inflict casualties and then withdrawal. They attempt to canalize UO forces and limit their ability to maneuver or mass. Small-arms weapons, sniper rifles, rocket-propelled grenades (RPG), mines, improvised explosive devices (IED), and booby traps are often the preferred weapons.

Enemy forces in conventional major combat operations oppose U.S. forces with a variety of means, including high technology capabilities built into mechanized, motorized, and light Infantry forces. These forces may be equipped with newer generation tanks and Infantry fighting vehicles and have significant numbers of antitank guided missile systems, Man-Portable Air Defense System (MANPADS) weapons, advanced fixed- or rotary-wing aviation assets, missiles, rockets, artillery, mortars, and mines. They may field large numbers of Infantry and robust military and civilian communications systems. In addition, they may possess weapons of mass destruction. Enemy forces in major combat operations may be capable of long-term resistance using conventional formations, such as divisions and corps. They may also conduct sustained unconventional operations and protracted warfare.

The enemy in unconventional small-scale contingency environments employs forces characterized by limited armor. Some are equipped with small numbers of early generation tanks, some with mechanized forces but most forces are predominately Infantry. Guerrillas, terrorists, paramilitary units, special-purpose forces, special police, and local militias are present in the environment. These forces are equipped primarily with antitank guided missile systems, MANPADSs, mortars, machine guns, and explosives. Their forces are expected to have robust communications, using conventional military devices augmented by commercial equipment, such as cell phones. These forces may not be capable of long-term, sustained, high-tempo operations. They can conduct long-term, unconventional terrorist and guerrilla operations.

Insurgents or Guerrillas

Insurgents are members of a political party who rebel against established leadership. Guerrillas are a group of irregular, predominantly indigenous personnel organized along military lines to conduct military and paramilitary operations in enemy-held, hostile, or denied territory.

Insurgents and guerrillas are highly motivated and can employ advanced communications; some precision weapons, such as guided mortar rounds and MANPADS missiles; and some ground-based sensors in varying combinations with conventional weapons, mines, and IEDs. They usually conduct psychological and other information warfare against the HN government and population, sometimes using assassinations, kidnappings, and other terrorist techniques. Because of this, the BCT should communicate clearly with the population and operate in support of HN government forces rather than act independently as the main security and combat force.

Under the conditions of insurgency within the urban environment, the commander should emphasize—

- Developing population status overlays showing potential hostile neighborhoods
- Developing an understanding of how the insurgent or guerrilla organization operates and its organization
- Determining primary operating or staging areas
- Determining mobility corridors and infiltration/exfiltration routes
- Determining most likely targets
- Determining where the enemy's logistic facilities are and how they operate
- Determining the level of popular support (active and passive)
- Determining the recruiting, command and control, reconnaissance and security, logistics (to include money), and operations techniques and methods.
- Locating neutrals and those actively opposing these organizations
- Using pattern analysis and other tools to establish links between the insurgent or guerilla organization and other organizations (to include family links)
- Determining the underlying social, political, and economic issues

Enemy Tactics

Adaptive urban enemies seek to modify their operations to create false presentations and reduce signatures to influence and disrupt accurate intelligence preparation of the battlefield (IPB). They also attempt to deceive the BCT by showing it exactly what it expects to see. Enemy forces and organizations position decoys and deception minefields in locations where the BCT expects to see them and emplace real mines where the BCT does not anticipate them.

In complex urban terrain, the enemy can close undetected with BCT forces and employ low-signature weapons against command posts (CP), communications nodes, sustainment units, and uncommitted forces. This makes the survivability of these elements and forces at the BCT level more difficult in an urban environment. The need to find, engage, and defeat the enemy must include an understanding that all forces within the BCT must be prepared to fight and secure themselves, their equipment, and their means to move and maneuver. This, combined with commercially available deception measures available to the enemy, raises the level of uncertainty and slows the pace of BCT maneuver, potentially making it more vulnerable.

Urban enemies seek to complicate BCT targeting by "hugging" BCT forces or through shielding their forces among civilian populations or within important cultural landmarks and social or religious structures. Enemy use of high technology systems also makes discerning the signatures of high-payoff systems more difficult, further confounding BCT targeting efforts. Differentiating between valid and invalid targets is time-consuming and impacts reconnaissance and security capabilities through enemy deception and dispersion.

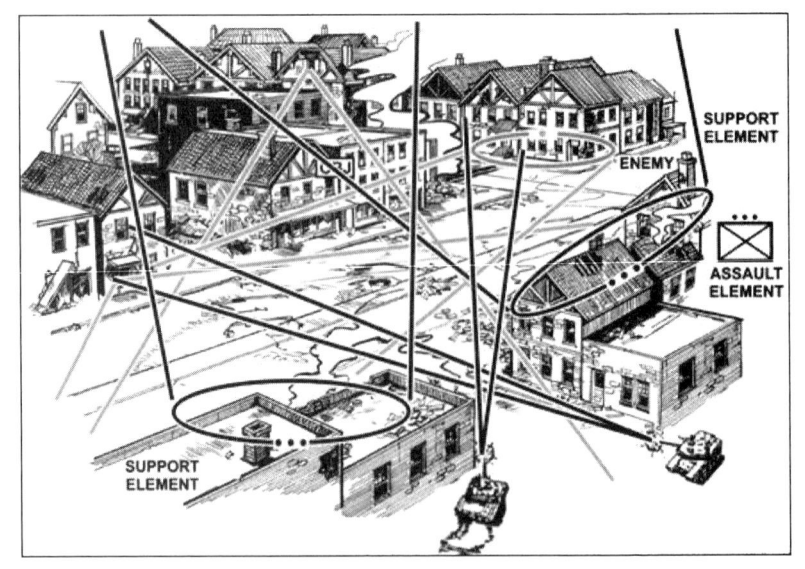

Cordon

A cordon is a line of troops or military posts that enclose an area to prevent passage. The Infantry platoon normally conducts a cordon as part of a larger unit. It is established by positioning one or more security elements on key terrain that dominates avenues of approach in and out of the objective area. The overall goal is the protection of the maneuver element, and to completely dominate what exits or enters the objective area. This requires a detailed understanding of avenues of approach in the area. There are many techniques used to facilitate isolation including, blocking positions, direct fire (precision and area), indirect fire, roadblocks, checkpoints, and observation posts. The same techniques can be used to cordon and search a small urban area (such as a village) surrounded by other terrain.

Ideally these positions are occupied simultaneously, but a sequential approach can also be useful. Limited visibility aids can be used in the establishment and security of the cordon. The security element can either surround the area while the maneuver element simultaneously moves in, or it can use a sequential technique in which they use stealth to get into position before the actual assault.

Plans should be developed to handle detained personnel. Infantrymen will normally provide security and accompany police and intelligence forces who will identify, question, and detain suspects. Infantry may also conduct searches and assist in detaining suspects, but their principal role is to reduce any resistance that may develop and to provide security for the operation. Use of force is kept to a minimum unless otherwise directed.

III. Assault a Building

Squads and platoons, particularly when augmented with engineers, are the best organized and equipped units in the Army for breaching protective obstacles; gaining access to buildings; and assaulting rooms, hallways, and stairways. Although there are specific drills associated with fighting in buildings, the overall assault is an operation, not a drill. During planning, the leader's level of detail should identify each window (aperture, opening, or firing port) in his sector fortifications. He should then consider assigning these as a specific TRP when planning fires.

Critical Tasks

There are a number of critical tasks that need emphasis for Infantry platoons assaulting a building:
- Isolate the building
- Gain and maintain fire superiority inside and outside the building
- Gain access to the inside of the building
- Move inside the building
- Seize positions of advantage
- Control the tempo

A. Entering the Building

After establishing suppression and obscuration, leaders deploy their subordinates to secure the near side and then, after gaining access, secure the far side. Gaining access to the inside of the building normally requires reducing protective obstacles.

Units gain access by using either a top or bottom entry. The entry point is the same thing as a point of penetration for an obstacle breach and as such is a danger area. The entry point will become the focus of fires for any enemy in a position to fire at it. It is commonly referred to as the "fatal funnel." Leaders ensure they have established measures to ensure the assault team has fire superiority when moving through the fatal funnel. Grenades (ROE determines fragmentation or concussion) are used to gain enough of a window of opportunity until the assault element can employ its small arms fire.

Top Entry

The top of a building is ordinarily considered a position of advantage. Entering at the top and fighting downward is the preferred method of gaining access to a building for a number of reasons. First, just as in operations on other types of terrain, it is easier to own the high ground and work your way down than it is to fight your way up when the enemy owns the high ground. Second, an enemy forced down to ground level may be tempted to withdraw from the building and expose himself to the fire of covering units or weapons. Third, the ground floor and basements are normally more heavily defended. Finally, the roof of a building is ordinarily weaker than the walls (and therefore easier to penetrate).

Top entry is only feasible when the unit can gain access to an upper floor or rooftop. Rooftops are danger areas when surrounding buildings are higher and forces can be exposed to fire from those buildings. Soldiers should consider the use of devices and other techniques that allow them upper level access without using interior stairways. Those devices and techniques include, but are not limited to, adjacent rooftops, fire escapes, portable ladders, and various Soldier-assisted lifts. For more information on top entry breaching, see FM 3-06.11.

Bottom Entry

Entry at the bottom is common and may be the only option available. When entering from the bottom, breaching a wall to create a "mousehole" is the preferred method because doors and windows may be booby-trapped and covered by fire from inside the structure. There are many ways to accomplish this, including employing CCMS, SLM, demolitions, hand tools, machine guns, artillery fire, and tank fire. The actual technique used depends on the ROE, assets available, building structure, and the enemy situation. If the assault element must enter through a door or window, it should enter from a rear or flank position after ensuring the entry point is clear of obstacles.

Secure the Near and Far Side of the Point of Penetration

Infantry platoons use the following drill for gaining access to the building. The steps of this drill are very similar to those drills described in Section IX to secure the near and far side of the point of penetration—

- The squad leader and the assault fire team move to the last covered and concealed position near the entry point
- The squad leader confirms the entry point
- The platoon leader or squad leader shifts the support fire away from the entry point
- The support-by-fire element continues to suppress building and adjacent enemy positions as required
- Buddy team #1 (team leader and automatic rifleman) remain in a position short of the entry point to add suppressive fires for the initial entry
- Buddy team #2 (grenadier and rifleman) and the squad leader move to the entry point. They move in rushes or by crawling.
- The squad leader positions himself where he can best control his teams
- Buddy team #2 position themselves against the wall to the right or left of the entry point.
- On the squad leader command of COOK OFF GRENADES (2 seconds maximum), the Soldiers employing the grenades shout, FRAG OUT, and throw the grenades into the building. (If the squad leader decides not to use grenades, he commands, PREPARE TO ENTER—GO!)
- Upon detonation of both grenades (or command GO), the buddy team flows into the room/hallway and moves to points of domination engaging all identified or likely enemy positions.
- Both Soldiers halt and take up positions to block any enemy movement toward the entry point.
- Simultaneously, buddy team #1 moves to and enters the building, joins buddy team #2, and announces, CLEAR.
- The squad leader remains at the entry point and marks it IAW unit SOP. He calls forward the next fire team with, NEXT TEAM IN.
- Once the squad has secured a foothold, the squad leader reports to the platoon leader, FOOTHOLD SECURE. The platoon follows the success of the seizure of the foothold with the remainder of the platoon.

When using a doorway as the point of entry, the path of least resistance is initially determined on the way the door opens. If the door opens inward, the Soldier plans to move away from the hinged side. If the door opens outward, he plans to move toward the hinged side. Upon entering, the size of the room, enemy situation, and obstacles in the room (furniture and other items) that hinder or channel movement become factors that influence the number one man's direction of movement.

B. Clearing Rooms

Ref: FM 3-21.8 (FM 7-8) The Infantry Rifle Platoon and Squad, pp. 7-40 to 7-42.

Although rooms come in all shapes and sizes, there are some general principles that apply to most room clearing tasks. For clearing large open buildings such as hangars or warehouses, it may be necessary to use subordinate units using a line formation while employing traveling or bounding overwatch. These methods can effectively clear the entire structure while ensuring security.

Room clearing techniques differ based on METT-TC, ROE, and probability of noncombatants inside the building. If there are known or suspected enemy forces, but no noncombatants inside the building, the platoon may conduct high intensity room clearings. If there are known or suspected noncombatants within the building, the platoon may conduct precision room clearings. High intensity room clearing may consist of fragmentation grenade employment and an immediate and high volume of small arms fire placed into the room, precision room clearing will not.

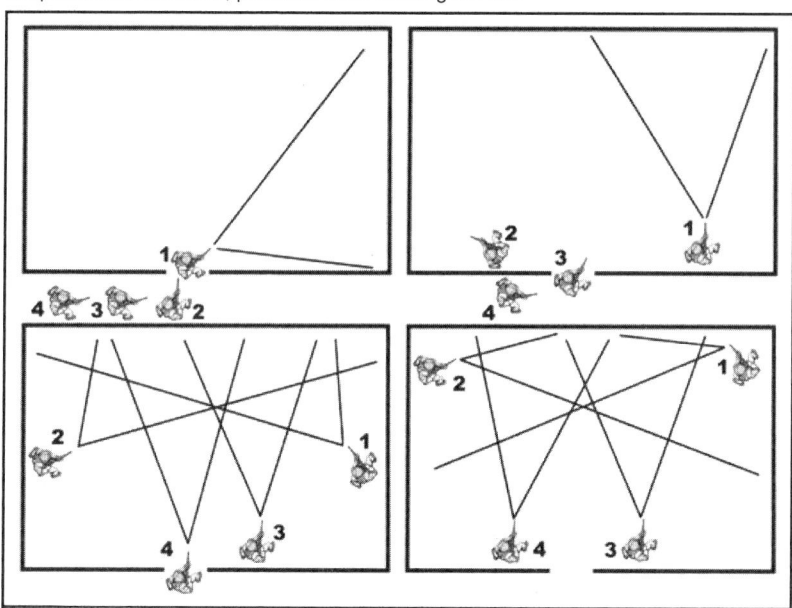

- **#1 Man**. The #1 man enters the room and eliminates any immediate threat. He can move left or right, moving along the path of least resistance to a point of domination—one of the two corners and continues down the room to gain depth.

- **#2 Man**. The #2 man enters almost simultaneously with the first and moves in the opposite direction, following the wall. The #2 man must clear the entry point, clear the immediate threat area, and move to his point of domination.

- **#3 Man**. The #3 man simply moves in the opposite direction of the #2 man inside the room, moves at least 1 meter from the entry point, and takes a position that dominates his sector.

- **#4 Man**. The #4 man moves in the opposite direction of the #3 man, clears the doorway by at least 1 meter, and moves to a position that dominates his sector.

Once the room is cleared, the team leader may order some team members to move deeper into the room overwatched by the other team members. The team leader must control this action. In addition to dominating the room, all team members are responsible for identifying possible loopholes and mouseholes. Cleared rooms should be marked IAW unit SOP.

(Urban & Regional Environments) I. Urban Operations 7-13

C. Moving in the Building

Ref: FM 3-21.8 (FM 7-8) The Infantry Rifle Platoon and Squad, pp. 7-42 to 7-46. See also pp. 2-32 to 2-33 for discussion of clearing as a tactical mission task.

Movement techniques used inside a building are employed by teams to negotiate hallways and other avenues of approach.

Diamond Formation (Serpentine Technique)

The serpentine technique is a variation of a diamond formation that is used in a narrow hallway. The #1 man provides security to the front. His sector of fire includes any enemy Soldiers who appear at the far end or along the hallway. The #2 and #3 men cover the left and right sides of the #1 man. Their sectors of fire include any enemy combatants who appear suddenly from either side of the hall. The #4 man (normally carrying the M249) provides rear protection.

Vee Formation (Rolling-T Technique)

The rolling-T technique is a variation of the Vee formation and is used in wide hallways. The #1 and #2 men move abreast, covering the opposite side of the hallway from the one they are walking on. The #3 man covers the far end of the hallway from a position behind the #1 and #2 men, firing between them. The #4 man provides rear security.

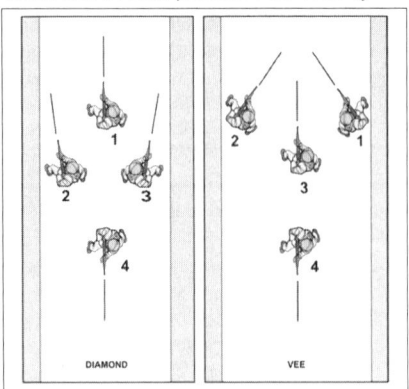

The unit is using the diamond (serpentine) formation for movement (Figure 7-13 A).

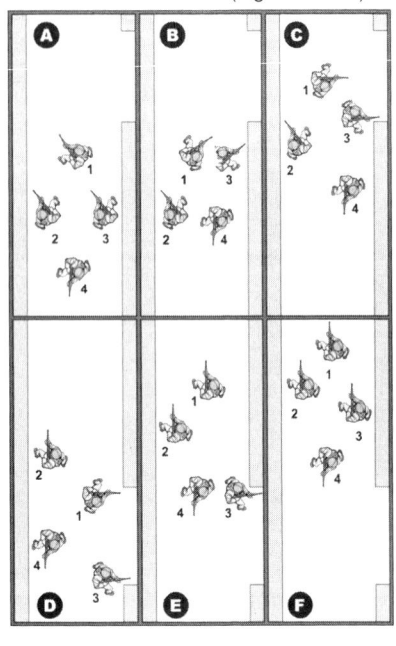

To clear a hallway—

- The team configures into a modified 2-by-2 (box) formation with the #1 and #3 men abreast and toward the right side of the hall. The #2 man moves to the left side of the hall and orients to the front, and the #4 man shifts to the right side (his left) and maintains rear security. (When clearing a right-hand corner, use the left-handed firing method to minimize exposure [Figure 7-13 B]).

- The #1 and #3 men move to the edge of the corner. The #3 man assumes a low crouch or kneeling position. On signal, the #3 man, keeping low, turns right around the corner and the #1 man, staying high, steps forward while turning to the right. Sectors of fire interlock and the low/high positions prevent Soldiers from firing at one another [Figure 7-13 C]).

1. Clearing Hallway Junctions

Hallway intersections are danger areas and should be approached cautiously. Figure 7-13 depicts the fire team's actions upon reaching a "T" intersection when approaching along the "cross" of the "T".

- The #2 and #4 men continue to move in the direction of travel. As the #2 man passes behind the #1 man, the #1 man shifts to his left until he reaches the far corner (Figure 7-13 D).
- The #2 and #4 men continue to move in the direction of travel. As the #4 man passes behind the #3 man, the #3 man shifts laterally to his left until he reaches the far corner. As the #3 man begins to shift across the hall, the #1 man turns into the direction of travel and moves to his original position in the diamond (serpentine) formation (Figure 7-13 E).
- As the #3 and #4 men reach the far side of the hallway, they, too, assume their original positions in the serpentine formation, and the fire team continues to move (Figure 7-13 F).

2. Clearing a "T" Intersection

Figure 7-14 depicts the fire team's actions upon reaching a "T" intersection when approaching from the base of the "T". The fire team is using the diamond (serpentine) formation for movement (Figure 7-14 A).

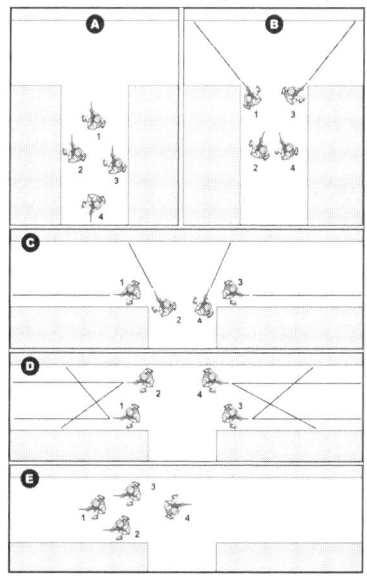

- The team configures into a 2-by-2 (box) formation with the #1 and #2 men left and the #3 and #4 men right. (When clearing a right-hand corner, use the left-handed firing method to minimize exposure [Figure 7-14 B]).
- The #1 and #3 men move to the edge of the corner and assume a low crouch or kneeling position. On signal, the #1 and #3 men simultaneously turn left and right respectively (Figure 7-14 C).
- At the same time, the #2 and #4 men step forward and turn left and right respectively while maintaining their (high) position. (Sectors of fire interlock and the low/high positions prevent Soldiers from firing at another [Figure 7-14 D]).
- Once the left and right portions of the hallway are clear, the fire team resumes the movement formation (Figure 7-14 E). Unless security is left behind, the hallway will no longer remain clear once the fire team leaves the immediate area.

3. Clearing Stairwells and Staircases

Stairwells and staircases are comparable to doorways because they create a fatal funnel. The danger is intensified by the three-dimensional aspect of additional landings. The ability of units to conduct the movement depends upon which direction they are traveling and the layout of the stairs. Regardless, the clearing technique follows a basic format:

- The leader designates an assault element to clear the stairs
- The unit maintains 360-degree, three-dimensional security in the vicinity of the stairs.
- The leader then directs the assault element to locate, mark, bypass, and or clear any obstacles or booby traps
- The assault element moves up (or down) the stairway by using either the two-, three-, or four-man flow technique, providing overwatch up and down the stairs while moving. The three-man variation is preferred.

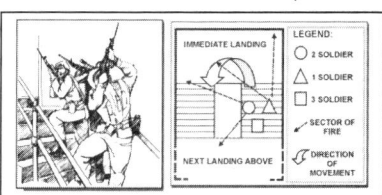

(Urban & Regional Environments) I. Urban Operations 7-15

IV. Follow Through

After securing a floor (bottom, middle, or top), selected members of the unit are assigned to cover potential enemy counterattack routes to the building. Priority must be given initially to securing the direction of attack. Security elements alert the unit and place a heavy volume of fire on enemy forces approaching the unit.

Units must guard all avenues of approach leading into their area. These may include—

- Enemy mouseholes between adjacent buildings
- Covered routes to the building
- Underground routes into the basement
- Approaches over adjoining roofs or from window to window

Units that performed missions as assault elements should be prepared to assume an overwatch mission and to support another assault element.

To continue the mission—

- Momentum must be maintained. This is a critical factor in clearing operations. The enemy cannot be allowed to move to its next set of prepared positions or to prepare new positions.
- The support element pushes replacements, ammunition, and supplies forward to the assault element
- Casualties must be evacuated and replaced
- Security for cleared areas must be established IAW the OPORD or TSOP
- All cleared areas and rooms must be marked IAW unit SOP
- The support element must displace forward to ensure that it is in place to provide support (such as isolation of the new objective) to the assault element

The compartmentalized nature of urban terrain, limited observation and fields of fire, and the vast amounts of potential cover and concealment mean that defenders can disperse and remain undetected. The origin of enemy gunfire can be difficult to detect, because distance and direction become distorted by structures. (Dept. of Army photo by Spc. Kieran Cuddihy).

II. Fortified Areas

Ref: FM 3-21.8 (FM 7-8) The Infantry Rifle Platoon and Squad, pp. 7-48 to 7-53.

Fortifications are works emplaced to defend and reinforce a position. Time permitting, enemy defenders build bunkers and trenches, emplace protective obstacles, and position mutually supporting fortifications when fortifying their positions. Soldiers who attack prepared positions should expect to encounter a range of planned enemy fires to include small arms fire, mortars, artillery, antitank missiles, antitank guns, tanks, attack aviation, and close air support. Attacking forces should also expect a range of offensive type maneuver options to include spoiling attacks, internal repositioning, counterattacks, and withdrawing to subsequent defensive positions. Spoiling attacks will attempt to disrupt the attacker's momentum and possibly seize key terrain. If driven out of their prepared positions, enemy troops may try to win them back by hasty local counterattacks or through deliberate, planned combined arms counterattacks. If forced to withdraw, the enemy forces may use obstacles, ambushes, and other delaying tactics to slow down pursuing attackers.

The attack of a fortified position follows the basic principles of tactical maneuver. However, greater emphasis is placed upon detailed planning, special training and rehearsals, increased fire support, and the use of special equipment.

The deliberate nature of defenses requires a deliberate approach to the attack. These types of operations are time consuming. Leaders must develop schemes of maneuver that systematically reduce the area. Initially, these attacks should be limited in scope, focusing on individual positions and intermediate terrain objectives. Leaders must establish clear bypass criteria and position destruction criteria and allocate forces to secure cleared enemy positions. Failure in this will likely result in enemy reoccupying the positions, isolating lead elements, and ambushes.

Characteristics

The intense, close combat prevalent in trench clearing is remarkably similar to fighting in built up areas. Comparable characteristics include:

- **Restricted Observation and Fields of Fire**. Once the trench is entered, visibilities may be limited to a few meters in either direction. This compartmentalization necessarily decentralizes the engagement to the lowest level.

- **Cover and Concealment**. The nature of a trench system allows covered movement of both friendly and enemy forces. To prevent being flanked or counterattacked, junctions, possible entry points, and corners should be secured.

- **Difficulty in Locating the Enemy**. The assault element may come under fire from multiple mutually supporting positions in the trench or a nearby position. The exact location of the fire may be difficult to determine. Supporting elements should be capable of locating, suppressing, or destroying such threats.

- **Close Quarters Fighting**. Because of the close nature of the trench system, Soldiers should be prepared to use close quarters marksmanship, bayonet, and hand-to-hand fight techniques.

- **Restricted Movement**. Trench width and height will severely restrict movement inside the system. This will ordinarily require the assault element to move at a low crouch or even a crawl. Sustainment including ammunition resupply, EPW evacuation, casualty evacuation, and reinforcement will also be hampered.

- **Sustainment**. The intensity of close combat in the trench undoubtedly results in increased resource requirements.

I. Find

Finding the enemy's fortified positions relates back to the position's purpose. There are two general reasons to create fortified positions. The first includes defending key terrain and using the position as a base camp, shelter, or sanctuary for critical personnel or activities. This type of position is typically camouflaged and difficult to locate. When U.S. forces have air superiority and robust reconnaissance abilities, enemy forces will go to great lengths to conceal these positions. Sometimes the only way to find these enemy positions is by movement to contact. When Infantry platoons or squads encounter a previously unidentified prepared enemy position, they should not, as a general rule, conduct a hasty attack until they have set conditions for success.

The second general purpose for fortified positions is to create a situation in which the attacker is required to mass and present a profitable target. This type of position normally occurs in more conventional battles. These positions can be relatively easy to find because they occupy key terrain, establish identifiable patterns, and generally lack mobility.

Attacking fortified positions requires thorough planning and preparation based on extensive reconnaissance.

II. Fix

An enemy in fortified defenses has already partially fixed himself. This does not mean he will not be able to maneuver or that the fight will be easy. It does mean that the objective is probably more defined than with an enemy with complete freedom of movement. Fixing the enemy will still require measures to prevent repositioning to alternate, supplementary, and subsequent positions on the objective and measures to block enemy counterattack elements.

III. Finish (Fighting Enemies in Fortifications)

Finishing an enemy in prepared positions requires the attacker to follow the fundamentals of the offense-surprise, concentration, tempo, and audacity to be successful.

The actual fighting of enemy fortifications is clearly an Infantry platoon unit function because squads and platoons, particularly when augmented with engineers, are the best organized and equipped units in the Army for breaching protective obstacles. They are also best prepared to assault prepared positions such as bunkers and trench lines. Infantry platoons are capable of conducting these skills with organic, supplementary, and supporting weapons in any environment.

Leaders develop detailed plans for each fortification, using the SOSRA technique to integrate and synchronize fire support and maneuver assets. Although there are specific drills associated with the types of fortifications, the assault of a fortified area is an operation, not a drill. During planning, the leader's level of detail should identify each aperture (opening or firing port) of his assigned fortification(s) and consider assigning these as a specific target when planning fires. Contingency plans are made for the possibility of encountering previously undetected fortifications along the route to the objective, and for neutralizing underground defenses when encountered.

A. Securing the Near and Far Side—Breaching Protective Obstacles

To fight the enemy almost always requires penetrating extensive protective obstacles, both antipersonnel and antivehicle. Of particular concern to the Infantrymen are antipersonnel obstacles. Antipersonnel obstacles (both explosive and nonexplosive) include, wire entanglements; trip flares; antipersonnel mines; field expedient devices (booby traps, nonexplosive traps, punji sticks); flame devices; rubble; warning devices; CBRN; and any other type of obstacle created to prevent troops from entering a position. Antipersonnel obstacles are usually integrated with enemy fires close enough to the fortification for adequate enemy surveillance by day or night, but beyond effective hand grenade range. Obstacles are also used within the enemy position to compartmentalize the area in the event outer protective barriers are breached. See Appendix F for more information on obstacles.

The following steps are an example platoon breach:

- The squad leader and the breaching fire team move to the last covered and concealed position near the breach point (point of penetration)
- The squad leader confirms the breach point
- The platoon leader or squad leader shifts the suppressing element away from the entry point
- The fire element continues to suppress enemy positions as required
- Buddy team #1 (team leader and the automatic rifleman) remains in a position short of the obstacle to provide local security for buddy team #2
- The squad leader and breaching fire team leader employ smoke grenades to obscure the breach point
- Buddy team #2 (grenadier and rifleman) moves to the breach point. They move in rushes or by crawling
- The squad leader positions himself where he can best control his teams.
- Buddy team #2 positions themselves to the right and left of the breach point near the protective obstacle
- Buddy team #2 probes for mines and creates a breach, marking their path as they proceed
- Once breached, buddy team #1 and buddy team #2 move to the far side of the obstacle and take up covered and concealed positions to block any enemy movement toward the breach point. They engage all identified or likely enemy positions.
- The squad leader remains at the entry point and marks it. He calls forward the next fire team with, "Next team in."
- Once the squad has secured a foothold, the squad leader reports to the platoon leader, "Foothold secure." The platoon follows the success of the seizure of the foothold with the remainder of the platoon.

B. Knocking Out Bunkers

The term bunker in this discussion covers all emplacements having overhead cover and containing apertures (embrasures) through which weapons are fired. The two primary types are reinforced concrete pillboxes, and log bunkers. There are two notable exploitable weaknesses of bunkers.

First, bunkers are permanent, their location and orientation fixed. Bunkers cannot be relocated or adjusted to meet a changing situation. They are optimized for a particular direction and function. The worst thing an Infantry platoon or squad can do is to approach the position in the manner it was designed to fight.

Second, bunkers must have openings (doors, windows, apertures, or air vents). There are two disadvantages to be exploited here. First, structurally, the opening is the weakest part of the position and will be the first part of the structure to collapse if engaged. Second, a single opening can only cover a finite sector, creating blind spots.

C. Assaulting Trench Systems

Ref: FM 3-21.8 (FM 7-8) The Infantry Rifle Platoon and Squad, pp. 7-51 to 7-53.

Trenches are dug to connect fighting positions. They are typically dug in a zigzagged fashion to prevent the attacker from firing down a long section if he gets into the trench, and to reduce the effectiveness of high explosive munitions. Trenches may also have shallow turns, intersections with other trenches, firing ports, overhead cover, and bunkers. Bunkers will usually be oriented outside the trench, but may also have the ability to provide protective fire into the trench.

The trench provides defenders with a route that has frontal cover, enabling them to reposition without the threat of low trajectory fires. However, unless overhead cover is built, trenches are subject to the effects of high trajectory munitions like the grenade, grenade launcher, plunging machine gun fire, mortars, and artillery. These types of weapon systems should be used to gain and maintain fire superiority on defenders in the trench.

The trench is the enemy's home, so there is no easy way to clear it. Their confined nature, extensive enemy preparations, and the limited ability to integrate combined arms fires makes trench clearing hazardous for even the best trained Infantry. If possible, a bulldozer or plow tank can be used to fill in the trench and bury the defenders. However, since this is not always feasible, Infantry units must move in and clear trenches.

1. Entering the Trenchline

To enter the enemy trench the platoon takes the following steps:

- The squad leader and the assault fire team move to the last covered and concealed position near the entry point
- The squad leader confirms the entry point
- The platoon leader or squad leader shifts the base of fire away from the entry point
- The base of fire continues to suppress trench and adjacent enemy positions as required
- Buddy team #1 (team leader and automatic rifleman) remains in a position short of the trench to add suppressive fires for the initial entry
- Buddy team #2 (grenadier and rifleman) and squad leader move to the entry point. They move in rushes or by crawling (squad leader positions himself where he can best control his teams).
- Buddy team #2 positions itself parallel to the edge of the trench. Team members get on their backs
- On the squad leader command of COOK OFF GRENADES (2 seconds maximum), they shout, FRAG OUT, and throw the grenades into the trench
- Upon detonation of both grenades, the Soldiers roll into the trench, landing on their feet and back-to-back. They engage all known, likely or suspected enemy positions.
- Both Soldiers immediately move in opposite directions down the trench, continuing until they reach the first corner or intersection
- Both Soldiers halt and take up positions to block any enemy movement toward the entry point
- Simultaneously, buddy team #1 moves to and enters the trench, joining buddy team #2. The squad leader directs them to one of the secured corners or intersections to relieve the Soldier who then rejoins his buddy at the opposite end of the foothold.
- At the same time, the squad leader rolls into the trench and secures the entry point.
- The squad leader remains at the entry point and marks it. He calls forward the next fire team with, NEXT TEAM IN
- Once the squad has secured a foothold, the squad leader reports to the platoon leader, FOOTHOLD SECURE. The platoon follows the success of the seizure of the foothold with the remainder of the platoon.

7-20 (Urban & Regional Environments) II. Fortified Areas

The leader or a designated subordinate must move into the trench as soon as possible to control the tempo, specifically the movement of the lead assault element and the movement of follow-on forces. He must resist the temptation to move the entire unit into the trench as this will unduly concentrate the unit in a small area. Instead,, he should ensure the outside of the trench remains isolated as he maintains fire superiority inside the trench. This may require a more deliberate approach. When subordinates have reached their objectives or have exhausted their resources, the leader commits follow-on forces. Once stopped, the leader consolidates and reorganizes.

The assault element is organized into a series of three-man teams. The team members are simply referred to as number 1 man, number 2 man, and number 3 man. Each team is armed with at least one M249 and one grenade launcher. All men are armed with multiple hand grenades.

The positioning within the three-man team is rotational, so the men in the team must be rehearsed in each position. The number 1 man is responsible for assaulting down the trench using well aimed effective fire and throwing grenades around pivot points in the trenchline or into weapons emplacements. The number 2 man follows the number 1 man closely enough to support him but not so closely that both would be suppressed if the enemy gained local fire superiority. The number 3 man follows the number 2 man and prepares to move forward when positions rotate.

While the initial three-man assault team rotates by event, the squad leader directs the rotation of the three-man teams within the squad as ammunition becomes low in the leading team, casualties occur, or as the situation dictates. Since this three-man drill is standardized, three-man teams may be reconstituted as needed from the remaining members of the squad. The platoon ldr controls the rotation between squads using the same considerations as the squad leaders.

2. Clearing the Trenchline

Once the squad has secured the entry point and expanded it to accommodate the squad, the rest of the platoon enters and begins to clear the designated section of the enemy position. The platoon may be tasked to clear in two directions if the objective is small. Otherwise, it will only clear in one direction as another platoon enters and clears in the opposite direction.

The lead three-man team of the initial assault squad moves out past the security of the support element and executes the trench clearing drill. The number 1 man, followed by number 2 man and number 3 man, maintains his advance until arriving at a pivot, junction point, or weapons emplacement in the trench. He alerts the rest of the team by yelling out, POSITION or, JUNCTION, and begins to prepare a grenade. The number 2 man immediately moves forward near the lead man and takes up the fire to cover until the grenade can be thrown around the corner of the pivot point. The number 3 man moves forward to the point previously occupied by number 2 and prepares for commitment.

If the lead man encounters a junction in the trench, the platoon leader should move forward, make a quick estimate, and indicate the direction the team should continue to clear. This will normally be toward the bulk of the fortification or toward command post emplacements. He should place a marker (normally specified in the unit TSOP) pointing toward the direction of the cleared path. After employing a grenade, the number 2 man moves out in the direction indicated by the platoon leader and assumes the duties of the number 1 man. Anytime the number 1 man runs out of ammunition, he shouts, MAGAZINE, and immediately moves against the wall of the trench to allow the number 2 man to take up the fire. Squad leaders continue to push uncommitted teams forward, securing bypassed trenches and rotating fresh teams to the front. Trenches are cleared in sequence not simultaneously.

3. Moving in a Trench

Once inside, the trench teams use variations of the combat formations to move. These formations are used as appropriate inside buildings as well. The terms hallway and trench are used interchangeably. The column (file) and box formations are self explanatory. The line and echelon formations are generally infeasible.

Ideally the team is able to destroy the bunker with standoff weapons and HE munitions. However, when required, the fire team can assault the bunker with small arms and grenades. A fire team (two to four men) with HE and smoke grenades move forward under cover of the suppression and obscuration fires from the squad and other elements of the base of fire. When they reach a vulnerable point of the bunker, they destroy it or personnel inside with grenades or other hand-held demolitions. All unsecured bunkers must be treated as if they contain live enemy, even if no activity has been detected from them. The clearing of bunkers must be systematic or the enemy will come up behind assault groups. To clear a bunker—

- The squad leader and the assault fire team move to the last covered and concealed position near the position's vulnerable point
- The squad leader confirms the vulnerable point
- The platoon leader/squad leader shifts the base of fire away from the vulnerable point
- The base of fire continues to suppress the position and adjacent enemy positions as required
- Buddy team #1 (team leader and the automatic rifleman) remain in a position short of the position to add suppressive fires for buddy team #2 (grenadier and rifleman)
- Buddy team #2 moves to the vulnerable point. They move in rushes or by crawling
- One Soldier takes up a covered position near the exit
- The other Soldier cooks off a grenade (2 seconds maximum), shouts, FRAG OUT, and throws it through an aperture
- After the grenade detonates, the Soldier covering the exit enters and clears the bunker
- Simultaneously, the second Soldier moves into the bunker to assist Soldier #1
- Both Soldiers halt at a point of domination and take up positions to block any enemy movement toward their position
- Buddy team #1 moves to join buddy team #2
- The team leader inspects the bunker, marks the bunker, and signals the squad leader
- The assault squad leader consolidates, reorganizes, and prepares to continue the mission

IV. Follow Through

The factors for consolidation and reorganization of fortified positions are the same as consolidation and reorganization of other attacks. If a fortification is not destroyed sufficiently to prevent its reuse by the enemy, it must be guarded until means can be brought forward to complete the job. The number of positions the unit can assault is impacted by the—

- Length of time the bunkers must be guarded to prevent reoccupation by the enemy
- Ability of the higher headquarters to resupply the unit
- Availability of special equipment in sufficient quantities
- Ability of the unit to sustain casualties and remain effective

As part of consolidation, the leader orders a systematic search of the secured positions for booby traps and spider holes. He may also make a detailed sketch of his area and the surrounding dispositions if time allows. This information will be helpful for the higher headquarters intelligence officer or if the unit occupies the position for an extended length of time.

Chap 7
III. Desert Operations

Ref: FM 90-3 (FMFM 7-27), Desert Operations (Aug '93).

I. Desert Environments

By definition, a desert region receives less than 10 inches (25cm) of rainfall annually. There are 22 deserts covering 30 percent of the earths' landmass. The largest non-polar desert is the Sahara that covers the northern half of the African continent, an area larger than the entire contiguous 48 states of America. The word desert comes from Latin desertum meaning "an abandoned place." Though admittedly sparse, both large and small cities are distributed across the deserts of the world. Human population in the desert centers on fresh water sources, oil reserves, and seaports.

Successful desert operations require adaptation to the environment and to the limitations its terrain and climate impose. Equipment and tactics must be modified and adapted to a dusty and rugged landscape where temperatures vary from extreme highs down to freezing and where visibility may change from 30 miles to 30 feet in a matter of minutes. Deserts are arid, barren regions of the earth incapable of supporting normal life due to lack of water. (Dept. of Defense photo by Chance Haworth).

A. Weather in the Desert

Storms in the desert are commonly refer to dust storms. Dust storms are a meteorological formation that may produce hurricane force wind delivering walls of thick dust several hundred feet (100m) high. Dust storms may last as long as several days and can produce dangerous levels of static electricity. Visibility closes to just 30 feet (10m) and travel by land-borne or airborne vehicle is extremely dangerous.

Typically, deserts also have a rainy season in which nearly all of the little precipitation falls. Rains commonly follow heavy dust storms. Flash flooding is a threat during rains because the desert floor has modest absorption properties.

During the rainy season it is not uncommon to see temperature lows at or near freezing, 32°F (0°C) during the night. Daily highs during the rainy season may not get much above 50°F (10°C).

However, rainy seasons are the exception in the desert. Most of the year the desert is blistering hot with daily high temperatures soaring up to and even beyond 130°F (54°C). Nightly lows can plummet 40°F (22°C) from the daily high in just a matter of a couple hours, causing high winds. This is due to the lack of substantial vegetation and the arid soil's inability to retain heat.

B. Terrain & Vegetation Characteristics

Consideration of OCOKA presents unique challenge in the desert – not because of significant impairment to mobility, but instead due to a lack of obstacles. In most desert environments key terrain is often unavailable and the avenue of approach may offer unlimited options.

The flipside of the open nature of the desert is that concealment and even adequate cover are elusive. Particularly when moving during daylight hours, the enemy is almost always privy to friendly maneuver.

Terrain Formations

Deserts may be broken into four generic classifications. That includes sand dune desserts, rocky deserts, mountain deserts, and arctic deserts. The last two desert types, mountain and arctic, are discussed in their associated sections of this manual. This section will focus on sand dune deserts and rocky deserts exclusively.

Sand dune desserts are denoted by wind drifting soils that form small to large dunes and valleys, called "slacks." A single dune can actually form a series of ridges that extend beyond 60 miles (100km) and crest as high as 1,600 feet (500m). Slacks, too, can be very small or very large measuring as much as 2 miles (3km) from dune peak to peak.

Rocky deserts are perhaps the most common and include very flat regions of sun-baked silt and even dry "salt lakes." More typically rocky deserts exhibit exhaustive networks of rock outcroppings and wind/water formed ravines called "waddis."

Vegetation

Vegetation in desert regions has little impact on considerations of OCOKA. What vegetation is available offers little concealment, no cover, and only rarely offers any practical advantage such as shade or relief from the heat. On the other hand, for purposes of primitive survival needs, the presence of vegetation indicates the potential of a water source.

There is little to no vegetation present in sand dune deserts. Vegetation is present in most rocky deserts. Sagebrush, grasses, and various cactus plants are common, particularly in the lowlands areas where seasonal rains pool.

C. Impact on Mobility

Roadways are more easily established in rocky deserts. Off-road travel by both vehicle and foot is easily achieved, the heat and lack of water not withstanding. Unfortunately the same cannot be said of sand dune deserts.

The lack of plant roots in the soil allows dunes to move. It is not uncommon for large dunes to move as much as 100 meters in five years time. This makes roadway construction through a sand dune desert all but impossible. Off road travel by vehicle is not feasible for tracked or wheeled vehicles, and there is considerable danger of a vehicle flipping as the sand dune gives way to the weight of the vehicle. Light 4x4 quad motorcycles are one of the few options.

Even foot travel quickly becomes exhausting as the sand slips with each step taken. For the most part, the only reasonable mobility option in regards to sand dune deserts is either to go around it, or to insert and extract vertically by air.

Visibility in the desert extends for miles. Given the tendency to employ armored vehicles for their relative speed and overwhelming firepower, the trade off is that stealth is difficult to achieve. Armored formations result in plumes of dust that can be seen over very long distances. (Dept. of Defense photo by Cecillio Ricardo).

II. Desert Effects on Personnel

The weather and terrain of the desert have a very tangible impact on operations, with the most significant factor being extreme heat. The human body's demand for water is enormous under these conditions.

Consider that working troops may need to as much as 4 gallons (15 liters) of drinking water per day. Combat units calculate another 2.5 gallons (9.5 liters) per individual for cooking and cleaning. At 6.5 gallons (24.5 liters) per individual, a company of 140 troops would consume a total of 910 gallons (3,444 liters) each day, at a weight of 7,600 lbs. Commanders must also consider the load weight imposed by the requirement for water during desert operations.

A. Injuries & Disease in the Desert

See the following page for discussion of injuries and disease in the desert.

B. Uniforms & Special Equipment

The extreme heat of desert operations demands specialized uniforms and items that extend the durability of the uniform, and therefore the comfort of the troops during extended missions. Since the troops must wear the uniform over the entire body to avoid severe sunburn, the uniforms are made of a very light, tough material such as Ripstop in either 65/35 nylon-cotton or 100 percent poplin. Uniforms are worn loosely. Specialty items are necessary for troop comfort in the desert. Without such items, routine injury can exacerbate to more severe injury. These items include:

- Sun and wind goggles to keep dust out of the eyes
- Suede-out leather mechanic type gloves to protect from hot surfaces
- Knee pads protect against injuries in rocky deserts
- Full brim "Boonie" hats keep the sun off the face and neck
- Scarves to cover the nose and mouth from inhaling dust
- Sweat rags or cooling bandanas kept wet around the neck

Injuries & Disease in the Desert
Ref: FM 90-3 (FMFM 7-27), Desert Operations (Aug '93), chap. 1.

Contaminated water supplies bring dysentery, skin infection and even malaria. This again puts great emphasis on a healthy, reliable water supply. Often water will need to be purified before use. Flies and rodents are prolific in the desert, and carry a variety of diseases. Units must take preventative measures to repel such animals. Protective screening in dining areas is a must.

Heat Exhaustion & Heat Stroke
The most immediate and pervasive dangers are heat exhaustion and the more severe heat stroke. Heat exhaustion is a medical condition of water and/or salt depletion in the body. Signs of dehydration include dark colored urine, thirst, fatigue and headache. Signs of salt depletion include muscle cramps, dizziness and vomiting. Left unattended heat exhaustion can quickly worsen into heat stroke.

Prevention and treatment for heat exhaustion is to hydrate with drinks that include electrolytes, and to take frequent rest breaks in shaded areas. Due to excessive sweating in jungle climates, fluid consumption should equal 1 gallon (3.8 liters) per day at rest; 1.5 gallons (5.7 liters) while at moderate work; and upwards of 2 gallons (7.5 liters) of fluids per day for heavy continuous work.

Heat stroke is a potentially deadly medical condition. It usually occurs due to dehydration, plus high temperatures and/or heavy physical exertion. When the bodily core temperature reaches 105°F (40.5°C) or higher, the central nervous system begins to malfunction and the brain and other organs are damaged.

Signs and symptoms of heat stroke mimic those of heat exhaustion, but later stages include a lack of sweating, hot red skin, seizures, and unconsciousness.

First aid includes any attempt to lower the subject's core bodily temperature. This most commonly involves pouring water over the subject while loosening restrictive clothing. If the subject is still conscious, have the subject sip water or liquids. An IV should be administered at once. Medevac promptly.

Hyponatremia
A condition exists in which electrolyte salts are flushed from the bodily fluids through excessive drinking in a short period of time. This results in human body cells expanding to accept more water from the bodily fluids. Such a state would be fine for most tissues of the body, but the human brain cannot expand in this manner and death is a potential result.

Drinking too much water risks hyponatremia. The exact amount of "too much" water is still debated, however even physically fit troops working in moderate to heavy labor may experience a dangerous loss of electrolytes at somewhere near 4 gallons (15 liters) of water intake per day.

Prevention involves drinking liquids mixed with powdered drinks. Electrolytes are salts found in the powdered drink mixed in the MRE.

Malaria
This is one of the more serious afflictions in the desert and it can be fatal in some cases. Early signs and symptoms include headache, fever, fatigue and back pain. Later symptoms include a dry cough, shivering while sweating, vomiting and seizures.

Prevention includes insect repellants such as DEET (N,Ndiethylmetatoluamide), insecticide sprays, and mosquito bar netting when sleeping.

Treatment requires medical evacuation. For uncomplicated malaria oral anti-malarial prescription drugs are used. For severe cases, the subject is brought to intensive care units to fight many of the symptoms such as high fever while using prescription-based drugs in tandem.

III. Mission Command Considerations

Maneuver in desert operations is a large affair, most often conducted with the battalion and even entire brigade moving toward a single tactical objective. It is not uncommon for battalion commanders to monitor and control the entire mission as it develops. Command is centered at battalion and brigade level, and reports to higher organizational command at division, corps, or joint task force.

There are few restrictions on command and control in desert environments, with the notable exceptions of mountain and arctic regions covered elsewhere in this manual. However control measures, communications, and weapons all have unique considerations in the desert.

A. Control Measures

Again the relatively flat, open, and sometimes featureless desert presents challenges for fire and maneuver coordination between various units. The boundaries for each unit on map overlays are ideally establish along prominent terrain or man-made features such as highways. Such features are rare.

This means successful coordination places enormous emphasis on proper navigation. Masterful use of map, compass, and Global Positioning Satellite (GPS), coupled with an effective communication plan are critical to mission accomplishment. This is particularly so when two or more units attempt a passage of lines in hours of darkness. Creative and perhaps unorthodox solutions may be employed as friend-or-foe identifiers to prevent fratricide.

B. Communications

In general, communications do well in desert conditions, in part due to a lack of vegetation and tall terrain features (mountain deserts excepted). However the constant dust and the extreme temperatures play havoc on electronic devices.

Electronic circuit boards have a maximum heat allowance by design. The use of the device increased heat output, yet the extreme heat from the hottest season in the desert can literally shut down or even damage circuit boards. Too, batteries will deplete much faster in hot temperatures. When possible electronics should be removed from direct sunlight and placed in well-ventilated containers for better performance.

Normal radio transmitting ranges may be cut in half during the day as the extreme heat causes anomalies that attenuate radio signal. Yet during nighttime operations in the desert, normal transmitting ranges can be exponentially increased. This causes communication security (COMSEC) issues when a unit is broadcasting much further than intended.

C. Weapons

Small arms perform reasonably well in desert environments. The dust, however, will be a constant factor. The use of thick grease or excessive oil on a weapon will result in dust build up that may cause immediate malfunction upon use. Lubricate only the surface of moving parts of the firearm in desert conditions.

Also, during periods of extreme heat small arms cool gradually. This makes extended employment of weapons in cyclical fire or rapid fire problematic. It risks overheating the weapon and causing malfunction. Short bursts are preferred in most situations in the desert.

Lastly, the long distances that can be viewed in desert environments cause target acquisition problems. An illusion referred to as "optical bending" creates a ghost image that makes the target look higher in elevation than it actually is. Overshooting enemy targets at mid to far distances is common in the desert.

Overall, small arms perform well in the dry desert environment. Routine maintenance keeps them functioning reliably.

IV. Tactical Considerations
Ref: FM 90-3 (FMFM 7-27), Desert Operations (Aug '93), chap. 3.

Desert operations establish highly centralized control from commands of higher unit formations. Missions take place in vast geographic areas, yet the command team in the Tactical Operation Center (TOC) commonly monitors and influences the battle as it progresses through a network of ISRC4 capital.

This means tasks to subordinate tactical units at the fireteam, squad, platoon, company, and even the battalion level are often little more than a series of well-executed battle drills. Subordinate units are given an identified objective, a distance, direction and time hack. Each unit will accomplish the mission by employing unit SOP.

A. Offense Operations
In the desert the offense is swift, powerful and violent. It combines mechanized forces with highly synchronized combat engineer, artillery and Close Air Support (CAS) assets. Engagements frequently begin at the maximum effective ranges of the weapon systems employed, well beyond the range of visual recognition.

In desert operations small tactical units perform battle drill often as a single large formation. Here US Soldiers practice vehicle dismount and attack drills. (Dept. of Army).

Deliberate Attack is the preferred method of the initial engagement. The deliberate attack requires excellent information regarding the enemy location, size, capabilities, and disposition. The lack of vegetation and prohibitive terrain in the desert allow excellent collection of such information ahead of the attack. Heavy aerial and artillery bombardment precede deliberate attacks. Under conditions of limited visibility, combat engineers pave the channels through enemy obstacles such as minefields, wire entanglements, and buttresses. Mechanized armor and infantry formations break through the breach to capitalize on the momentum of violence and destroy targeted enemy nodes of command and control.

Movement to Contact (MTC), in desert operations, almost invariably follows the breach caused by a deliberate attack. In this way, local tactical gains appreciate larger operational and strategic success.

The MTC is conducted either by a larger main body behind the attacking force, or by a reserve force held in store during the deliberate attack. MTC continues to push the retreating enemy force back, and seeks to seize key terrain or defeat the enemy in detail.

Exploitation is the means by which an offensive force gains key terrain after a successful deliberate attack. This may be achieved by the MTC that pushes through and beyond the breech, or it may be achieved through the shaping operations of an enveloping force.

Pursuit is the means by which an offensive force destroys the enemy force after a successful deliberate attack. Pursuits are conducted as single, double, or vertical envelopments by forces conducting shaping operations in conjunction with the decisive operation of the deliberate attack and MTC. The pursuit sometimes tasks the enveloping force to initially conduct an exploitation to seize key terrain along the enemy's route of egress, and then fix the enemy in battle. With the enemy disrupted and committed to battle in small pockets of resistance along their line of egress, the MTC is then able to effectively defeat the enemy in detail using classic hammer and anvil maneuver.

Raids are commonly used in desert operations as a means of shaping operations. While raids seek very limited tactical goals such as destroying enemy ISRC4 resources or enemy Air Defense Artillery (ADA) assets, these missions are critical.

Ambushes are somewhat rare in offensive desert operations. However the ambush remains a viable component of the defense, even in desert operations.

B. Defense Operations

The defense, too, is a large affair in desert operations. On rare occasions an outpost can be established as a strongpoint defense of key terrain. Otherwise, an area defense may extend continuously in breadth over hundreds of miles or kilometers. Alternatively, a mobile defense or retrograde may extend an equal number of miles in depth. For the largest task forces in desert operations, both an area defense and a retrograde are developed simultaneously.

The desert presents few flanking obstacles with which a defense may tie into. If impassable terrain features – the sea, plateau, mountain canyon, sand dunes, or the deep silt of a dried salt lake – cannot be established on the flanks then flanks may be secured with massive minefields to deny the enemy maneuver.

Tactical units in the defense spend an inordinate amount of time conducting surveillance and screening operations forward of the engagement area. During nighttime hours or periods of limited visibility, a network of ambushes may be employed to disrupt enemy reconnaissance.

C. Enabling Operations

The defense requires coordinated planning and sustainment. Combat support and service support units shape the success of any operation. Tactical units must be prepared to provide security for supporting units.

Reconnaissance in Force (RIF) is the primary form of ground reconnaissance in the desert. RIF swaps stealth for speed, and values timely information over catching the enemy unaware. Thus, RIF is typically conducted by mechanized or motorized mounted patrols. And unlike other operational environments that might dispatch a squad or fireteam to conduct reconnaissance, RIF typically involves platoons or entire companies for each reconnaissance mission.

Screening patrols and local area security patrols are conducted as a means of counter-reconnaissance to disrupt enemy activity. This is particularly effective when supporting sustainment operations or a relief-in-place.

Lastly, tactical commanders must become competent at effectively conducting passage of lines in a wide variety of weather and light conditions. This is a critical enabling task for extended offensive operations common to desert warfare.

Army operations are ideally suited to desert environments. Its thrust of securing and retaining the initiative can be optimized in the open terrain associated with the desert environments of the world. In that environment, the terrain aspect of METT-T offers the potential to capitalize on the four basic tenets of the doctrine initiative, agility, depth, and synchronization. (Dept. of Army photo).

D. Maneuver

Maneuver must be at the maximum tactical speed permitted by the terrain, dust conditions, and rate of march of the slowest vehicle, using whatever cover is available. Even a 10-foot sand dune will cover and conceal a combat vehicle. Air defense coverage is always necessary as aircraft can spot movement very easily due to accompanying dust clouds. In some situations movement may be slowed to reduce dust signatures. Rapid movement causes dramatic dust signatures and can reveal tactical movements.

To achieve surprise, maneuver in conditions that preclude observation, such as at night, behind smoke, or during sandstorms. In certain circumstances, there may be no alternative to maneuvering in terrain where the enemy has long-range observation. Then it is necessary to move at the best speed possible while indirect fires are placed on suspected enemy positions. Speed, suppressive fires, close air support, and every other available combat multiplier must be brought to bear on the enemy.

Tactical mobility is the key to successful desert operations. Most deserts permit good to excellent movement by ground troops similar to that of a naval task force at sea, Use of natural obstacles may permit a force to establish a defensive position that theoretically cannot be turned from either flank; however, these are rare. Desert terrain facilitates bypassing enemy positions and obstacles, but detailed reconnaissance must be performed first to determine if bypassing is feasible and will provide an advantage to friendly forces.

Dismounted infantry may be used to clear passes and defiles to eliminate enemy ATGM positions prior to the mounted elements moving through.

Avenues of approach of large forces may be constrained due to limited cross-country capability of supply vehicles coupled with longer lines of communications. The limited hard-surface routes that do exist are necessary for resupply.

Chap 7
IV. Cold Region Operations

Ref: ATTP 3-97.11/MCRP 3-35D, Cold Region Operations (Jan '11).

I. Cold Regions

Cold regions are unique environments that together make up about a quarter of the earth's landmass at the northernmost and southernmost locales, and are found seasonally in high altitude mountains. In cold regions the annual snowfall exceeds two feet (0.6 meters); the average temperature of the year is below freezing; large ice formations remain for half the year or longer; and a permanent sheet of ice called permafrost lies just beneath the topsoil. In addition to characteristic weather, cold regions have distinctive flora and fauna.

In addition to hauling equipment, the Small Unit Support Vehicle (SUSV) can pull troops cross-country on skis, referred to as skijoring. (Dept. of Army photo).

The **arctic** is a cold region of low coastal and interior plains. Grasslands void of trees, with little snowfall and even less evaporation denote the arctic. The arctic gets so little precipitation that it is classified as a desert, yet in the warmest season the surface ice may melt forming a humid, muddy marsh due to water trapped in the topsoil above the permafrost layer. The arctic has very few human settlements. Military operations tend to focus on maintaining seaports and roads.

The **subarctic** is a cold region below the Arctic Circle that includes low costal and interior plains, high interior plains, and mountain ranges. Vast boreal forests of coniferous pine trees and the most extreme temperature changes on earth denote the subarctic. There are more significant human settlements, road infrastructure, and natural resources in the subarctic that render military operations all the more likely, but similarly as daunting as the arctic.

(Urban & Regional Environments) IV. Cold Region Operations 7-31

A. Weather in Cold Regions

The weather in cold regions includes a combination of temperature, precipitation, and visibility conditions. These weather factors often compound upon each other to create extreme operating conditions.

Temperatures

Cold region temperature varies greatly, most notably in the subarctic which can see summer temperatures range as high as 90°F (32°C) and then drop down to low extremes of -60°F (-51°C) during winter months. There are five categories of cold recognized by the US military:

- Wet Cold is defined as 52°F (11°C) down to 20°F (-6°C)
- Dry Cold is defined as 20°F (-6°C) down to -5°F (-20°C)
- Intense Cold is defined as -5°F (-20°C) down to -25°F (-31°C)
- Extreme Cold is defined as -25°F (-31°C) down to -40°F (-40°C)
- Hazardous Cold is defined as below -40°F (-40°C)

Precipitation

Arctic precipitation is less than 10 inches of water per year, while the subarctic typically sees just 15 inches of water per year. Rainstorms can occur during the warm summer months, but seldom produces substantial rain accumulation. Whereas rain is more commonly measured in water accumulation per hour, snow is more typically measured in terms of visibility by distance.

- Rain = light, moderate, with heavy rain more than .5in/hour (1.25cm)
- Light Snow = visibility at 1,000m or greater, up to 1in/hour (2.5cm)
- Moderate Snow = visibility at 400m to 1,000m, 1-3in/hour (7.5cm)
- Heavy Snow = visibility at 400m or less, more than 3in/hour

Visibility

Vision presents yet another hazard. The arctic and subarctic receive sunlight for long periods of time, and then enter twilight and nighttime conditions for equally long periods of time. These light conditions create visual effects that may confuse or disorient individual troops. They may also mask or disclose friendly positions to enemy observation, and visa versa.

- Blizzard occurs in heavy snowfall or snow blown by high winds resulting in low visibility distances
- White Out occurs in bright sunlight through dense low clouds to obscure the horizon resulting in a loss of depth perception
- Gray Out occurs during twilight giving all snow surfaces the same gray appearance resulting in a loss of depth perception
- Ice Fog occurs at extreme cold temperatures in still air with a vapor source present resulting in a fog that obscures visibility

B. Terrain & Vegetation Characteristics

Cold regions entail a variety of environmental characteristics that impact tactical considerations of OCOKA as well as safety considerations for each mission.

Terrain Formations

Less commonly known terrain characteristics include:

- Ice Caps – permanent expansive sheets of ice without vegetation
- Glaciers – accumulated snow compressed into slow eroding rivers of ice
- Crevasse – a deep and sometimes sudden crack in an ice cap or glacier
- Permafrost – a semi-permanent sheet of ice just below or at the soil line
- Muskeg – a grassland marsh created by melting snow and poor drainage

All five characteristics above impact mobility. They present slippery, low-traction driving conditions that may exclude the use of vehicles. Furthermore, the potential for deep crevasse in ice caps and glaciers render even dismounted patrols a treacherous option across these ice formations, unless units are trained and properly equipped.

Vegetation

Cold regions present unique flora, in part due to the bitter cold but also due to remarkably low annual rainfall. Though a large amount of fresh water is present in cold regions, it takes the form of frozen ice all or most of the year. These conditions result in robust but often sparse vegetation:

- Tundra – vast expanses of various grasses with little or no trees
- Boreal Forests – dense coniferous pine trees that grow in low elevations

Cold region tundra offers no concealment for moving patrols. For halted patrols tundra offers only modest concealment, but no cover from enemy fires. Boreal forests on the other hand offer fantastic concealment for moving or halted patrols, yet again it offers very little in the way of cover. Furthermore, friendly observation in a boreal forest is as equally hindered as the enemy's observation.

C. Impact on Mobility

Cold regions present numerous weather and terrain challenges for both vehicle and personnel mobility. Furthermore, tactical operations are not feasible in cold regions during winter months without heating stoves and warming tents. Without these resources combat troops will quickly succumb to the cold. As such, patrols must include tents, stoves, and fuel as part of their essential load-out items.

Vehicles

In general it is more feasible to conduct vehicle patrols and combat action during the winter months when the ground is frozen solid. In the warm summer months open terrain often turns into grassland marshes that are impassable for most land vehicles.

Typically wheeled vehicles are restricted to movement by roads in cold regions. In addition to off road hazards in the seasonal marshland bogs, wheeled vehicles such as the HMMWV, MRAP and Stryker are more quickly and significantly impaired in as little as 12 inches (30cm) of snow depth during the winter months.

Tactically speaking, this renders wheeled vehicles of all types as follow on forces during offensive operations. To attack a determined enemy with wheeled vehicles channeled on roads leaves those vehicles and troops vulnerable to destruction by enemy ambush, spoiling attack, and counterattack by air.

Tracked vehicles appreciate much greater movement capabilities during colder months in cold regions. The ground must be significantly frozen to support the weight of tracked armor vehicles such as the M1 Abrams, M2 Bradley, M113 (Gavin) series, and M88 recovery vehicle. Even with frozen ground, during wet cold temperatures the crews of these vehicles may frequently stop to remove ice and snow build up from road wheels. Otherwise they risk throwing a track or similar types of impediment.

Large offensive actions are only practical for airmobile forces, or heavy tracked vehicles during periods of dry cold and intense cold temperatures with frozen ground conditions. In dry powdered snow conditions heavy tracked vehicles can maneuver in snow depths of 5-foot (1.5m) across uneven terrain.

Whether mounted or dismounted, all cold region patrols include tents, stoves, and fuel as part of their essential load-out items. The squad sled or SUSV is required to transport such heavy equipment.

(Urban & Regional Environments) IV. Cold Region Operations 7-33

Personnel

Dismounted patrols must consider the additional weight required of extreme cold weather uniforms and equipment. Multiple layers of clothing and special equipment such as skis, snowshoes, and sleds force patrols to slow down or risk overheating and exhaustion, two potentially deadly factors. Also, thick clothing inhibits dexterity with weapons, radios, and targeting systems.

II. Cold Weather Effects on Personnel

Cold region operations have been compared to warfare in deep space. Simply getting dressed for a patrol in extreme cold temperatures has similarities to dressing for a walk on the dark side of the moon. And because so few humans live in cold regions, the environment appears alien and hostile.

Indeed there are two opponents in cold region operations – the weather and the enemy. In some instances, simply outlasting your enemy in harsh weather may present a victory. History is full of such examples: Napoleon's defeat in Russia circa 1812; Germany's defeat in Russia circa 1941; and the American route by Chinese Communist forces in the Chosin Reservoir, Korea circa 1950.

To combat cold weather effects on personnel it will be necessary to understand how the human body reacts to cold environments, to identify and prevent cold weather injuries, and to include nutrition and rest cycles into mission planning.

A. Bodily Reaction to Cold Temperatures

The human body attempts to maintain a constant core temperature of 98°F (37°C) through metabolic conversion of energy to increase heat, and respiratory breathing plus sweating to cool down.

For operations in cold regions, the primary concern will be heat loss. The human body looses heat through radiation, convection, conduction, respiration and perspiration (sweating).

Radiation & Convection

An estimated 60 percent of all body heat is lost through radiation and convection, particularly through thin-skinned areas such as the head, wrist, and ankles. The body radiates heat from natural metabolic activity, but it is the convectional transfer of body heat into the cold air that is the most common cooling factor of body temperature.

Perspiration & Respiration

Sweating further speeds the process of convection as the skin attempts to cool down rapidly by creating a wet surface. Similarly, breathing pushes warm wet air out of the lungs and draws cold dry air into the body's core. This, too, intensifies the cooling of the human body.

Conduction

The process of transferring heat from one object to another cooler object can occur at an alarming pace. When a combat troop lies on the frozen ground to take up a good firing position, or falls asleep on the cold metal surface of a vehicle platform, conduction is taking place. Great care must be taken to ensure that individuals in laying or sitting positions for significant lengths of time are appropriately insulated from cold surfaces.

B. Nutrition and Rest Requirements

Due to respiration, perspiration, and metabolic demand to constantly heat the body, individuals working in temperatures at or below 20oF (-6oC) should drink about 1.5 gallons (5.7 liters) of water per day, and increase calorie intake by as much as a 40 percent daily. That's roughly 4,500 calories every 24 hours.

The Meal Ready to Eat (MRE) includes approximately 1,200 calories. Each combat troop would require four MRE daily. Yet four MRE adds only ¼ gallon (less than 1 liter) of water to the diet. By comparison, the Meal Cold Weather (MCW) contains approximately 1,400 calories and each MCW requires ¼ gallon of water to reconstitute the contents of the MCW. This means the requirement to reconstitute meals encourages hydration, ensuring roughly half of the daily requirement of water is consumed with the essential three MCW per day.

Combat troops struggle to drink enough water in cold regions in part due to a "whaling effect" in which the extremities redirect blood flow to the organs in the body core. This not only results in higher output of urine, called diuresis, but also makes it uncomfortable for individuals to drink cold liquids.

The consequence is dehydration that can ironically manifest as heat casualty. Heat casualties in cold temperatures routinely become hypothermic, presenting further danger. There are numerous techniques commanders and combat troops can employ to prevent dehydration.

The human body's circadian rhythm, or sleep cycle, is regulated by light. Light suppresses the brain's pineal gland production of melatonin. In hours of darkness, the pineal gland produces more melatonin causing drowsiness. Thus, long periods of darkness that are common to cold regions cause combat troops to become lethargic, especially when stationary.

III. Mission Command Considerations

The principles of command and control do not change for cold region operations, yet there are numerous additional considerations to weigh. The bitter cold has an identified effect on the human body, equipment, and logistics. No sooner does the cold give way to warm seasons and the emergence of grassland marshes create a whole new set of command and control issues.

A. Control Measures

Fire and maneuver control measures in cold region operations are somewhat similar to desert operations in that the terrain is often flat and featureless. This makes establishing boundaries a challenging task. It's easy to see such control measures on a map; it is difficult to reference those controls on flat terrain.

In the offense the general rule of thumb is to mass the attacking forces in close proximity, thereby negating distant coordination between units. In the defense each subordinate unit is assigned a sector, most commonly along the lines of communication. This eases command and control at least during periods of good visibility. In darkness or inclement weather, commanders must weigh the risk of unconventional means of fire and maneuver control measures such as lights and visual markers.

B. Communications

Communication equipment in cold region operations suffers from magnetic storms, freezing temperatures, and ice.

Magnetic Storms

Magnetic storms, called Aurora Borealis, involve highly charged particles from the sun that are subsequently caught in the earth's upper atmosphere and then pulled by the earth's magnetic field toward the poles. These storms can disrupt electronic communications, particularly AM ranges. Ironically, the Aurora Borealis has been known to sometimes transmit FM ranges further distances.

Temperature

Temperature effects battery power. Nickel-Cadmium (NiCad) batteries operate well in hot climates, but in high-drain communication electronics begin to encounter

IV. Tactical Considerations

Ref: ATTP 3-97.11/MCRP 3-35D, Cold Region Operations (Jan '11), chap. 2.

Warfare in cold regions is similar to a slow game of chess, arguably even more so than other operational environments. Considerable time is spent positioning and posturing, with only the occasional lightening-fast attack.

Operations in cold regions routinely focus on gaining, securing, and expanding the lines of communication – most notably seaports, airports and roadways internal to the area of operations.

Large-scale offensive actions are commonly limited to specific seasons in cold regions. For long periods during the year defensive outposts characterize the battlespace with only small-scale combat actions and patrols.

A. Offense Operations

The element of surprise is remarkably difficult to achieve in cold region operations. Any movement across the open plain is easily detected during the long periods of excellent visibility. Too, noise travels better in cold temperatures, disclosing the presence of vehicle formations at considerable distances.

This is not to say that surprise is impossible, and indeed it is valued in cold regions. However surprise tends to be limited to dismounted patrols in areas with dense boreal forest or, during summer months, higher elevation tundra. Higher elevation tundra appreciates better drainage and evaporation during seasonally warm periods and usually doesn't transform into marshy bogs. It is much more feasibly traversed by foot and vehicle patrols.

Cold region offensive operations are characterize by mass, speed, and tempo.

Mass is the concentration of combat power. In the tactical offense massing includes the combination of integral troop strength and unit firepower, plus coordinated engineer, artillery, and Close Air Support (CAS) assets. Mass is focused on a single point of the enemy's defense rather than attempting to attack along the entire front of the enemy's defensive line. The intent is to overwhelm enemy forces at a single location.

Speed refers to the amount of time it takes to close the distance between friendly and enemy forces. Again while in very specific instances the element of surprise is gained through a stealthy surreptitious approach toward enemy positions, for larger scale offensive action speed is the means of catching the enemy unprepared. This almost invariably involves a combination of artillery, CAS and mechanized or air assault.

Tempo is the ability to deliver numerous offensive actions simultaneously or in rapid succession. An enemy may capably repel a single attack, however multiple attacks along the enemy's defenses and/or repeated attacks at a single point in the enemy's defense reduce the enemy's combat power and cause logistical chokepoints as the enemy attempts to resupply and reinforce those positions.

To achieve mass, speed and tempo combat units execute carefully planned and swiftly conducted deliberate attack. This requires accurate intelligence on enemy locations and relative combat power. It also demands permissive weather conditions.

Mechanized assault is ideal over frozen ground during heavy snowfall that limits enemy visibility. Air assault is ideal in any season during nighttime hours and clear sky conditions. Both types of assault are best conducted against enemy objectives that are on or near roadways. This allows follow on forces to exploit, sustain, or reinforce the deliberate attack.

A special purpose attack is exceptionally risky at temperatures below -5°F (-20°C) for two reasons. First, both the raid and ambush require combat troops to lie prone on the frozen ground, undetected for long periods of time. Conduction of body heat to the frozen ground quickly results in cold weather injuries. This further places the patrol in danger of death, detection and capture by enemy.

Second, both the raid and near ambush are initiated with unusually high volumes of weapons fire. The close proximity of the enemy during the raid and near ambush demands overwhelming violence and high rates of fire. Otherwise, the momentum is lost and the enemy will repel or destroy the raid or ambush patrol.

Yet at temperatures below -5°F (-20°C) small arms and especially machineguns are susceptible to catastrophic failures when operating at rapid or cyclical rates of fire. Weapons must be warmed prior to performing these types of fires.

These limitations essentially eliminate the option to conduct the raid or near ambush in intense cold and extreme cold conditions. However, due to greater distances between forces and slower rates of fire, the far ambush is a feasible option at temperatures below -5°F (-20°C).

B. Defense Operations

Defense is the normal state of affairs in cold region operations. Combat is limited in warm seasons that flood grasslands into marshes and can swallow armored vehicles. Even in seasons of bitter cold roadways are often closed due to snow and ice, frustrating ground-based reinforcement and resupply.

Perhaps fortunately, warfare in this environment is rarely focused on influencing or controlling the civil populace, in part because cold regions are so sparsely populated. And because there exists so little infrastructure, there is little need to take and hold vast expanses of land. There are too few roadways to sustain such ambitious operational goals in any case.

This creates an environment in which a mobile defense stages a network of outposts to control critical lines of communication. The mobile defense has a specific intent of repulsing the enemy through a series of counterattacks rather than a single, decisive engagement.

As such, the two most critical tactical aspects of the mobile defense in cold region operations are the designating alternate fighting positions, and the coordination of supporting artillery, armor, and air defense fires. These factors give the mobile defense the flexibility and depth necessary to defeat enemy offensive action. Still it is worth noting that a mobile defense presents greater challenges for sustainment, and any defense will fail if it cannot be sustained.

Mobile defense outposts tend to be small enough not to attract enemy air attack or artillery resources, but large enough to rotate units through the ongoing tasks of surveillance/security watch, local screening and reconnaissance patrolling, and rest cycles.

Typically this means that each rifle company is separated into platoon outposts with mutually overlapping observation to cover down on a fairly large defensive sector. In this manner, at least one squad (1/3rd of the platoon) is in a rest cycle at all times and the outpost maintains security 24 hours per day.

C. Enabling Operations

In any operating environment enabling operations cover a wide array of issues and tasks. However in cold region environments tactical units are regularly committed to enabling tasks of reconnaissance and engineering projects.

Reconnaissance is deliberate and ongoing. It is carefully managed to provide continuous tiered coverage. This is the primary means of gaining and retaining situational awareness both inside and outside the area of operations.

Intelligence, Reconnaissance and Surveillance (ISR) assets include manned and Unmanned Aerial Vehicles (UAV) as well as human intelligence gathered from outpost surveillance and local area patrols. Reconnaissance patrols may include relatively short missions of zone reconnaissance, route reconnaissance, and area reconnaissance. In specific instances the situation may dictate a need for reconnaissance in force, which compromises stealth for speed. Reconnaissance in force is most often conducted with vehicles for immediate reporting of observed information.

Engineering projects are the means by which the offense and defense are sustained in cold region operations. Roadways must be cleared, reinforced, and expanded to connect a network of outposts for resupply and reinforcement..

Small unit commanders consider security detail assignments when weighing force protection and economy of force planning.

failure at temperatures even below 50°F (10°C). Alkaline batteries and rechargeable Lithium Ion batteries fair better than NiCad at lower temperatures, however they both experience similar failures for high-drain electronics below freezing temperature 32°F (0°C).

In general, the circuitry of electronic equipment is not designed for low thresholds of temperature. The presence of moisture complicates the situation further as circuit boards may simply freeze under these conditions. It is commonly the case that communication and information electronics require warming to above freezing temperatures before they will function as intended.

Ice negatively impacts communications in more ways than freezing circuit boards. It also creates a surface so hard and so thick that antenna stakes cannot be driven into the earth. Also, sheets of thick ice have low conductivity and make for a poor electronic ground.

Combat units in cold regions may be spread over significant distances defending lines of communication. Establishing a larger communication network is a challenge. Communication systems fail with regularity in cold climates. All communication plans should be reinforced with tandem networks of radio, satellite phone, computer, landline, audio-visual, and/or courier.

C. Weapons

The impact of cold temperatures on small arms is of considerable tactical concern. Small arms encounter a series of malfunctions below -5°F (-20°C) with surprising frequency. Smaller parts of the weapon such as rods and pins freeze rigid and snap, or seize. Springs lose flexibility. In full-auto weapons, stiffly frozen springs can cause a "run away gun" when the bolt carrier does not recoil far enough back to catch the disconnector.

Dry lubricants will help, but are not a cure for brittle metal. The remedy is fire discipline. When combat troops maintain a sustained rate of fire for the first minute or two of an engagement, the weapon naturally heats up even the smallest parts involved in the mechanical operation of the weapon system.

Once warmed by a sustain rate of fire, the weapon is then capable of being employed at a rapid or cyclical rates of fire. However, the complete cooling of a weapon system can take as little as ten minutes in extreme cold temperatures. The warming process must be repeated before higher rates of fire are used.

D. Maneuver

Mounted and airmobile maneuver are preferred in cold regions as these vehicles have the means to carry all necessary life-sustaining equipment, and deliver the troops rested and prepared to fight.

Land-borne vehicles have significant limitations in cold regions. Generally, tracked vehicles are suitable in dry cold (below 20°F/-6°C) or intense cold (below -5°F/-20°C) conditions when the ground is frozen solid enough to support the weight of such vehicles. Yet even tracked vehicles can quickly become sunk in the grassland marshes of the warm season.

Wheeled vehicles are commonly restricted to roadways throughout the year. They are frequently bogged down in summertime marshes, and trapped in wintertime snowdrifts when they travel off road.

Airmobile transportation is a suitable maneuver option year-round. It is best achieved during hours of darkness under clear sky conditions. However a note of caution is merited. If a unit is deployed well beyond resupply or extraction by roadway, this presents a potential risk during the cold season. Enemy anti-aircraft capabilities and inclement weather can delay resupply and extraction by air. Over a significant period of time in extreme cold weather with no resupply the friendly unit's combat readiness may degrade with disastrous results.

Chap 7
V. Mountain Operations

Ref: FM 3-97.6 (90-6), Mountain Operations (Nov '00).

I. Mountain Environments

Mountain weather can vary sharply and has enormous implications for tactical missions. Erratic weather conditions from extreme cold to hot temperatures, extremely arid to periods of considerable precipitation, and from calm to violent winds dictate a wider variety of combat troop equipment and uniforms, placing greater demand on unit logistical requirements.

Talus boulders present challenges and opportunities in mountain warfare. Ascent is relatively easy but descent requires greater care. (Dept. of Army photo).

A. Temperatures

Generally, mountain temperatures decrease approximately 5 degrees for every one thousand feet (300 meters) increased in altitude. This varies of course depending on air humidity. Furthermore, during periods of cloud overcast the temperatures may counter-intuitively become warmer than during periods of clear sky. This phenomenon is referred to as inversion.

Normal daytime to nighttime temperatures will vary 40 degrees in the mountains. Where there are no forest canopies, the temperature variation in the mountains can vary at even greater extremes.

Extreme temperatures have marginal effect on weapon ballistics, but create a greater demand on weapon maintenance as the moving parts can become frozen with ice and snow. In rare but specific weather and humidity conditions, small arms fire will leave frozen vapor trails behind the bullet, disclosing the firing position of both friendly and enemy troops.

(Urban & Regional Environments) V. Mountain Operations 7-39

B. Weather Impact on Movement

Tactical commanders must consider the additional weight required of extreme cold weather uniforms and equipment versus the option of surrendering high ground during periods of extreme cold while on patrol.

Additional patrolling considerations include the understanding that extreme cold temperatures can freeze the ground solid enough to make vehicle movement feasible, yet with precipitation in the form or ice and snow this same ground may be impassible for vehicle, animal or human.

Airborne vehicles such as helicopters have finite capabilities at high altitude. Thinner air and weather conditions mean that helicopters cannot carry normal loads beyond 13,000 feet (3,900 meters). Medical evacuations are also rare above that limit, although there have been a few daring rescues at elevations reaching 18,000 feet (5,400 meters).

Even during periods of moderate weather, what would otherwise be routine rain and wind gusts may become a mission hazard in mountainous terrain. Wind gusts are channeled by mountain ridges and may winds capable of damaging, destroying or sweeping away critical equipment and clothing items. The steep terrain and ravines may also become filled with impassable flash floods during a mountain rainstorm.

Inclement weather indicators include:

- Increasing humidity forming a halo effect around the sun or moon
- Snow blowing from mountain peaks
- Gradual cloud density and lowering of cloud overcast
- Rapid cloud movement especially when merging from multiple altitudes
- Rapid decrease or increase of temperatures
- Rapid onset of high winds

II. Effects on Personnel

Mountain operations subject combat troops to harsh, unforgiving environmental conditions that often include decreased oxygen, rapid loss of water, and greater caloric demand. Because of these compounding factors, there is an increased risk for and to injured or ill troops.

A. Water

Altitude, temperature, wind and physical exertion all take effect on the evaporation of water through the skin and respiratory system. Simply put, tactical operations at higher elevation require greater water intake.

On a typical day at altitudes below 5,000 feet (1,500 meters) the human body will lose 0.7 gallons (2.5 liters) of water through a combination of urination and respiration at rest or light duty. At higher elevations the human body attempts to cope with decreased oxygen levels by increasing respiratory breathing, blood pressure, and pulse rate. Increased water is necessary for all of these bodily functions.

Water consumption requirements of 1.5 gallons (5.5 liters) per day at sea level increase to 3 gallons (11 liters) per day at an altitude of just 6,000 feet. Higher altitudes, arid climates, extreme temperatures and intensified physical activity will further increase water consumption demands.

Yet there is also a maximum of daily water intake each person can consume. Hyponatremia presents a condition in which electrolyte salts are flushed from the bodily fluids. This results in human body cells expanding to accept more water from the bodily fluids. This is fine for most tissues of the body, but the human brain cannot expand in this manner and death is a potential result.

C. Geologic & Vegetation Characteristics
Ref: FM 3-97.6 (90-6), Mountain Operations (Nov '00), chap. 1.

Mountain environments contain a variety of geological characteristics that impact tactical considerations of OCOKA for each mission. Major and minor terrain features and even micro terrain are easily anticipated. However, less commonly known characteristics include:

- **Talus** – large boulder rock formations
- **Scree** – small loose rocks

The size of **talus** renders slopes covered with these boulders relatively easy in the ascent, but require considerable effort to navigate in the descent. On the other hand, talus offers excellent cover from enemy fire and acceptable concealment from enemy observation.

As opposed to talus, the loose rock nature of a slope covered in **scree** make the ascent challenging as scree is known to give away under foot and cause rock slides. Whereas such slides make the descent a relatively easy proposition, though with some injury risk due to falling rock. Scree does not usually offer any value in regard to either cover or concealment from enemy fires and observation.

CLASS	TERRAIN	MOBILITY	SKILL LEVEL
1	Gentle Slopes & Trails	Walking	Unskilled
2	Steep Rugged Terrain	Walking & Use of Hands	Basic Mountaineer
3	Low Grade Climbing	Walking, Use of Hands & Fixed Rope	Basic Mountaineer
4	Steep Grade Climbing	Fixed Rope	Assisted by Assault Climbers
5	Vertical Grade Climbing	Alpine Rock Climbing	*Only* Assault Climbers

Ref: FM 3-97.6, Mountain Operations, chap. 1, fig. 1-3.

In addition to geological characteristics, mountains have unique flora and fauna that must be considered during OCOKA:

- **Tundra** – high altitude short grasses
- **Tussocks** – lower altitude long, clumps of grass
- **Deciduous Forests** – trees with thick trunks and seasonal broad leaves
- **Coniferous Forests** – trees with thin trunks and year-round evergreen

The short characteristic of **tundra** grasses offers little in the way of concealment. However when clumps of **tussocks** are plentiful, concealment from enemy observation can easily be achieved.

Coniferous forests offer trees with dense evergreen pine and thinner trunks. There are exceptions to this rule, of course, such as the Ponderosa pine trees unique to mountain forests in the American southwest. However, as a general rule of thumb coniferous forests are more difficult to navigate due to dense evergreen vegetation that grows low to the ground.

Pine trees offer little in the way of cover from enemy fires. In a mountain coniferous forest the micro terrain offers more practical solutions for cover. Yet pine trees offer excellent concealment from enemy observation all year around because unlike the seasonal leaves of deciduous trees, pine trees do not shed their pines.

Mountains may be covered in either deciduous or coniferous forests, mixed forests, or none at all in the most arid conditions. Furthermore, trees do not generally grow above altitudes of 10,000 feet (3,050 meters).

Deciduous forests offer trees with thick trunks that are more sparsely populated. This means that navigation tends to be easier with better visibility. It also means that older, thicker deciduous tree trunks offer better protection from enemy small arm fire.

B. Altitude

Ref: FM 3-97.6 (90-6), Mountain Operations (Nov '00), chap. 1.

Elevations in altitude correlate to lower saturations of oxygen in the air and blood. Rapid changes in altitude do not allow for the human body to acclimate to different oxygen levels. It is important then that troop commanders consider a trade off between speed of movement for stealth of movement. Slower ascent and descent on the mountain may result in more effective tactical operations.

CATEGORY	ALTITUDE	EFFECT ON LABOR	ACCLIMATION TIME
Low	Below 5,000 ft (1,500 m)	None up to 8 hours of moderate labor.	A day or less.
Moderate	5,000 to 8,000 ft (1,500 to 2,400 m)	Mild AMS symptoms typical for *moderate* labor up to 8 hours.	70% in 5 days 90% in 15 days
High	8,000 to 14,000 ft (2,400 to 4,200 m)	Increased risk of illness for *moderate* labor up to 6 hours.	70% in 7 days 90% in 21 days
Very High	14,000 to 18,000 ft (4,200 to 5,400 m)	Increased risk of illness for *moderate* labor up to 4 hours.	70% in 10 days 90% in 30 days
Extreme	Above 18,000 ft (5,400 m)	Increased risk of illness for *light* labor up to 30 minutes.	3 additional days, once acclimated to very high altitude
	At 22,000 ft (6,700 m)	Light labor at 10-minute increments.	Oxygen supplement required
	At 25,000 ft (7,600 m)	Light labor at 5-minute increments.	Oxygen supplement required
	At 28,000 ft (8,500 m)	Light labor at 3-minute increments.	Oxygen supplement required

Ref: FM 3-97.6, Mountain Operations, chap. 1, fig. 1-7.

Hypoxia

Hypoxia is a condition whereby limited oxygen is transferred to the muscle and tissues of extremities of the arms, legs, and even head. All troops experience hypoxia, though some more acutely than others, and a lack of water, nutrition plus extreme temperatures exasperate the condition of hypoxia.

Acute Mountain Sickness (AMS)

Headache, nausea, fatigue and dizziness. Rest, hydration and proper nutrition will alleviate AMS.

High Altitude Pulmonary Edema (HAPE)

Difficulty breathing often accompanied by pinkish foamy saliva. Coughing or wheezing obstructions of the airway are common symptoms. Over time the victim will slip into a coma or die if left untreated.

High Altitude Cerebral Edema (HACE)

Increasingly extreme symptoms of AMS, including mental confusion, hallucinations, and an unnatural swaying of the upper body may signal the onset of HACE. If left untreated, the victim will slip into a coma and die.

Drinking too much water risks hyponatremia. The exact amount of "too much" water is still debated, however even physically fit troops working in moderate to heavy labor may experience a dangerous loss of electrolytes at somewhere near 4 gallons (15 liters) of water intake per day. Therefore rest and nutrition are equally important for sustained mountain operations.

Finally, at 8.35 pounds per gallon of water (2.2 pounds per liter), commanders must consider the load weight imposed by the requirement for extra water during high altitude missions. Water weight and water resupply are critical.

C. Nutrition

High altitudes require additional calories per day due to the human body's attempt to cope with low oxygen through increased respiration, blood pressure and pulse. Without adequate nutrition, combat troops become lethargic and there is a greater chance of altitude sickness.

In preparation for mountain operations, troops must consume a balanced diet of proteins, fats, and carbohydrates. Fresh vegetables and fruits are an excellent source of necessary vitamins and minerals, but in high altitude environments such foods are difficult to procure. When necessary vitamin and mineral supplements should be consumed.

Yet combat troops commonly lose their appetites at high altitudes. This is problematic because, like water, food is best consumed before the human body feels the energy sapping effect of hypoglycemia. Otherwise, if food is only eaten when hunger is present, that food may take a couple or more hours before it can be used as energy.

Consider that a 22-year-old male at 180 pounds would require 2,800 calories per day for light duty and exercise at 6,000 feet (1,800 meters). Yet during mountain patrolling operations that same person would require 4,000 calories per day to sustain health and energy, compared to just 3,600 calories at sea level.

The point here is that additional food is required per day, per combat troop at higher altitudes. Of course, it is also believed that as the human body slowly becomes fully acclimated it also begins to adapt to higher altitudes and uses oxygen more efficiently. But such adaptation can take months or even years.

D. Casualties

As discussed previously in the section regarding oxygen changes and implications for altitude sickness, higher elevations invariably come with much lower temperatures. This means that cold weather injuries are second in occurrence only to altitude sickness in mountain operations. Frostbite, hypothermia, immersion foot, snow blindness and even severe sunburn are common yet preventable weather injuries. Troops must pack protective uniforms and equipment accordingly given the erratic weather conditions on the mountain.

Higher elevations also have a negative impact on illness and injury recovery. Recovery is slow due to the harsh demands on the human body and the lower oxygen saturation. Oxygen is necessary for cell recovery. As such, it is often advisable to return injured or ill troops back down to sustainable elevations.

Finally it has been noted in various after action reports that combat injuries due to small arms, explosives and blunt force trauma in mountainous elevations produced unanticipated results. At high altitudes wounded troops experience hypothermia and shock much quicker and often with fatal results due to extreme cold and lower oxygen levels.

This places greater burden on field medics and unit logistics when carrying casualty warming bags and oxygen tanks. It also precipitates a need for faster casualty evacuation – a difficult feat given the altitude limitations of helicopters.

III. Tactical Considerations

Ref: FM 3-97.6 (90-6), Mountain Operations (Nov '00), chap. 4.

Tactical missions in mountainous terrain embody a blurring of the line between offensive, defensive and enabling operations. Mountain operations are fought to gain control of key terrain with the objective of establishing and securing the passes, ridges, and natural chokepoints along the lines of communications.

The intent may be to facilitate large follow-on offensive formations, or to allow logistical transport of supply, or even to influence the indigenous civil population by separating them from enemy insurgent and guerilla forces.

A. Offense Operations

The principle element of surprise dominates the mountain battlespace. Surprise requires either lightning fast maneuver or slow stealthy movement to achieve unexpected advantage in positioning combat power. Such operations regularly demand a lengthy planning and preparation period.

Offensive action in mountain operations commonly assumes the Movement to Contact (MTC), ambush or raid. These forms of offense do not require follow-up missions such as pursuit or exploitation, and indeed rarely are these conducted. That is, the MTC and special purpose attacks such as the raid and ambush are intended to achieve tactical success for limited operational advantage.

MTC is a process of finding the enemy, fixing the enemy, and finishing the enemy by defeat in detail. While MTC may be employed universally in various battlespace environments (discussed chapter two of this manual) in mountain warfare the attacking force includes the main body and an additional blocking element that is pre-positioned before the main body initiates contact with the enemy. The attacking force pushes the retreating enemy back into the blocking elements attack-by-fire in a classic "hammer and anvil" maneuver.

Once the vehicles occupy a designated overwatch position and gain appropriate weapon angles to conduct attack by fire, the dismounted patrols initiate their attack against the enemy. As the enemy retreat, they are suppressed in their lines of egress by heavy weapons fire from the blocking position. This

The highest elevations are not necessarily required in mountain combat operations, yet the dominant terrain should always be obtained. Dominant terrain permits high angles of attack and maximum effective range of small arms. (Dept. of Army photo).

permits the dismounted attacking force to defeat the enemy in detail while the enemy is separated and pinned in groups of disorderly retreat.

Arguably, ambush is perhaps the most effective small unit tactic in mountain operations. Whereas MTC and raid require remarkably accurate intelligence to gain situational awareness and pinpoint fixed enemy resources, the ambush casts a broader net in order to catch enemy targets of opportunity.

Other than high altitude considerations on personnel and weapons, plus the steep elevation impact on marksmanship, there is little further consideration for ambush techniques in the mountains. What is noteworthy about ambush in mountainous terrain is that the steeply walled ravines and canyons make ambush countermeasures difficult, while making networks of ambushes all the more feasible.

Infiltration almost invariably precedes any offensive action in mountain warfare. In specific circumstances a logistical cache may pave the way for an infiltration, although great care must be taken not to disclose the cache site or rendezvous timing to the enemy. To do so would risk a counter-ambush by enemy forces upon the friendly patrol.

B. Defense Operations

The primary intent of defense in mountain operations is to deny enemy use of key terrain while maintaining friendly lines of communication.

This presents an apparent blurring of the offensive-defensive dynamic in that the establishment of an outpost defense in order to secure friendly lines of communication is a form of operational offense into the enemy's territory. In short, key terrain assets denied to the enemy may be viewed simultaneously as operational offense and defense.

Mountain defense is established through a series of strongpoint defensive nodes. Each defensive strongpoint maintains 360-degree physical security, yet also overlaps its primary defensive fires with other strongpoint defensive nodes. This is typically achieved along the peaks of a single ridgeline running parallel to the engagement area, or along multiple high points of various terrain features that face into the engagement area – along the lines of communication.

Coordinated fires are essential. Otherwise the enemy will simply maneuver to a better firing angle and bring their weapons to bear on the defense. This includes the need to integrate artillery and mortar fires, plus Close Air Support (CAS).

Artillery will ideally be far enough back to take advantage of increased angles of fire, and target reference points such as along defilades should be established and confirmed within the engagement area. Yet it should be noted that the steep mountainous terrain might also mitigate the effectiveness of artillery radar during counter-battery fires.

Mortars are well suited to mountain warfare due to their high angle of fire, rapid fire, and portability. Mortars can deliver effective fires to dead space and reverse slopes of terrain.

With the defense establish, fires coordinated and communication plus logistic plans in place, the primary activity of the mountain defensive operations is screening patrols.

Screening patrols deny enemy patrols access to the defensive nodes through defilades and gaps that naturally form in mountainous terrain. Screening patrols may ambush enemy targets of opportunity or conduct spoiling attacks against massing enemy forces.

C. Enabling Operations

Enabling operations in the mountains are as numerous and diverse as any other environment. As discussed previously, high altitude weather conditions and the impact on personnel and equipment must be weighed carefully.

However it is the mission of reconnaissance that is paramount for all offensive and defensive planning. Reconnaissance may be conducted by screening patrols, and often this is the case for gaining situational awareness of the local area. However, for situational awareness beyond the engagement area and rear areas, reconnaissance assets must be planned, managed and employed continuously.

Unmanned Aerial Vehicles (UAV) are effective for looking into defilade canyons for enemy presence, and all the more so when operating above the high altitude tree line. At the small unit level, vehicles like the 4x4 quad motorcycle aid reconnaissance teams in rugged but low-grade terrain conditions. But for the most difficult terrain dismounted patrols are necessary. Scout teams qualified in assault climbing compose valuable reconnaissance asset in such case.

Assault Climbers obtain classification through specialized training and are invaluable when assisting combat patrols with basic mountaineering skill levels, or when conducting autonomous high elevation mountain reconnaissance. (Dept. of Army photo).

IV. Mission Command Considerations

While all principles of command and control remain as valid for mountain operations as they do in other environments, the steeply channeled terrain presents limitations unique to mountain operations. Large troop formations moving swiftly across the battlespace are extremely rare in the mountains. Instead, mountain operations are more often characterized as stealthy maneuver and careful coordination of supporting fires and logistics.

A. Control Measures

Fire and maneuver control measures are remarkably intuitive in mountain environments. This is due in part to a significant advantage of observation afforded by higher elevations of ridgelines and mountain peaks. It is also the case that such large terrain features are used as easily identified boundary limits.

Fires should be interlocked with consideration given to vertically elevated or plunging fires. Again the steep terrain offers excellent backdrops for such coordinated fires, yet also creates natural but dangerous dead spaces that must be considered.

B. Communications

Communications in mountain operations suffer significantly due to inclement weather, freezing temperatures, greater distances between combat elements, and difficulty in establishing line-of-sight due to precipitous terrain.

Since combat units are commonly spread over greater distances in mountain operations, establishing a larger communication network is a daunting challenge. Yet all of the communications that can be brought to bear elsewhere in other environments are equally viable in the mountains.

C. Navigation

An excessive reliance on Global Positioning Satellite (GPS) navigational devices in the mountains is problematic. The National Geospatial-Intelligence Agency (NGA) insists GPS signals may be disrupted by sun activity impact on satellite services, less than ideal satellite positioning, topography complications, heavy cloud overcasts, blizzard conditions, sand storms, and thick forest canopy.

While all of these limiting factors apply in mountainous terrain, topography complication is perhaps the most common limitation. The walls of deep canyons and ravines frequently block adequate satellite signals for those patrols at lower elevations.

In general, mountain troops require extensive training for land navigation in the mountains where tasks such a range estimation and pace count are all the more challenging. Complicating the matter further is a common trend for topographical maps to be incorrect and with outdated marginal data. Too, training with an altimeter is uniquely necessary for land navigation in the mountains.

D. Maneuver

Mounted maneuver in either land-borne vehicle or airmobile assault in helicopter presents the most desirable means of maneuver in mountainous terrain. This type of mounted maneuver delivers the troops to dominant terrain with all necessary equipment, rested and prepared to fight.

Yet both land-borne vehicle and airmobile helicopters have significant limitations and vulnerabilities. Thinner atmospheric conditions mean helicopters struggle to carry their full load capacities even in high altitudes (8,000 to 14,000 ft.) let alone very high or extremely high altitudes. And vehicles, particularly heavy armored vehicles cannot traverse the steep slope grades of the precipitous terrain. Dismounted foot patrols are frequently the only plausible option.

VI. Jungle Operations

Ref: FM 90-5, Jungle Operations (Aug '82).

I. Jungle Environments

The word jungle comes from Sanskrit jangala, meaning "wasteland." Today the word jungle encompasses any climate with year-round warm temperatures, moderate to extreme rainfall, and dense vegetation. This includes three distinct climate zones – the equatorial tropics, tropical regions, and subtropical regions.

In general small arm weapons perform very well in jungle environments. Weapons require regular maintenance and a constant coat of protective oil. Combat troops should be trained in close quarter marksmanship in anticipation of enemy engagement at very short distances. (Dept. of Defense photo by Brian J. Slaght).

Jungle environments present unique challenges for military operations and yet because such a large percentage of the world's population lives in and around the jungle, an ability to conduct military operations in the jungle is essential. Fortunately, the US military has a long history of jungle operations all over the globe from which to pull lessons learned.

A. Weather in the Jungle

Generally speaking jungle environments enjoy consistently warm temperatures throughout the year. Indeed, of all the operational environments jungle climates experience the least fluctuation in daily and yearly temperatures.

Nonetheless there are noticeable differences between equatorial tropic, tropical, and subtropical climates.

Equatorial Tropics

This region is comparatively small at less than 5 percent of the earth's landmass. It includes land at or near the equatorial belt that circles the globe. The equatorial tropics are known for having the most consistent weather patterns and mild, pleasant temperatures.

Temperatures at or near the equatorial tropics remain remarkably constant with daily highs at 87°F (30°C) and nightly lows dropping just slightly to 75°F (24°C). Very seldom do temperatures rise above 92°F (33°C).

Precipitation at or near the equatorial tropic is also consistent at 2-3 inches (5-7cm) of rainfall each month, for a total of about 30 inches (75cm) annually.

Tropical Regions

This region of the world includes less than 20 percent of the earth's landmass, but an estimated 40 percent of the human population. It also accounts for all rainforests.

Tropical climates experience more fluctuation in temperatures than equatorial tropics. At extremes the daily high can exceed 100°F (38°C) temperatures, and the nightly lows may occasionally dip to 65°F (18°C). Farther away from the equator, tropical regions often have a cool dry season and a wet hot season.

Constant rains characterize precipitation in tropical regions, often year-around. Tropical rainforests experience rainy seasons up to 250 days each year and accumulate between 80-400 inches (200-1,000cm) of annual rainfall.

Subtropical Regions: This region accounts for another 20 percent of the earth's landmass and includes those areas just north and south of the tropical zone. The subtropics have at least two distinct seasons, a hot wet summer and a cool dry winter. Yet for much of the year it enjoys pleasant weather conditions.

Temperatures in subtropical regions witness greater variation with cooler dry seasons and warmer wet seasons called monsoons. At extremes the daily high can reach 106°F (41°C) temperatures during the warm season, and the nightly lows can drop to 50°F (10°C) during the cool season.

Precipitation in subtropical regions is usually characterized by a rainy season and dry season that can each last for months. Subtropical deciduous forests typically receive between 30-80 inches (75-200cm) of rainfall annually.

C. Impact on Mobility

The dense forests and swamplands of jungle regions greatly limit vehicle mobility to existing roadways. Airmobile and dismounted patrols dominate maneuver in the jungle environment.

Vehicles: Jungle operations are dominated by airmobile vehicles and dismounted patrolling. Land-borne vehicles of all types play a vital supporting role in jungle operations. While such vehicles are limited almost exclusively to roadways, they are tasked with the sustainment of continuous supply delivery.

Roadways are vulnerable to enemy ambush and counterattack. Gun trucks, up-armored wheeled vehicles, and even tracked combat vehicles are often required to function as escort for sustainment convoys.

In addition, while armored tanks are significantly limited in jungle operations, they are exceptionally valued for their firepower on outpost duty. In the event jungle operations deviate to urban combat, here again tank firepower is highly prized.

Foot and vehicle patrols also traverse tropical savannahs with relative ease. Yet while visibility is terrible in grasses over 5 feet (1.5m) high, regrettably grasslands provide little concealment from airborne observation and attack.

B. Terrain & Vegetation Characteristics

Ref: FM 90-5, Jungle Operations (Aug '93), chap. 1.

Tactical considerations of OCOKA present unique challenges in the dense jungle canopy and tropical grasslands. Visibility in the jungle is greatly diminished making concealment all but guaranteed for both friend and foe alike.

Dismounted patrolling remains the only viable option for imposing force in jungle operations. US Marines patrol the jungles of Columbia. (Dept. of Defense photo by Brian J. Slaght).

Terrain Formations

Farmable plains in jungle regions are regarded at a premium, especially when low altitude mountains are present. As such flat land is commonly cultivated into expansive open fields of rice paddies, cane fields, and fruit tree plantations or other similar produce. And because jungle regions are highly populated, villages, townships and cities also vie for lowland plains.

Swamps are the exception to habitual use of lowland terrain for farms and cities. Swamps restrict movement for vehicles and often even for dismounted patrols. In such cases travel through swamps may require small, flat-bottom boats.

Swamps fall into two general categories:
- **Mangrove swamps** – costal region swamps often with brackish water and dense shrub trees that grow only 3-15 feet (1-5m) high, as are common in the Everglades of Florida.
- **Palm swamps** – freshwater inland swamps set in mature forests trees that grow 65 feet (20m) or taller with dense upper canopies, as is common in the Bayous of Louisiana.

Hilly and mountainous terrain is commonly void of human development in jungle regions, leaving pristine jungle forests covering the high ground.

Vegetation

The jungle climate is covered with dense vegetation. This includes rainforests and deciduous forest, but also includes grasslands called savannahs.

- **Rainforests** – mature forests with trees growing as tall 200 feet (60m) and multiple canopy layers that leave a dark rotten floor with no undergrowth. The canopy ceiling can be 25 feet (8m) above the ground.
- **Deciduous Forests** – new growth or mature forests with trees commonly reaching 65 feet (20m) high. These forests typically have just one upper canopy that allows sunlight to the forest floor that may in spots produce a thick underbrush canopy.
- **Savannahs** – treeless grassland with broadleaf grasses that grow 3-15 feet (1-5m) high. These grasslands can appear as small meadows in the jungle, or grow as vast open prairies across flats and mountain foothills.

The more canopies present in the jungle, the less sunlight can break through to the forest floor. Yet the lack of undergrowth in dark jungle forests makes for easier movement when conducting dismounted patrols.

II. Jungle Effects on Personnel

The extreme conditions of jungle environments are often exaggerated. The reality is that humans in general prefer the mild warm temperatures of the tropics and subtropics. Indeed, jungles regions are popular vacation destinations.

High humidity and difficult off-road travel can create an appreciable level of discomfort. It takes most people just a few days to acclimate to jungle climates. However, in temperatures above 90°F (32°C) the combination of heat and humidity present specific operational hazards. Precautions are necessary.

A. Injuries & Disease in the Jungle

Jungle environments present potential weather injuries and bacterial infections due to the warm, moist climate. Oddly, the jungle humidity can cause equal discomfort in seasonal cool wet conditions. Though cases of hypothermia in the jungle are exceedingly rare, individuals must prepare according to the weather.

The human body must maintain a constant core temperature of 98°F (37°C) by cooling through radiation, convection, respiration and perspiration. Some 60 percent of all body heat is lost through radiation and convection, particularly through thin-skinned areas such as the head, wrist, and ankles. In jungle climates this is difficult because the air is almost as warm as the body's core temperature. Yet the act of sweating and breathing aid the process of convection as the skin attempts to cool down rapidly by creating a wet surface.

The most immediate and pervasive dangers are heat exhaustion and the more severe heat stroke.

Constant moisture pressed against the skin can cause anything from mild irritation to more serious medical conditions such as jungle rot and immersion foot.

Heat rash comes from clogged sweat pores usually at the point of continual contact with wet clothing. It appears to be a series of small red pimples but may also present as a solid pink blemish. This is common skin under the arm and inner thigh. It may swell and is often irritated and itchy. Prevention and treatment include the removal of wet clothing and allowing the skin to air dry.

A mosquito-borne disease of the liver, malaria is one of the more serious afflictions in jungle regions and it can be fatal in some cases. Early signs and symptoms include headache, fever, fatigue and back pain. Later symptoms include a dry cough, shivering while sweating, vomiting and seizures.

Various bacterial infections may invade internally from contaminated water, or topically to the skin from open wounds or leeches. Prevention primarily includes the identification and consumption of clean water, or alternatively for surface water sources the use of a combination of iodine tablets and commercial-off-the-shelf water purification filters. When collecting surface water, rainfall runoff is most desirable. But allow 20 minutes of rain before collecting water. This reduces the amount of bacteria because a great deal of bacteria is washed away in the first 20 minutes of heavy rainfall.

Leeches are not poisonous, but their bites can become infected. To remove leeches do not pull the leech. Part of the leech may be left behind in the wound and it will become infected.

Attacks by animals in the jungle against humans are remarkably rare. Still, jungle climates are host to a wide variety of animal life including snakes and crocodilians – alligators, crocodiles and caiman.

B. Uniforms & Special Equipment

Long patrols away from the comfort of the Forward Operating Base (FOB) or outpost require some special equipment in jungle operations. Of particular use are sleeping equipment, water purification equipment, and uniform items.

The basha (a.k.a. "shelter" or "hootch") is a thick PVC vinyl sheet in camouflage color patterns with grommet eyelets. They are employed as an open-air tent. In the jungle small conventional tents and bivy shelters don't allow enough airflow. They are too hot and retain too much water. Bashas keep rain and bugs off the troops. They are typically large enough to accommodate two or three individuals. Bashas are very light, pack small, and can be set just 18 inches (45cm) high. They are quickly set up and taken down with only six bungee cords.

The mosquito bar (a.k.a. mosquito netting) is a tightly woven but breathable mesh material that is hung over a sleeping position to keep out mosquitoes and other insects. It requires a few minute of fussing to set up, but is excellent for preventing malaria.

The jungle hammock is a netted sleeping material that is tied between two trees. The jungle version of this hammock includes mosquito netting to prevent bites and often comes with an internal overhead basha. When patrolling through swampland it is frequently difficult to find dry ground. The jungle hammock is an agreeable solution, although it does have the disadvantage of elevating the profile of the patrol base to at least waist high.

Water purification comes in the form of iodine tablets and commercial filters. Unit supply typically issues these items prior to conducting patrols in jungle regions.

Jungle uniforms are made of lightweight, fast-drying, water-wicking materials. These requirements hold true for then entire uniform:
- Shirt and blouse in Ripstop material
- Full brim "Boonie" hat in Ripstop material
- Sleeveless tank-top in wicking synthetic or natural fibers
- Tropical boot with drain holes and leather-nylon material upper

The recent prevalence of Ripstop material in 65/35 percent nylon-cotton blend was expected to satisfy the US military's intent for an all-season, all-climate fabric. Yet while this version of Ripstop material has proven substantially lighter and very robust, recent testing by the US military (USMC) suggests that the lighter and more breathable Ripstop in 100% cotton-poplin weave is far better suited to jungle climates as it dries much faster.

III. Mission Command Considerations

The conduct of command and control is unique in the jungle. Dense vegetation and terrain can disrupt normal communications. Even when airborne vehicles are used overhead for command and control, it is more often only as a relay for electronic communication between tactical units. This is because the dense vegetation also greatly reduces visual observation from the air.

The bulk of tactical missions in jungle operations are commanded at the rifle company, platoon, or squad level. Command centers at battalion and brigade levels are relegated to synchronizing and sustaining the various efforts because they so seldom can see or meaningfully influence the operation as it unfolds.

Instead, higher command at battalion and brigade in jungle operations become very adept at ensuring their subordinate commanders have all the necessary resources to effectively accomplish their missions. This requires not just efficient management of resources, but also brilliant leadership in influencing and mentoring subordinate commanders. In this way, the battalion and brigade commanders are shape each engagement even without being physically present.

A. Control Measures

Fire and maneuver control measures in jungle operations develop plans that include major terrain features as boundaries, effective use of scouts and indigenous guides.

IV. Tactical Considerations
Ref: FM 90-5, Jungle Operations (Aug '93), chap. 5.

Jungle operations are frequently decentralized, involving numerous on-going and semi-autonomous tactical missions. Battalion and brigade-size maneuver is rare. Instead missions in the jungle are typically conducted at the company, platoon and squad level of command.

A. Offense Operations
In the jungle, the primary means of achieving the tactical offense is the ambush. This cannot be overstated. The impact of a successful ambush carried out against an enemy force is immediate and widespread. It breaks enemy morale, undermines logistical efforts, and destroys the enemy's combat power. The ambush is used in various forms and is commonly networked together as an area ambush, and may even be employed as shaping operations for larger missions.

Historically, deliberate attacks are problematic in jungle operations for three established reasons. First, dense vegetation makes gathering specific information about the enemy's position very difficult. Second, even when intelligence is accurate regarding the enemy's disposition, the dense vegetation greatly inhibits targeting systems of Close Air Support (CAS) and artillery assets. Third, command and control is challenging in the dense vegetation, and all the more so for the highly synchronized fire and maneuver of a deliberate attack.

Instead, Movement to Contact (MTC) is favored over deliberate attacks for offensive missions at the company and battalion level. Yet the MTC also struggles with targeting systems of CAS and artillery. For this reason Forward Observers (FO) are commonly tasked to each platoon in jungle operations.

Yet at the platoon level, the tactical offense in the jungle remains almost invariably a form of ambush, an occasional MTC, and the scarce raid.

B. Defense Operations
While decentralized tactical missions are the norm for offensive action, battalion or company-sized strongpoint defenses more commonly characterize the defense. Placed on the dominant terrain with a 360-degree orientation, these defense positions are called Forward Operating Bases (FOB) or outposts.

Here the company, platoons and squads plan and prepare for upcoming missions as well as rest and resupply. The FOB is a staging area, often positioned well within or near enemy territory, from which to press the tactical offense.

When conducting patrols, temporary overnight defenses are established. When using a deliberate stealth method, the patrol base is used to evade enemy detection. When using a rapid tempo method of patrolling, the more substantial Nighttime Defensive Perimeter (NDP) is employed, complete with an improvised Landing Zone (LZ), developed fighting positions, forward wire obstacles, and anti-personnel mines.

C. Enabling Operations
At the tactical level of jungle operations, small units are tasked to reconnaissance patrols, counter-reconnaissance (screening) patrols, and local area security patrols. Reconnaissance becomes all the more valuable because technical advantages such Unmanned Aerial Vehicles (UAV) are mitigated by dense vegetation in the jungle. Screening and security disrupt the enemy's efforts to use the concealment of dense vegetation to approach friendly forces.

Relief in place is also common when assuming duties at an outpost, however this is typically achieved by vertical redeployment using helicopters. And officers and NCOs must become competent at effectively conducting passage of lines in a wide variety of weather and light conditions.

A typical progression of airmobile MTC in jungle operations:

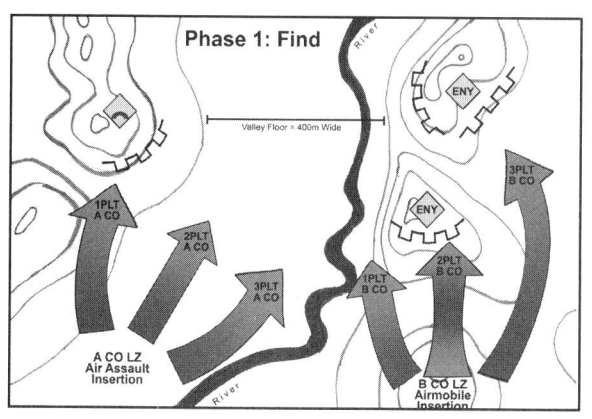

Phase 1: Find
Two rifle companies are inserted to separate LZ and then tasked to platoon lanes to seek the enemy.

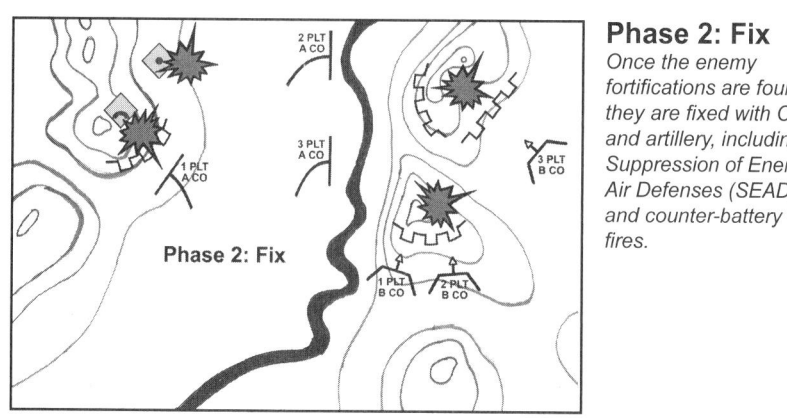

Phase 2: Fix
Once the enemy fortifications are found, they are fixed with CAS and artillery, including Suppression of Enemy Air Defenses (SEAD) and counter-battery fires.

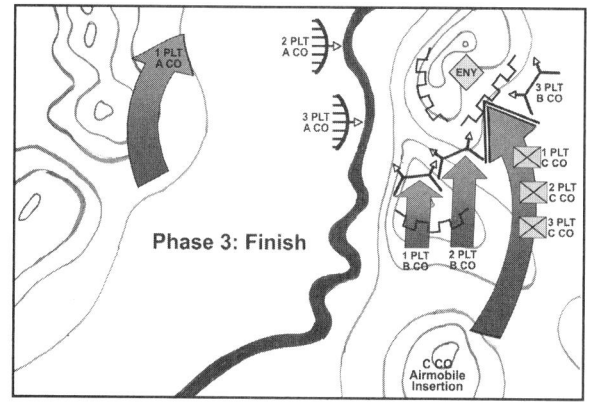

Phase 3: Finish
Platoons assume support-by-fire positions or ambushes in the enemy's route of egress as the third rifle company vertically inserts and masses on the main enemy objective. The enemy is defeated in detail.

Fire and maneuver control graphics that mark a map overlay are virtually impossible to identify in the jungle. Because of this it is best to use major terrain features for boundaries to the flanks and limit of advance. Ridgelines, roads, rivers, and plantations or villages are all easily recognized by combat patrols.

B. Communications

Perhaps the single greatest obstacle to communication in jungle operations is the humidity that can corrode batteries and even short electronic circuit boards. Rubber seals, moisture collecting desiccants, and regular maintenance go a long way in preventing such decay.

Furthermore, dense vegetation attenuates radio signals of all types. Absorption of radio signal is more acute in the wet humid environment of the jungle. The presence of steep terrain complicates radio signals even further.

Transmitting ranges are greatly diminished. Establishing a larger communication network is a challenge. Communication systems fail with regularity in jungle climates. All communication plans should be reinforced with tandem networks of radio, satellite/airborne systems, computer, landline, audio-visual, and/or courier.

C. Weapons

In jungle climates humidity forms on un-protected metal and quickly turns to rust, particularly when carbon or mud is present on the metal surface. Small arms require regular cleaning and relentless protecting with oil lubricant.

That noted, modern small arms with metal alloys and polymer stocks perform well in jungle environments. Routine maintenance will keep them reliable.

The distances of combat are greatly reduced due to the dense vegetation of the jungle. It is common to engage the enemy at as little as 10 meters or even less. However, open-bolt weapons such as light, medium and heavy machineguns experience an abnormally high rate of failure-to-fire on the first shot. For this reason crew served weapons should be protected by positioning them at the mid-section of patrolling formations.

Additionally, combat troops must be trained on close-quarter-marksmanship techniques for dismounted patrols in jungle operations.

D. Maneuver

Airmobile and light infantry formations dominate maneuver in jungle operations.

There are two established and effective methods to penetrate enemy defenses in jungle environments – deliberate stealth or rapid tempo. Both methods involve dismounted patrolling and airmobile assets.

The method of deliberate stealth is achieved when dismounted patrols insert into the jungle, normally by helicopter, and then surreptitiously enter enemy territory to conduct combat against enemy forces.

The method of rapid tempo is achieved by simply out-walking the enemy. A steady march will defeat the enemy's ability to keep pace, and often defeats the enemy's mortar and artillery assets by walking beyond weapon range. Patrols using this method routinely cover 10 to 20 kilometers per day, and may sustain such patrolling operations for up to a month at a time.

A rapid tempo reflects the original concept of light infantry formations. It is as viable a concept today as it has been over the past 2,000 years. The idea is that "light infantry" carry only a half-day (12-hour) ration of munitions, batteries, water and food. Carrying a maximum load of 45lbs (20kg) or less, and at a pace of just 1 to 2 kilometers per hour, combat troops are fully capable in aggressive battle.

Chap 8: Patrols & Patrolling

Ref: FM 3-21.8 (FM 7-8) The Infantry Rifle Platoon and Squad, pp. 9-2 to 9-9 and The Ranger Handbook, chap. 5.

The two categories of patrols are combat and reconnaissance. Regardless of the type of patrol being sent out, the commander must provide a clear task and purpose to the patrol leader. Any time a patrol leaves the main body of the unit there is a possibility that it may become engaged in close combat.

Patrol missions can range from security patrols in the close vicinity of the main body, to raids deep into enemy territory. Successful patrolling requires detailed contingency planning and well-rehearsed small unit tactics. The planned action determines the type of patrol. (Dept. of Army photo by Sgt. Ben Brody).

Combat Patrols
Patrols that depart the main body with the clear intent to make direct contact with the enemy are called combat patrols. The three types of combat patrols are raid patrols, ambush patrols (both of which are sent out to conduct special purpose attacks), and security patrols.

Reconnaissance Patrols
Patrols that depart the main body with the intention of avoiding direct combat with the enemy while seeing out information or confirming the accuracy of previously-gathered information are called reconnaissance patrols. The most common types reconnaissance patrols are area, route, zone, and point. Leaders also dispatch reconnaissance patrols to track the enemy, and to establish contact with other friendly forces. Contact patrols make physical contact with adjacent units and report their location, status, and intentions. Tracking patrols follow the trail and movements of a specific enemy unit. Presence patrols conduct a special form of reconnaissance, normally during stability or civil support operations.

Note: See also p. 4-12 for discussion of patrols in support of stability operations.

I. Organization of Patrols

A patrol is organized to perform specific tasks. It must be prepared to secure itself, navigate accurately, identify and cross danger areas, and reconnoiter the patrol objective. If it is a combat patrol, it must be prepared to breach obstacles, assault the objective, and support those assaults by fire. Additionally, a patrol must be able to conduct detailed searches as well as deal with casualties and prisoners or detainees.

The leader identifies those tasks the patrol must perform and decides which elements will implement them. Where possible, he should maintain squad and fire team integrity.

Squads and fire teams may perform more than one task during the time a patrol is away from the main body or it may be responsible for only one task. The leader must plan carefully to ensure that he has identified and assigned all required tasks in the most efficient way. (Dept. of Army photo by Sgt. Ben Brody).

A patrol is sent out by a larger unit to conduct a specific combat, reconnaissance, or security mission. A patrol's organization is temporary and specifically matched to the immediate task. Because a patrol is an organization, not a mission, it is not correct to speak of giving a unit a mission to "Patrol."

The terms "patrolling" or "conducting a patrol" are used to refer to the semi-independent operation conducted to accomplish the patrol's mission. Patrols require a specific task and purpose.

A commander sends a patrol out from the main body to conduct a specific tactical task with an associated purpose. Upon completion of that task, the patrol leader returns to the main body, reports to the commander and describes the events that took place, the status of the patrol's members and equipment, and any observations.

If a patrol is made up of an organic unit, such as a rifle squad, the squad leader is responsible. If a patrol is made up of mixed elements from several units, an officer or NCO is designated as the patrol leader. This temporary title defines his role and responsibilities for that mission. The patrol leader may designate an assistant, normally the next senior man in the patrol, and any subordinate element leaders he requires.

A patrol can consist of a unit as small as a fire team. Squad- and platoon-size patrols are normal. Sometimes, for combat tasks such as a raid, the patrol can consist of most of the combat elements of a rifle company. Unlike operations in which the Infantry platoon or squad is integrated into a larger organization, the patrol is semi-independent and relies on itself for security.

Every patrol is assigned specific tasks. Some tasks are assigned to the entire patrol, others are assigned to subordinate teams, and finally some are assigned to each individual. An individual will have multiple tasks and subtasks to consider and carry out.

1. Pointman, Dragman, and Security Team

Security is everyone's responsibility. Having noted that, every patrol has a troop walking in front. This troop is called the pointman. He is responsible for making sure the patrol does not walk into enemy ambushes, minefields, or similar. The pointman has forward security. Sometimes a patrol will send the pointman with another patrol member to walk a short distance forward of the patrol.

Also, every patrol has someone who is last in the formation. This troop is called the dragman. He is responsible for making sure that no patrol members are left behind. He also makes sure that the enemy doesn't surprise the patrol from the rear unnoticed.

The security team is responsible for specifically pulling security to the left and right of the patrol. This is a critical task when crossing danger areas, so a specific team is identified to conduct this task.

2. Clearing Team

The clearing team crosses the danger area once the security team is in place. The clearing team has the specified responsibility of visually clearing and physically securing the far side of a danger area. It's important so another team is designated to conduct this task.

3. Compass & Pace Team

Obviously someone needs to make sure the patrol is headed in the right direction and that we don't travel too far. This is the job of the compassman and paceman. Typically the compass and pace team is positioned immediately behind the pointman. Additionally, a secondary compass and pace team is usually located in the back half of the patrol.

4. Command Team

The PL and a radio operator (RTO) make up the command team for most patrols. Doctrinally speaking, the APL is also part of this team but the APL is normally positioned near the very rear of the formation to help the dragman and ensure no patrol member is left behind.

5. Aid & Litter Team

Someone has to help pull wounded buddies out of harms way. There are usually two members of each fire team designated as aid and litter teams. These teams are spread throughout the patrol and have the responsibility of carrying and employing extra medical aid gear.

6. Enemy Prisoner of War (EPW) Search Team

EPW teams are responsible for controlling enemy prisoners IAW the five S's and the leader's guidance. These teams may also be responsible for accounting for and controlling detainees or recovered personnel.

7. Tracking Team

There are many different specialty teams that might be assigned to a patrol. Trackers are just one such resource. Explosive ordinance details (EOD) are another. Trackers are unique, however, because they are generally positioned just ahead of the pointman on the patrol.

8. Support Team

The support team is outfitted with heavy, crew-served weapons on the patrol. Of course, reconnaissance patrols usually do not make use of a support team. But when a support team is required, it will be positioned to the center of the patrol.

9. Assault & Breach Team

Reconnaissance rarely ever needs an assault team. The assault team may be dispersed throughout the patrol, but ideally is situated toward the rear. This is because the assault team is typically placed on the objective last.

Patrols & Patrolling 8-3

II. Planning & Conducting a Patrol

Leaders plan and prepare for patrols using troop leading procedures and an estimate of the situation. They must identify required actions on the objective, plan backward to the departure from friendly lines, then forward to the reentry of friendly lines.

The patrol leader will normally receive the OPORD in the battalion or company CP. Because patrols act semi-independently, move beyond the direct-fire support of the parent unit, and often operate forward of friendly units, coordination must be thorough and detailed.

Patrol leaders may routinely coordinate with elements of the battalion staff directly. Unit leaders should develop tactical SOPs with detailed checklists to preclude omitting any items vital to the accomplishment of the mission.

Items coordinated between the leader and the battalion staff or company commander include:

- Changes or updates in the enemy situation
- Best use of terrain for routes, rally points, and patrol bases
- Light and weather data
- Changes in the friendly situation
- The attachment of Soldiers with special skills or equipment (engineers, sniper teams, scout dog teams, FOs, or interpreters)
- Use and location of landing or pickup zones
- Departure and reentry of friendly lines
- Fire support on the objective and along the planned routes, including alternate routes
- Rehearsal areas and times. The terrain for the rehearsal should be similar to that at the objective.
- Special equipment and ammunition requirements
- Transportation support, including transportation to and from the rehearsal site
- Signal plan—call signs frequencies, code words, pyrotechnics, and challenge and password

As the patrol leader completes his plan, he considers the following elements.

- Essential and supporting tasks. The leader ensures that he has assigned all essential tasks to be performed on the objective, at rally points, at danger areas, at security or surveillance locations, along the route(s), and at passage lanes.
- Key travel and execution times. The leader estimates time requirements for movement to the objective, leader's reconnaissance of the objective, establishment of security and surveillance, compaction of all assigned tasks on the objective, movement to an objective rally point to debrief the patrol, and return through friendly lines.
- Primary and alternate routes. The leader selects primary and alternate routes to and from the objective. Return routes should differ from routes to the objective.
- Signals. The leader should consider the use of special signals to include arm-and-hand signals, flares, voice, whistles, radios, visible and nonvisible lasers.
- Challenge and password outside of friendly lines. The challenge and password from the SOI must not be used when the patrol is outside friendly lines.
- Location of leaders. The leader considers where he, the platoon sergeant, and other key leaders should be located for each phase of the patrol mission. The platoon sergeant is normally with the following elements for each type of patrol:
  - In a raid or ambush, he normally controls the support element.
  - On an area reconnaissance, he normally supervises security in the objective rally point (ORP).
  - On a zone reconnaissance, he normally moves with the reconnaissance element that sets up the link-up point.
- Actions on enemy contact. The leader's plan must address actions on chance, to include:
  - Handling of seriously wounded & KIAs
  - Prisoners captured as a result of chance contact who are not part of the planned mission

The following provides an overview of considerations when conducting a patrol:

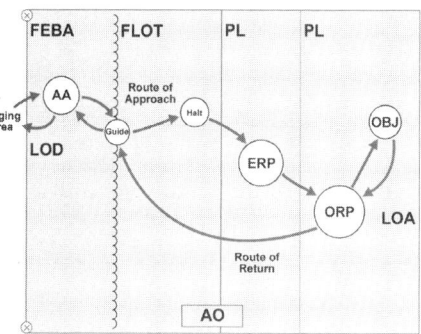

[Diagram showing: FEBA, FLOT, PL, PL, Staging Area, AA, Route of Approach, LOD, Guide, Halt, ERP, ORP, OBJ, LOA, Route of Return, AO]

Patrol planning starts in a staging area. At the AA, a passage lane is coordinated through the FLOT, and the patrol conducts a listening halt past the FEBA. Movement to and from the OBJ is reported via phase lines. (Ref: FM 7-8, chap 3, section I).

1. Occupy Assembly Area (AA)
The patrol marches to and gathers in a designated AA behind the forward line of troops (FLOT). This area allows final preparation—everyone must be accounted for and in the correct uniform, with the correct gear, having the correct information, etc.

2. Coordinate the Passage of Lines
Once all troops are accounted for, the FLOT guide links up with the PL. This coordination includes the normal contingency plans, including the near and far recognition signals, and how long the guide should wait on the far end of the obstacles should the patrol need to return unexpectedly.

3. Depart the Forward Line of Troops (FLOT)
The PL calls the assistant patrol leader (APL) forward to count the troops out through the forward obstacles with the guide. Once through the FLOT, the APL takes up a position in the rear of the patrol's formation. En route rally points (ERP) are identified as a fallback point if there is trouble along the route.

4. Establish a Listening Halt
The PL pauses the patrol and forms a 360-degree security. When it's safe to continue, the patrol will resume its march.

5. Occupy the Objective Rally Point (ORP)
The last stop prior to the objective is the ORP. This is where all plans are finalized and execution of actions on the objective begins. Maintain noise and light discipline.

6. Move through the Release Point (RP)
A security team is positioned forward where the patrol will begin their recon. This is the release point. From this point the PL releases authority to the subordinate leaders to achieve their tasks.

7. Conduct Actions on the Objective (OBJ)
Actions on the objective will vary depending on whether the patrol is a reconnaissance or a combat patrol. These actions should be clearly delineated and rehearsed by team members.

8. Reoccupy the ORP
The mission is almost over. Back in the ORP is where the patrol members disseminate any necessary information about the mission and make final coordination for the route back to the AA behind the FLOT.

9. Reenter the FLOT
The PL conducts a far recognition signal with the FLOT. This typically involves a simple radio call, but it can also be a flare, gong, or smoke signal. The patrol lets the FLOT know the patrol is returning so that the patrol is not mistaken for enemy activity. The PL stops the patrol in a security halt and moves forward with a security team to give the near recognition signal.

Once the near recognition is established, the security team returns to the security halt and leads the patrol to the guide. The PL counts each member in by name.

10. Reoccupy the AA
The PL should render his patrol report to the commander. This report may be verbal or written, simple, or elaborate depending on the situation and the commander's requirements. Continue the mission; reorganize and reconstitute.

III. Elements of a Combat Patrol

There are three essential elements for a combat patrol: security; support; and assault. The size of each element is based on the situation and the analysis of METT-TC.

1. Assault Element

The assault element is the combat patrol's decisive effort. Its task is to conduct actions on the objective. The assault element is responsible for accomplishing the unit's task and purpose. This element must be capable (through inherent capabilities or positioning relative to the enemy) of destroying or seizing the target of the combat patrol. Tasks typically associated with the assault element include:

- Conduct of assault across the objective to destroy enemy equipment, capture or kill enemy, and clearing of key terrain and enemy positions
- Deployment close enough to the objective to conduct an immediate assault if detected
- Being prepared to support itself if the support element cannot suppress the enemy
- Providing support to a breach element in reduction of obstacles (if required)
- Planning detailed fire control and distribution
- Conducting controlled withdrawal from the objective

Additional tasks/special purpose teams assigned may include search teams, prisoner teams, demolition teams, breach team, and aid and litter teams.

2. Support Element

The support element suppresses the enemy on the objective using direct and indirect fires. The support element is a shaping effort that sets conditions for the mission's decisive effort. This element must be capable, through inherent means or positioning relative to the enemy, of supporting the assault element. The support force can be divided into two or more elements if required.

The support element is organized to address a secondary threat of enemy interference with the assault element(s). The support force suppresses, fixes, or destroys elements on the objective. The support force's primary responsibility is to suppress enemy to prevent reposition against decisive effort. The support force—

- Initiates fires and gains fire superiority with crew-served weapons and indirect fires
- Controls rates and distribution of fires
- Shifts/ceases fire on signal
- Supports the withdrawal of the assault element

3. Security Element

The security element(s) is a shaping force that has three roles. The first role is to isolate the objective from enemy personnel and vehicles attempting to enter the objective area. Their actions range from simply providing early warning, to blocking enemy movement. This element may require several different forces located in various positions. The patrol leader is careful to consider enemy reserves or response forces that, once the engagement begins, will be alerted. The second role of the security element is to prevent enemy from escaping the objective area. The third role is to secure the patrol's withdrawal route.

There is a subtle yet important distinction for the security element. All elements of the patrol are responsible for their own local security. What distinguishes the security element is that they are protecting the entire patrol. The security element is organized to address the primary threat to the patrol—being discovered and defeated by enemy forces prior to execution of actions on the objective. To facilitate the success of the assault element, the security element must fix or block (or at a minimum screen) all enemy security or response forces located on parts of the battlefield away from the raid.

Chap 8
Patrols & Patrolling
I. Traveling Techniques

Ref: FM 7-92 Infantry Reconnaissance Platoon and Squad (Airborne, Air Assault, Light Infantry), chap 3; and FM 3-19.4 Military Police Leader's Handbook, chap 7.

Traveling techniques can be used with any of the attack formations. In essence, these techniques are concerned with the distances between troops and units while moving. The critical factor of any movement technique is that the patrol leader (PL) can see the subordinate leaders and vise versa. This is because most of the communication and coordination is achieved through hand and arm signals—which requires line of sight.

The first technique, traveling, is used primarily for walking a patrol down a road or path in fairly secured areas. Patrols use the traveling technique when enemy contact is unlikely.

The second technique, traveling overwatch, is the most common technique employed when moving troops in unsecured areas. Patrols use the traveling overwatch when enemy contact is likely.

The third technique, bounding overwatch, is the preferred technique when security is the most important factor. Patrols use the bounding overwatch when enemy contact is expected.

Note: See following pages (pp. 8-8 to 8-9) for further discussion.

The squad is the essential fire and maneuver element. US Army squads include two fire teams, while Marine squads include three fire teams made up of a rifleman, automatic rifleman, grenadier, and team leader. A squad with as few as seven troops can be split into two fire teams and a squad leader. (Photo by Jeong, Hae-jung).

Traveling Techniques

I. Traveling

1. The patrol is massed together as one entity for ease of command and control (C2). This technique allows for speed of movement.

2. Troops are spaced five meters apart. If marching on a road, two lines are formed with troops staggered left and right. This creates a distance of ten meters between the troops on one side of the road, but still only five meters behind or in front of the troop to the opposite side of the road.

3. The patrol disperses to the left and right in the event of attack. This technique permits very little deterrence to the effectiveness of mass-casualty producing weapons, but does concentrate the troops for a massed assault in the event of a near ambush.

The traveling technique is used when enemy contact is unlikely. Marching troops by road is often the most efficient means of travel. As such, this technique mitigates our vulnerability if attacked. (Photo by Jeong, Hae-jung).

II. Traveling Overwatch

1. The patrol is separated into two or more elements. This technique is also fast. It has considerably more security and the flexibility for each element to maneuver in support of another if attacked. However, the PL losses some of the control in that each element is now commanded by a subordinate leader. The PL maintains contact with these leaders.

2. There is still five meters between troop, and the troops are staggered in two lines when roads are used. However, a distance of *at least* 20 meters is maintained between each element.

3. The patrol disperses left and right in the event of attack. This technique has improved security in its ability to deter the effect of mass-casualty producing weapons and has further advantages in regard to its ability to disperse and overwhelm a near or far ambush.

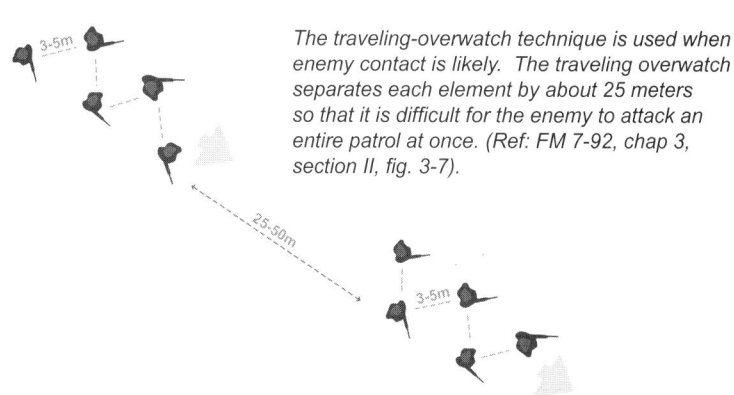

The traveling-overwatch technique is used when enemy contact is likely. The traveling overwatch separates each element by about 25 meters so that it is difficult for the enemy to attack an entire patrol at once. (Ref: FM 7-92, chap 3, section II, fig. 3-7).

III. Bounding Overwatch

1. The patrol is separated into two elements. This technique compromises speed for greater security and control.

2. The forward element halts in a position that offers the best observation of the terrain in front of the patrol. This element becomes the "overwatch" position. The position must offer some cover or concealment.

3. The trail element (behind the forward element) then bounds forward, either slightly left or right of the overwatch position.

4. Once the bounding element has successfully passed through the terrain, they take up a position that offers the best observation of the terrain in front of them. The bounding element now becomes the overwatch position and the old overwatch bounds forward.

5. This process is repeated until the patrol reaches its objective, or the PL selects another movement technique due to an improved security situation.

6. If the patrol comes under fire, the bounding overwatch becomes quick and violent. The overwatch position conducts suppressive fires while the PL directs the bounding element to either conduct a hasty attack against the enemy or break contact.

The bounding-overwatch technique is used explicitly when enemy contact is expected. The effort is to allow maximum use of combat power in the direction of movement, while exposing our smallest force to any potential enemy. (Photo by Jeong, Hae-jung).

On Point

Avoid confusing movement with maneuver. Maneuver is defined as "Movement supported by fire to gain a position of advantage over the enemy." At company level, the two overlap considerably. Tactical movement differs from maneuver, however, because maneuver is movement while in contact, but tactical movement is movement in preparation for contact. The process by which units transition from tactical movement to maneuver is called "actions on contact."

Troops must move. That's the nature of warfare. The trick is employing the appropriate technique to the present level of danger. Contact with the enemy is either unlikely, likely, or expected. Each technique allows for defensive fires in reaction to the enemy. The PL must consider the coordination of these fires in the planning phase.

The PL takes into consideration the factors of speed, control, and security. So, while bounding overwatch offers excellent control and security, it can be slow. On the other end of the spectrum is traveling, which maintains top speed and control, but dangerously lacks security. Somewhere in the middle of that continuum is the traveling overwatch, which offers adequate speed and security, but somewhat compromises the PL's control over the formation.

Keeping this in mind, the PL assesses the situation to determine the emphasis on speed, control, and security. In essence, the PL considers the appropriate distance between troops and subordinate elements.

Many issues factor into this consideration. Is the patrol moving troops in a secured area? Is it crossing rough or heavily vegetated terrain? Is the patrol moving at nighttime or daytime light conditions? How soon does the patrol need to be at its destination? How many troops are in this patrol—a fairly large unit, or a smaller one?

The most important consideration the PL takes into account is whether or not enemy contact is unlikely, likely, or expected.

Subordinate elements must not become so spread apart that they can no longer support each other. Weapon systems have finite ranges. In truth, weapon systems are more commonly limited by rugged or heavily vegetated terrain. In these cases, the distances between subordinate elements is even less. The PL chooses the appropriate technique for the given situation. Subordinate leaders enforce strict adherence to the assigned interval distances between troops and elements.

It is unrealistic to always move troops with the highest level of security, the bounding overwatch. The troops become easily fatigued and movement is slowed to an unreasonable pace. Conversely, it's unrealistic to always use the least amount of security, the traveling technique, in order to capitalize on speed.

Chap 8: Patrols & Patrolling
II. Attack Formations

Ref: FM 3-21.8 (FM 7-8) The Infantry Rifle Platoon and Squad, chap. 3 and 9; and FM 7-92 Infantry Reconnaissance Platoon and Squad (Airborne, Air Assault, Light Infantry), chap 3.

Squad formations include the squad column, the squad line, and the squad file. These formations are building blocks for the entire element. What that means is that the smallest element, the fire team, may be in a wedge while the larger element, such as the squad or platoon may be in another formation. It is quite possible to have fireteams in wedges, squads in columns, and the platoon in line—all at the same time. Additionally, this section will discuss a couple variations including the diamond and the staggered column.

Leaders attempt to maintain flexibility in their formations. Doing so enables them to react when unexpected enemy actions occur. (Dept. of Army photo by Senior Airman Steve Czyz).

Fire Team Formations

The term fire team formation refers to the Soldiers' relative positions within the fire team. Fire team formations include the fire team wedge and the fire team file. Both formations have advantages and disadvantages. Regardless of which formation the team employs, each Soldier must know his location in the formation relative to the other members of the fire team and the team leader. Each Soldier covers a set sector of responsibility for observation and direct fire as the team is moving. To provide the unit with all-round protection, these sectors must interlock.

The team leader adjusts the team's formation as necessary while the team is moving. The distance between men will be determined by the mission, the nature of the threat, the closeness of the terrain, and by the visibility. As a general rule, the unit should be dispersed up to the limit of control. This allows for a wide area to be covered, makes the team's movement difficult to detect, and makes them less vulnerable to enemy ground and air attack. Fire teams rarely act independently. However, in the event they do, they use a perimeter defense to ensure all-around security.

The squad leader adjusts the squad's formation as necessary while moving, primarily through the three movement techniques. The squad leader exercises command and control primarily through the two team leaders and moves in the formation where he can best achieve this. The squad leader is responsible for 360-degree security, for ensuring the team's sectors of fire are mutually supporting, and for being able to rapidly transition the squad upon contact.

The squad leader designates one of the fire teams as the base fire team. The squad leader controls the squad's speed and direction of movement through the base fire team while the other team and any attachments cue their movement off of the base fire team. This concept applies when not in contact and when in contact with the enemy.

Weapons from the weapons squad (a machine gun or a Javelin) may be attached to the squad for the movement or throughout the operation. These high value assets need to be positioned so they are protected and can be quickly brought into the engagement when required. Ideally, these weapons should be positioned so they are between the two fire teams.

Attack Formation Considerations

The decision on which formation to use comes down to four considerations:
- C2
- Maneuverability
- Firepower forward
- Protection of the flanks

Each formation offers distinct advantages over the other and employing the right formation for the situation allows the patrol to be very aggressive. However, stealth is still the preferred mode, allowing the unit get as close as possible.

A squad forms into a file with fireteams in wedges. This formation places the smallest footprint forward, while still maximizing the fire team's combat power. (Photo by Jeong, Hae-jung).

I. Attack Formations - The Line

The line formation places excellent firepower forward, employing virtually 100 percent of the unit's weapon systems to the front. Additionally, C2 is easily achieved along a line formation, making the line an excellent choice for frontal assaults against the enemy.

To execute the squad line, the squad leader designates one of the teams as the base team. The other team cues its movement off of the base team. This applies when the squad is in close combat as well. From this formation, the squad leader can employ any of the three movement techniques or conduct fire and movement.

The disadvantages include a lack of maneuverability, difficulty in changing direction, and an almost complete inability to protect the flank. Regardless of the interval distance between each troop in the line formation, they are literally lined up in a side-by-side fashion. This means that only the last troop on either flank can engage an enemy force to the sides of this formation.

Troops stand abreast of each other to form the line—with key leaders situated in the middle or just behind the formation. This formation is effective when we expect to gain fire superiority to the immediate front. (Photo by Jeong, Hae-jung).

1. All troops are formed into a rank, side-by-side. Each troop faces forward and has essentially a 90° sector of fire. Subordinate leaders maintain control over the formation, careful not to allow any portion of the line formation to get ahead of the others. This could risk fratricide or at the very least, mask the fires of friendly troops.

2. Hand and arm signals are the preferred method of communication. Communication is passed left and right along the formation. This means that every sixth step of the left foot, each troop should look left and right to see if any information is being passed along the line formation.

3. When any one member of the formation stops, every member halts. Each member takes a knee upon the formation's halt facing forward. The troop on the far left and right face out accordingly. After five minutes, each patrol member drops their rucksack and assumes a prone position until the signal to move out is given.

(Patrolling) II. Attack Formations 8-13

II. Attack Formations - The File

The squad file has the same characteristics as the fire team file. In the event that the terrain is severely restrictive or extremely close, teams within the squad file may also be in file. This disposition is not optimal for enemy contact, but does provide the squad leader with maximum control. If the squad leader wishes to increase his control over the formation he moves forward to the first or second position. Moving forward also enables him to exert greater morale presence by leading from the front, and to be immediately available to make key decisions. Moving a team leader to the last position can provide additional control over the rear of the formation

The file formation lends great ease of C2, maneuvers almost as easily as the individual troop, and can employ virtually every weapon to either flank.

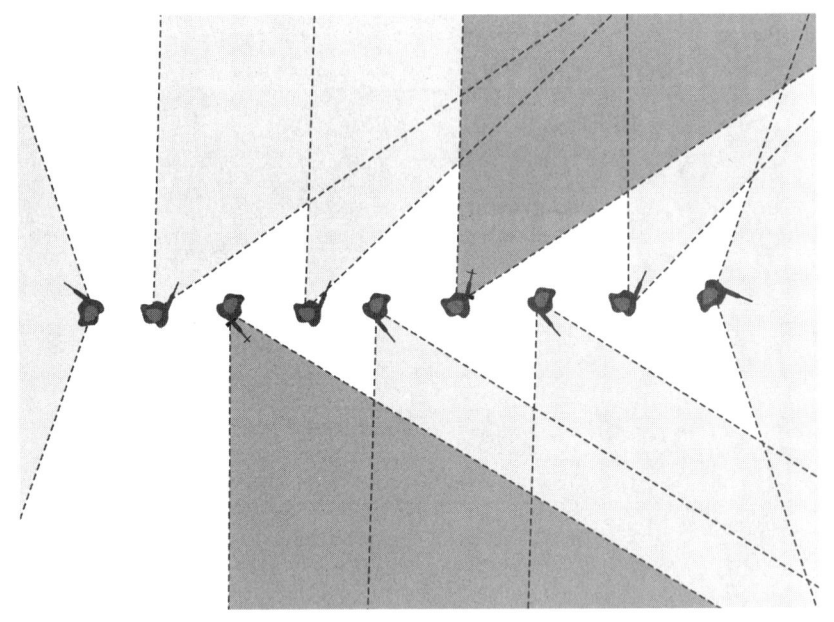

The file formation allows for the best command and control of troops. It is also the easiest formation to maneuver. And while it has ample security to the flanks, it makes poor use of combat power forward or to the rear—where it is very vulnerable. (Ref: FM 7-92, chap 3, sect II, fig. 3-5).

The file is an excellent choice for moving through difficult terrain. Because C2 is communicated so easily, the file is also ideal for moving in times of limited visibility, such as nighttime. In battle, the file also has advantages. The file is a difficult formation to ambush because it permits the use of virtually every single weapon system to either flank. Additionally, this formation is ideal for penetrating or flanking an enemy position because, as the file comes perpendicular to the enemy position, a left or right turn allows every troop to employ their weapon against the enemy. In this case, the file transforms into a line formation—which is excellent for attacking forward.

A disadvantage of the file formation is its inability to place adequate fires forward or backward of the formation. Troops behind the pointman cannot fire forward without the risk of hitting their own troops. If the enemy is able to place significant fires upon the file formation, this can prove to be disastrous.

Execution

1. This formation is constructed by having each troop follow the pointman in single file.

2. The pointman's sector of fire is the 120° field of view to his front. The second man in line must monitor a 90° sector of fire to the left of the formation. The third man in the line must monitor a 90° sector of fire to the right of the formation, and so on. The sectors of fire are staggered left and right for every member of the patrol except the dragman. The dragman's sector of fire is a 120° field of view to the rear of the formation.

3. Hand and arm signals are the preferred method of communication. Communication is passed up and down the formation. This means that every sixth step of the left foot, each troop should turn around to see if any information is being passed UP the column formation.

4. When any one member of the formation stops, every member halts. Typically, each member takes a knee upon the formation's halt. After three minutes, each patrol member takes a couple steps in the direction they are facing (their sector of fire). This clears the center path for leaders and key teams to use. After five minutes, each patrol member drops their rucksack and assumes a prone position until the signal to move out is given.

Each troop follows behind the pointman to form the file—like ducks in a row. Each troop is assigned a sector of observation alternating left and right. The dragman watches rearward. This formation places excellent fires to the flank. (Photo by Jeong, Hae-jung).

Variation: The Staggered Column

When the patrol uses a road or developed path, they will form two lines, one on each side of the road. This is achieved by alternately assuming a position based on the opposite side of the road for the man in front of you. More simply, if the pointman takes the left side, the next troop takes the right, and the next troop takes the left, and so on. This forms two columns, one to the right side of the road and to the left side of the road. Otherwise, the staggered column functions exactly like the column file.

(Patrolling) II. Attack Formations 8-15

III. Attack Formations - The Wedge

Offsetting each troop behind the pointman forms the wedge. The wedge permits excellent firepower forward and to either flank. A version known as the modified wedge places the last man in the formation all the way back, and positioned behind the pointman. This is also called the diamond. (Photo by Jeong, Hae-jung).

The wedge is the basic formation for the fire team. The interval between Soldiers in the wedge formation is normally 10 meters. The wedge expands and contracts depending on the terrain. Fire teams modify the wedge when rough terrain, poor visibility, or other factors make control of the wedge difficult. The normal interval is reduced so all team members can still see their team leader and all team leaders can still see their squad leader. The sides of the wedge can contract to the point where the wedge resembles a single file. Soldiers expand or resume their original positions when moving in less rugged terrain where control is easier.

In this formation the fire team leader is in the lead position with his men echeloned to the right and left behind him. The positions for all but the leader may vary. This simple formation permits the fire team leader to lead by example. The leader's standing order to his Soldiers is: "Follow me and do as I do." When he moves to the right, his Soldiers should also move to the right. When he fires, his Soldiers also fire. When using the lead-by-example technique, it is essential for all Soldiers to maintain visual contact with the leader.

The wedge formation is somewhat of a compromise between the line and column formations. The wedge formation scores high in terms of firepower forward and protection of the flanks. It also scores moderately in maneuverability.

Using the wedge formation, the patrol can still employ almost all of the weapons forward against an enemy force. Additionally, since about half of all weapons can be instantly brought to bear to either flank, this formation proves to be very difficult to ambush or flank. While pivoting the formation is a bit difficult—especially in steep or heavily vegetated terrain—it is far easier to maneuver the wedge than the line formation.

Execution

1. This formation is constructed by offsetting each troop to the left and right of the pointman. This forms a wide, inverted "V".

2. The pointman's sector of fire is, again, the 120° field of view to his front. The next two members in the formation, behind the pointman's position and offset to the left and right, monitor a 90° sector of fire that begins directly forward and covers their immediate left or right, respectively. This is also true for the last troop(s) in the formation, being offset one more time to the left or right of the troops in front of them. There is no rear sector of fire because subsequent fireteams follow behind.

3. Hand and arm signals are the preferred method of communication. Communication is passed up and down the formation. This means that every sixth step of the left foot, each troop should turn inward to see if any information is being passed along the wedge formation.

4. While it is acceptable to temporarily halt the wedge formation—in which each troop takes a knee—it is inadvisable to halt troops for any length of time in this formation. That is because C2 is very difficult to achieve and soon troops and subordinate leaders lose situational awareness. If a long stop is required, the PL designates another attack formation before halting the patrol or the PL rallies the patrol into a security halt.

The squad wedge with the fireteams in wedge (when three fireteams are used) makes maximum use of the squad's firepower forward and to the flanks. (Photo by Jeong, Hae-jung).

Variation: The Diamond

The diamond formation, also known as the "modified wedge", is an acceptable alternative to the wedge. If there are four members of the fire team, simply place the fourth troop, last in line, directly behind the pointman. If there are five members of the fire team, place the fire team leader in the very middle of the formation…also in line with the point and dragman.

Be warned that the diamond formation will not allow a maximum deployment of the fire team's weaponry against targets forward of the patrol. However, it still allows an acceptable percentage of the weapons to be brought to bear against an enemy force in front and to the flanks of the formation. The trade-off is that with the diamond formation, the fire team may move with more speed, change directions with more ease, and provide 360° of security for itself.

On Point

Combat formations are composed of two variables: lateral frontage, represented by the line formation; and depth, represented by the column formation. The advantages attributed to any one of these variables are disadvantages to the other. Leaders combine the elements of lateral frontage and depth to determine the best formation for their situation. In addition to the line and column/file, the other five types of formations—box; vee; wedge; diamond; and echelon—combine these elements into varying degrees. Each does so with different degrees of emphasis that result in unique advantages and disadvantages

Attack formations are designed to allow the maximum use of the patrol's weaponry, while limiting the patrol's exposure to the enemy. Every troop in the formation knows their sector of fire according to their position within the formation. The PL selects the appropriate formation based on considerations of C2, maneuverability, firepower forward of the formation, and protection of the formation's flanks.

Each tactical situation is unique and the patrol is not restricted to just one formation or another. Employ all of them if necessary. Generally, it is better to use the attack formation that allows optimal command and control to maneuver within striking distance of the enemy. At that time the patrol may need to change attack formations in order obtain the greatest security and make maximum use of the patrol's firepower.

The final consideration might be called "follow through." It is important that the patrol is not exhausted to the point that they cannot continue the mission. Use the right formation for the given situation.

Security Checks While on Patrol

Patrol members must assist their patrol leader by applying basic patrolling techniques consistently. This gives the team leader more time to concentrate on assisting the patrol leader in the conduct of the patrol. Team members should concentrate on maintaining spacing, formation, alertness, conducting 5 and 20 meter checks and taking up effective fire positions without supervision.

5 and 20 Meter Checks

Every time a patrol stops, it should use a fundamental security technique known as the 5 and 20 meter check. The technique requires every patrol member to make detailed, focused examinations of the area immediately around him, and looking for anything out of the ordinary that might be dangerous or significant. Five meter checks should be conducted every time a patrol member stops. Twenty meter checks should be conducted when a patrol halts for more than a few minutes.

Soldiers should conduct a visual check using unaided vision, and by using the optics on their weapons and binoculars. They should check for anything suspicious, and anything out of the ordinary. This might be as minor as bricks missing from walls, new string or wire run across a path, mounds of fresh dirt, or any other suspicious signs. Check the area at ground level through to above head height.

When the patrol makes a planned halt, the patrol leader identifies an area for occupation and stops 50 meters short of it. While the remainder of the patrol provides security, the patrol leader carries out a visual check using binoculars. He then moves the patrol forward to 20 meters from the position and conducts a visual check using optics on his weapon or with unaided vision.

Before actually occupying the position, each Soldier carries out a thorough visual and physical check for a radius of 5 meters. They must be systematic, take time and show curiosity. Use touch and, at night, white light if appropriate.

Any obstacles must be physically checked for command wires. Fences, walls, wires, posts and the ground immediately underneath must be carefully felt by hand, without gloves.

8-18 (Patrolling) II. Attack Formations

III. Crossing a Danger Area

Ref: FM 3-21.8 (FM 7-8) The Infantry Rifle Platoon and Squad, pp. 3-33 to 3-37 and FM 7-93 Long-Range Surveillance Unit Operations, appendix J.

Crossing danger areas can be achieved through one of a series of battle drills designed to get the patrol to the far side of the danger area with the very least amount of exposure, and the maximum amount of necessary firepower positioned to deflect an enemy attack. In essence, the patrol will be moving from one concealed position to another, getting through the danger area as safely and as quickly as possible.

Types of Danger Areas

Danger areas fall into two categories, linear and open. Each category has two sub-categories, big and small. The numerous types of danger areas require that patrols have multiple methods in their bag of tricks to get safely across the danger area.

Roads, paths, creeks, and open fields present opportunities for ambush and sniping missions. Natural and man-made obstacles allow for fairly long sectors of fire because they are relatively clear.

A patrol leader (PL) must assess is which type of danger area the patrol is presented with. Ideally, the patrol circumvents a danger area—that is, the patrol goes around. However, linear danger areas rarely leave that option. Instead, the patrol must traverse these danger areas by crossing them.

So, the PL has to assess the type and relative size of the danger area. Also, the PL has to assess the likelihood of enemy contact. It's a pretty quick mental checklist:

- Linear vs. Open
- Big vs. Small
- Time Constraints

If the patrol is moving through territory with significant enemy presence, hopefully the PL allotted a realistic amount of time to conduct the mission. The patrol employs a more deliberate method of crossing the danger area, one that offers maximum protection to the front and flanks. For a linear danger area, this might mean the heart-shaped method. For an open danger area, this might mean the box method.

If, on the other hand, the patrol is moving quickly through territory with sparse enemy presence and time is of a high priority, then the patrol employs a method of crossing the danger area that makes maximum use of speed as a form of security, with minimal protection to the front and flanks. For a linear danger area, this might mean the patch-to-the-road method. For an open danger area, this might mean the bypass method.

The size of the danger area must also be considered. Even in the case of patrolling through territory with sparse enemy presence, if the danger area is too large to use speed as a form of security...it may be best to use a method that offers a greater form of security.

The platoon leader or squad leader decides how the unit will cross based on the time he has, size of the unit, size of the danger area, fields of fire into the area, and the amount of security he can post. An Infantry platoon or squad may cross all at once, in buddy teams, or one Soldier at a time. A large unit normally crosses its elements one at a time. As each element crosses, it moves to an overwatch position or to the far-side rally point until told to continue movement.

(Patrolling) III. Crossing a Danger Area 8-19

I. Patch-to-the-Road Method

Using this method, a nine-man squad should be able to cross the danger area in ten seconds or less. *Speed is a form of security.* This method also allows the column formation to be maintained, which means greater control and communication for the PL.

1. The point man brings the patrol to a halt and signals that he has come upon a danger area. The PL comes forward to view the danger area, assesses the situation, and selects a method of negotiating the danger area.

2. If the patch-to-the-road method is selected, the PL communicates this to the team with the appropriate hand and arm signal. The entire patrol closes the intervals between members shoulder-to-shoulder. The patrol members must actually touch each other. This is done even during daylight hours. This will allow a very fast pace when crossing and prevent a break in contact.

3. The two-man security team moves from the rear of the formation up to the front. At the PL's signal, the first security troop steps up to the danger area only as far as he needs to look left and right. If the road is clear of enemy presence, the troop takes a position so he can view down the road to his right. In this position, his unit patch (on the upper part of his left arm sleeve) will be facing toward the middle of the road. Thus, the method is called "patch-to-the-road."

This method uses speed as the primary form of security. A left and right security overwatch is provided locally. At the patrol leader's signal, the rest of the patrol move in file across the danger area. (Photo by Jeong, Hae-jung).

8-20 (Patrolling) III. Crossing a Danger Area

4. As soon as the security troop on the near side of the danger area levels his weapon down the road, the second member of the security team immediately rushes across the danger area and takes up a position to view down the opposite direction of the road. At this point, both team members have their unit arm patches facing toward the middle of the road and they are pointing in the *opposite direction*.

5. As soon as the security troop is on far side if the danger area levels his weapon down the road, this signals the PL to stand the remaining patrol members and RUN across the danger area. This is done literally by holding onto the gear of the troop to the front.

6. As the last troop passes the near side security troop, he firmly says, "Last man." An acceptable alternative is to tap the security troop on the shoulder. In either case, this indicates to the security troop to stand up and run across the danger area behind the patrol.

7. The security troop will say firmly, "Last man", to the far side security troop or tap him on the shoulder. This lets that troop know to follow behind.

8. Now the entire patrol is back in its original marching order on the far side of the objective.

It is important that as the pointman initially crosses the danger area, that he makes a quick dash into the tree line to visually inspect the space the patrol will occupy. The *only reason to stop the patrol in the danger area* is if the pointman determines the far side tree line is booby-trapped. Even if the enemy has set up a near ambush, the patrol must assault through. No one stays in the danger area.

The potential danger here is that the security team troops become distracted from the mundane task of overwatching their sector. This is especially true if some snag holds up the process and the security team is forced to stand overwatch down the road for more that the allotted ten seconds.

It takes considerable discipline and lots of rehearsals to keep troops facing down a linear danger area, partially exposing themselves and generally feeling vulnerable when there is a hold-up such as another member tripping while running across the road, or getting caught on a fence wire, or dropping an unsecured piece of equipment and then doubling back to retrieve it. What generally happens at that point is that one or both of the security team members become agitated and turns to look to see what's going on in middle of the road instead of maintaining a vigilant overwatch of their sector.

Contingency Plan

Ideally, if the enemy does show up when the patrol is crossing a danger area, the security team will fire first. Or if there is on-coming traffic, the security team will shout a warning to the other patrol to momentarily halt and hide. This signal means no one else should attempt to cross the danger area. So it is imperative that the security team realizes they are to keep a vigilant overwatch of the danger area until:

- The patrol successfully traverses the danger area
- They are directed to hide from on-coming traffic
- Or the patrol becomes engaged in a firefight

If there is a break in contact due to traffic or contact with the enemy, each patrol must establish a method of link-up. Typically, if the patrol becomes separated, the patrol will rendezvous at the last designated en-route rally point (ERP).

II. Heart-Shaped Method

If the patrol has to pass through a linear danger area in territory with significant enemy activity, or if the linear danger area is simply too large to cross quickly with the patch-to-the-Road method, then the PL needs to select a method with the greatest amount of security the patrol can mass. The Heart-shaped method takes about three to five minutes even for a squad-sized patrol. It also has a tendency to scramble the order of march and requires a great command and control. But, if rehearsed thoroughly, these issues can be mitigated.

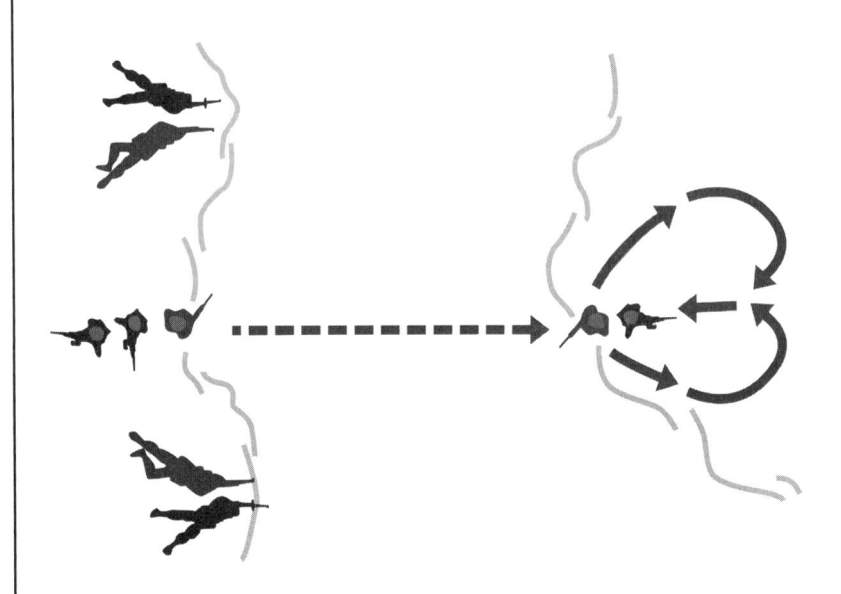

Crossing danger areas with this method makes maximum use of security and combat power. However, the heart-shaped method is time consuming. (Ref: FM 7-8, chap 2, section III, fig. 2-27).

1. The point man brings the patrol to a halt and signals he has come upon a danger area. The PL comes forward to view the danger area, assesses the situation, and selects a method of negotiating the danger area.

2. If the heart-shaped method is selected, the PL physically places a security team approximately 20~50 meters down to the right and places another security team 20~50 meters down to the left. The exact distance depends on the terrain and visibility of the danger area. The PL returns to the main body of the patrol.

3. The PL now sends a third security team to the far side, across the danger area. The PL does not go with the far-side security team. Instead, the PL points out an easily recognized object on the far side that is in line with the patrol's direction of movement.

8-22 (Patrolling) III. Crossing a Danger Area

4. This team will cross the danger area as the situation dictates—perhaps at a run, perhaps at a crawl. They have immediate fire support from the left and right side security teams.

5. Once concealed within the far side tree line, the security team conducts a quick listening halt to determine if the enemy is in the immediate area. If enemy are detected, the security team carefully makes its way back to the patrol and informs the PL of the situation.

6. If no enemy is detected, the far side security team physically inspects an area large enough for the entire patrol to fit. This is achieved by walking a designated distance into the tree line. Once they have walked the designated distance, members turn away from each other and pace off a determined distance to check the flanks. The security members then move back toward their original listening halt position. When looking at the path from a bird's eye perspective, it looks as though the security team has cut a heart-shaped path into the tree line. Thus it earned the name "heart-shaped."

7. When the security team has reassembled and determined the far-side security team free from enemy presence, they give the PL the "thumbs up" hand & arm signal. This lets the PL know the far side is secure and that the far-side security team is monitoring the danger area.

8. At this point, the PL leads the remainder of the patrol, minus the left and right security teams, across the field using the same path as the far-side security team took earlier. The left and right security teams continue to monitor the danger area.

9. When the patrol is safely on the far side of the danger area, the PL will signal by hand or by radio for the left and right security teams to cross the danger area, using the same path.

Contingency Plan

If the patrol is compromised while crossing the danger area, the patrol will rendezvous back at the last ERP. However, the patrol cannot simply run away and leave elements of the patrol still in the danger area. Without support, these troops would be killed or captured. Smoke canisters are employed to screen withdrawing troops and the left and right security teams place suppressive fires on the enemy until all patrol members have withdrawn. Once the patrol has withdrawn, the left and right security teams withdraw.

(Patrolling) III. Crossing a Danger Area 8-23

III. Bypass Method

The previously mentioned methods, patch-to-the-road and the heart-shaped, are all fine and well. But what if it is simply too dangerous to cross an open danger area? The patrol doesn't want to unnecessarily expose the patrol to enemy observation or fire. That could bring the mission to a quick end; especially if the patrol isn't suppose to make contact in the first place. In these cases, it's best to use the bypass. The bypass takes considerable time, but offers the greatest degree of stealth.

The bypass method is used for isolated danger areas, such as open meadows. The patrol takes several 90-degree turns until coming back on azimuth. The lateral distance is not added to the route pace count. (Photo by Jeong, Hae-jung).

1. After halting the patrol and signaling a danger area, the pointman and the PL confirm the patrol's direction of advance using a prominent feature as a point of reference on the far side of the danger area. This may be an easily recognized terrain feature, such as a rise or dip in the terrain, or it may be an easily recognized landmark, such as a tall tree, or a large boulder.

2. The PL estimates the distance to the far side of the danger area using visual techniques or the map. That distance is added to their present pace count.

3. Then, ignoring the pace count and compass bearing, the patrol simply follows the pointman as he skirts the danger area, keeping safely inside the tree line until the patrol gets to the designated feature on the far side of the danger area.

4. Here the pointman assumes the previous direction of advance and the patrol takes up the new pace count.

If the open danger area is so incredibly large that the patrol cannot even see the far side, one option is to deal with this terrain as "significantly thinning vegetation" instead of as a danger area. In such cases, the patrol assumes a wedge formation and significantly increases the interval between patrol members and subordinate teams. The patrol continues to move along the direction of advance in this manner until the terrain changes.

8-24 (Patrolling) III. Crossing a Danger Area

IV. Box Method

An alternative plan is the "box method" which is really a type of navigation technique that is closely related to the bypass. This method is more scientific in its execution and employs dead-reckoning skills.

1. After the pointman halts the patrol, the PL moves forward and confirms that the danger area is too large to see any prominent features on the far side.

2. The PL either adds 90° to the current direction of advance if he wants to turn the patrol to the right, or subtracts 90° from the current direction of advance if he wants to turn the patrol to the left. The new direction is issued to the lead team and pointman.

3. The patrol continues on the new direction of advance being careful to keep a pace count to record the distance traveled in this new direction. The patrol halts when the danger area is no longer visible.

4. The PL now assumes the old direction of advance for a distance that is greater than the length of the danger area. This information is confirmed on the map—since the far side of the danger area could not be visibly observed.

5. Again, the lead team and pointman pay careful attention to the distance traveled. Once the patrol has covered the prescribed distance, the pointman halts the patrol.

6. The PL now does the reverse of the earlier left or right turn. That is, the PL either adds 90° to the current direction of advance if he wants to turn the patrol back to the right, or subtracts 90° from the current direction of advance if he wants to turn the patrol back to the left.

7. When the patrol has traveled the same lateral distance as their first turn, mathematically speaking, the patrol is back on the original direction of advance. The patrol assumes the old direction of advance and takes up the new pace count. The entire danger area has been bypassed.

This method takes the patrol off route. This option can be selected only if the patrol has ample time to conduct the Box method and if the diverging route does not take the patrol out of the AO.

V. Crossing Large Open Areas

If the large open area is so large that the platoon cannot bypass it due to the time needed to accomplish the mission, a combination of traveling overwatch and bounding overwatch is used to cross the large open area. The traveling overwatch technique is used to save time. The squad or platoon moves using the bounding overwatch technique at any point in the open area where enemy contact may be expected. The technique may also be used once the squad or platoon comes within range of enemy small-arms fire from the far side (about 250 meters). Once beyond the open area, the squad or platoon re-forms and continues the mission.

On Point

When analyzing the terrain (in the METT-TC analysis) during the TLP, small unit leaders may identify danger areas. When planning the route, the leader marks the danger areas on his overlay. The term danger area refers to any area on the route where the terrain could expose the platoon to enemy observation, fire, or both. If possible, the platoon leader plans to avoid danger areas, but sometimes he cannot. When the unit must cross a danger area, it does so as quickly and as carefully as possible. During planning, the leader designates near-side and far-side rally points. If the unit encounters an unexpected danger area, it uses the en route rally points closest to the danger area as far-side and near-side rally points.

Examples of danger areas include—

- **Open Areas**. Conceal the platoon on the near side and observe the area. Post security to give early warning. Send an element across to clear the far side. When cleared, cross the remainder of the platoon at the shortest exposed distance and as quickly as possible.
- **Roads and Trails**. Cross roads or trails at or near a bend, a narrow spot, or on low ground.
- **Villages**. Pass villages on the downwind side and well away from them. Avoid animals, especially dogs, which might reveal the presence of the platoon.
- **Enemy Positions**. Pass on the downwind side (the enemy might have scout dogs). Be alert for trip wires and warning devices.
- **Minefields**. Bypass minefields if at all possible, even if it requires changing the route by a great distance. Clear a path through minefields only if necessary.
- **Streams**. Select a narrow spot in the stream that offers concealment on both banks. Observe the far side carefully. Emplace near- and far-side security for early warning. Clear the far side and then cross rapidly but quietly.
- **Wire Obstacles**. Avoid wire obstacles (the enemy covers obstacles with observation and fire).

Each danger area is unique and the PL will determine the manner in which the patrol overcomes each obstacle. The situation on the ground can change dramatically from what we see on a map. For instance, a linear danger area on the map might actually turn out to be a massive open danger area. Similarly, open danger areas on the map may actually be so overgrown that they present no danger area at all.

Enemy Contact at Danger Areas

An increased awareness of the situation helps the platoon leader control the platoon when it makes contact with the enemy. If the platoon makes contact in or near the danger area, it moves to the designated rally points. Based on the direction of enemy contact, the leader still designates the far- or near-side rally point. During limited visibility, he can also use his laser systems to point out the rally points at a distance. If the platoon has a difficult time linking up at the rally point, the first element to arrive should mark the rally point with an infrared light source. This will help direct the rest of the platoon to the location. During movement to the rally point, position updates allow separated elements to identify each other's locations. These updates help them link up at the rally point by identifying friends and foes.

IV. Establishing a Security Halt

Ref: FM 7-8 (FM 7-8) The Infantry Rifle Platoon and Squad, chap. 3 and 9.

Regardless of the mission, every patrol must halt at different locations along the route. This type of security halt is called the en route rally point (ERP). In addition to the ERP, every patrol makes a final stop prior to the assigned objective. This is done to coordinate between elements and make final preparations for actions on the objective. This type of security halt is called the objective rally point (ORP).

When the patrol halts, take a knee. Standing upright may draw fire. Troops face in the opposite directions forming 360-degree security. (Photo by Jeong, Hae-jung).

The ORP and ERP are security halts that afford 360° of security for the patrol as the patrol stops along its route toward the objective. Security halts provide concealment from enemy observation while plans and equipment are adjusted. The operative description for the security halt is *disciplined*. Noise and light discipline is strictly enforced.

The ERP is either occupied or at least designated along the route. Ideally, it is designated at easily recognized terrain or landmarks that offer cover and concealment. Again, the ERP does not need to be occupied…but it is still designated en route. The ERP may also be pre-designated in the plan.

The ORP is always pre-designated in the plan. It is placed far enough away from the enemy objective that the patrol can conduct final preparations and planning before conducting actions on the objective.

When halted longer than several minutes, take a step or two out and go to a prone position. This automatically forms a 'cigar-shaped' perimeter with the PL in the center. (Photo by Jeong, Hae-jung).

The patrol needs to stay safely concealed from the enemy's view and far enough that the noise of the patrol's final preparations won't be heard.

A rule of thumb is to remain 300 meters away or one terrain feature. Keeping one sizable terrain feature between the ORP and the objective significantly reduces the chance that the patrol will be detected. However, if the terrain is rather open, the ORP is kept at least 300 meters away from the objective. In practice, this decision really depends more on the type of terrain, the size of the patrol, and the nature of the patrol's final preparations.

As for the ERP, the only consideration regarding when and where to stop would be concealment.

I. Cigar-Shaped Method

Security halts are either taken by force or a leader's recon is sent forward to assess the anticipated site. The ERP is almost exclusively taken by force because there are so many unpredictable reasons for establishing an ERP. The drill that is most suitable for the ERP security halt is the "cigar-shaped" method.

The ORP, on the other hand, is a planned security halt and generally requires a leader's recon of the site to determine its suitability. In this case, the drill that is most suitable for the ORP is the "wagon wheel" method.

A small patrol often walks directly up to the ERP or ORP security halt. This practice is known as occupying by force. The implications are that if the enemy is detected near the security halt, the patrol either engages the enemy or quietly withdraws to a new position.

1. Upon reaching a desired location for the security halt, the patrol leader (PL) halts the patrol. During halts, all troops automatically take a knee.

2. The PL indicates to the patrol members that they are in a security halt. Each patrol member faces either left or right in an alternating pattern and takes two steps outward to form a cigar-shaped perimeter. The pointman, dragman, and PL are exempt from this maneuver and remain kneeling where they initially stopped. This leaves the PL in the center, the pointman at 12 o'clock, and the dragman at the 6 o'clock position.

3. Subordinate leaders then ensure each man is behind adequate cover and assigned a sector of fire. At a minimum, the 3, 6, 9, and 12 o'clock positions must be maintained and covering their sector of fire.

4. The PL pulls any necessary leaders to the center in order to confirm or adjust plans. If this security halt is an ORP, the PL begins work priorities for the ORP.

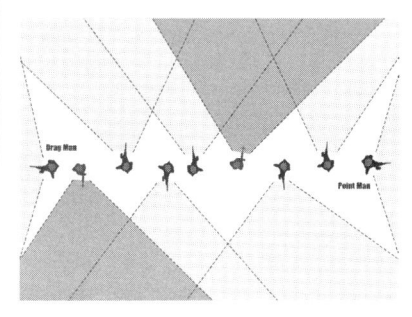

All troops should take a kneeling position upon a halt to lower their profile. They continue to face in their primary direction for security purposes—affording 360-degree security. (Ref: FM 7-8, chap 5, annex B, appendix 5, para 2-a).

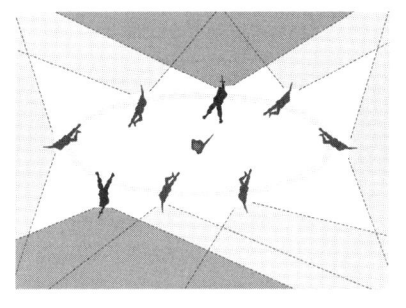

The 'cigar shape' is formed when troops take a step forward in the direction they are facing. They look for cover such as a thick tree or stone for frontal protection and assume a prone position. (Ref: FM 7-8, chap 5, annex B, appendix 5, para 2-b).

II. Wagon Wheel Method

This drill is used almost exclusively for occupying the ORP security halt. Developed in the jungles of Southeast Asia, this drill works well in areas of dense vegetation because it allows the PL to see where each troop is located. The effort is to get the entire patrol in a circular formation and this takes a bit more work than the cigar-shaped method. Care must be taken in selecting the ORP site, as well as occupying it with the least amount of noise and commotion.

This method of forming the ORP is one of the most simple. After a security team is placed as the anchor, the entire patrol plays "follow the leader" in a big circle. Once the circle is complete, the PL adjusts the circle evenly.

The wagon wheel method is conducted by having the patrol leader simply walk in a circle around the area of the intended ORP. Troops are adjusted once the circle is complete. Typically a machinegunner sits at 12 and 6 o'clock. (Photo by Jeong, Hae-jung).

1. The patrol assumes an ERP security halt approximately 100 meters out from the planned ORP site. A leader's recon is then conducted forward at the ORP site to make certain the terrain is appropriate for use.

2. The leader's recon typically involves four members of the patrol—the PL, a compass man, and a two-man security team. Before leaving the ERP, the PL issues the assistant patrol leader (APL) a contingency plan and coordinates for their return.

3. Once the leader's recon has reached the designated ORP site, the PL determines if the site is appropriate or selects another site nearby.

4. The PL places the two-man security back to back at the 6 o'clock position of the ORP. The security team will be left at the ORP to watch the objective and to guide the remainder of the patrol into position. The PL will leave a contingency plan with the security team before the PL and the compass man return to pick up the rest of the patrol.

5. The PL and the point man return to the rest of the patrol back in the security halt. The patrol resumes its marching order, and since the compass man has already been to the ORP and back, he can lead the patrol right to the security team at the 6 o'clock position. The patrol halts once the compass man links up with the security team and the PL moves forward.

6. From the 6 o'clock position the PL leads the patrol in a large circular path around the perimeter of the ORP. This forms the patrol into a large circle through an exercise of "follow the leader."

7. Once the circle has been completed, the subordinate leaders adjust the exact positions of the members to offer the best cover and to provide 360° sectors of fire.

8. The PL pulls subordinate leaders to the center in order to confirm or adjust plans. The PL sets work priorities in motion for the ORP.

This variation of the wagon wheel method requires a bit more coordination and is typically only used for larger patrols. Three elements make up three legs of the ORP—from 2 to 6 o'clock, from 6 to 10 o'clock, and from 10 to 2 o'clock. Machinegunners are typically placed at each apex. (Photo by Jeong, Hae-jung).

III. Priorities of Work at the Objective Rally Point

Regardless of the method used, once the ORP has been established, there is work to do. The ORP is merely a security halt where the patrol can finalize its preparations for the mission. Some of the work priorities are conducted concurrently; some may not be necessary at all.

1. Security is always the first priority. The ORP never falls lower than 50 percent security. That means that whatever task is necessary, half of the patrol maintains a vigilant guard of their sectors of fire.

2. A leader's recon is conducted. The leader's recon is optional and depends on the nature of the mission. The leader's recon team leaves and re-enters through the 12 o'clock position. It is coordinated with the patrol members in the ORP. Typically, the APL remains in the ORP while the PL, the security team leader, and a security team conduct the recon.

3. All special equipment is prepared. The patrol doesn't prepare explosives, anti-armor weapons, or conduct radio checks while sitting in position on the objective. That is too much noise and activity and the enemy would certainly see or hear the patrol.

4. Plans are finalized or altered. The leader's recon may come back with information that either slightly or dramatically alters the plan of the mission. Or, sometimes the failure of special equipment may require some improvising. In any case, these adjustments are made in the ORP and every member must be informed.

5. Weapons are prepared. Weapons are cleaned if the patrol made contact while en route to the ORP, or if the movement to the ORP took considerable time and moved through a notably dirty environment—such as fording a river or being inserted onto a sandy beach. Still, no more than 50 percent of the patrol members do this at one time. The other 50 percent pulls security.

6. Sleep and eating plans are initiated. If the situation dictates, the ORP may implement an eating and sleeping schedule. Fifty percent security is maintained, always.

(Patrolling) IV. Establish a Security Halt 8-31

On Point

Units conducting tactical movement frequently make temporary halts. These halts range from brief to extended periods of time. For short halts, platoons use a cigar-shaped perimeter intended to protect the force while maintaining the ability to continue movement. When the platoon leader decides not to immediately resume tactical movement, he transitions the platoon to a perimeter defense. The perimeter defense is used for longer halts or during lulls in combat.

There are many reasons a patrol may stop. The patrol leader might need to confirm the patrol's position on the map, the patrol periodically needs to listen and observe new surroundings, and the patrol may need a rest or to make final coordination prior to actions on the objective.

Security halts offer 360° of security when the patrol stops along its route toward the objective. Security halts provide concealment from enemy observation while plans and equipment are adjusted.

En Route Rally Point (ERP)

The ERP is a security halt along the route. Ideally, it is quickly recognizable on the ground in case the patrol needs to return to it. The ERP is rarely pre-designated, but is assigned or occupied along the route. When the ERP is occupied, it is almost always taken by force. The cigar-shaped method is well suited for this.

Objective Rally Point (ORP)

The ORP is the last stop prior to the objective, where the patrol can come together to coordinate with other friendly elements, to finalize the mission plans, and to rest prior to conducting actions on the objective. The ORP may be occupied by force—in which case the cigar-shaped method is used. More commonly, the ORP is first inspected by a leader's recon and then the patrol is carefully maneuvered into position. In this case, the wagon-wheel method is used.

The ORP is a point out of sight, sound, and small-arms range of the objective area. It is normally located in the direction that the platoon plans to move after completing its actions on the objective. The ORP is tentative until the objective is pinpointed. Actions at or from the ORP include—

- Issuing a final FRAGO
- Disseminating information from reconnaissance if contact was not made
- Making final preparations before continuing operations
- Accounting for Soldiers and equipment after actions at the objective are complete
- Reestablishing the chain of command after actions at the objective are complete

If the ORP is occupied under limited visibility, the method of occupation is rehearsed prior to the patrol's movement. Noise and light discipline are paramount.

Which method choosen to occupy a security halt depends on the terrain, the mission, and the number of troops available. Every combat mission will include security halts. Security is paramount.

V. Establishing a Hide Position

Ref: FM 7-93 Long-Range Surveillance Unit Operations, app. E.

Hide positions are primarily used for surveillance teams during reconnaissance operations. The recon team uses the hide position to rest troops while keeping them concealed. The troops can then be rotated to the surveillance position during their shift. Often the hide and surveillance site are combined. Hide positions may be subterranean or above ground.

However, hide positions are also used by small patrols operating for extended periods of time beyond the forward edge of battle area (FEBA). Such hide positions can be temporarily employed for many reasons. Perhaps the patrol entailed a march that could not be achieved in a single day or, more commonly, the patrol itself is of such a small size that a patrol base is neither feasible nor necessary. In these cases, the patrol can opt to implement a hide position in order to plan and rest.

Considerations

The hide position is similar to the patrol base in that it is a security perimeter that uses concealment for its primary defense. However, the hide position is not a patrol base. It is only a place to rest or observe.

There are several key differences between the hide position and the patrol base:

- No one departs or re-enters the hide position (No missions are conducted)
- It is intended for no more than 12 hours of use and is vacated
- Hide positions are never re-used due to risk of detection by the enemy
- Fighting positions are not built up

The hide position offers 360° security in that the entire patrol is positioned in a tight formation facing outward. The patrol is so close they act as a single fighting position, as few as two men can easily maintain security. The hide position does not require any communication system other than word of mouth and visual contact with other troops.

The hide position is established ideally no closer that 300 meters or a terrain feature away from an enemy force. Hide positions are intentionally placed in the most inhospitable terrain, such as in thick patches of thorn bushes or jagged rock formations. Although a bit uncomfortable, this terrain discourages enemy patrols.

The patrol leader (PL) designates the approximate location of the hide position on a map, or the PL may designate a condition under which the patrol establishes a hide position—such as after patrolling for a set number of hours or days. If the patrol is small, the PL may occupy the hide position by force.

For a medium sized patrol, the patrol establishes a security halt and sends a leader's recon forward to identify the hide position. At any rate, the PL indicates to the patrol that they are in hide position.

Establishing a Hide Position

I. Back-to-Back Method

This method is more practical for wooded and heavily vegetated terrain. When seated, the security team can observe of the likely avenues of approach or escape. This could not be achieved if the security team were laying in the prone in heavy vegetation. And frankly, if the patrol is exhausted enough that it has to use the hide position, placing the watch team on their bellies is just asking for trouble. An exhausted troop is much more likely to fall asleep lying down than sitting up—no matter how disciplined.

When a hide position is placed in a heavily vegetated area without decent observation, the two troops pulling security will sit back-to-back to form a 360-degree security position. The rest of the troops rest head-to-toe until it is their turn. (Photo by Jeong, Hae-jung).

1. The patrol members come shoulder to shoulder, take a knee and face left and right in an alternating pattern. All patrol members drop their rucksacks.

2. Half of the patrol members are designated to ready their sleeping bags and mats while the other half pull security. Once the first members have readied their sleeping positions as comfortably as possible, they sit on their equipment and pull security while the other members ready their sleeping bags and mats. No tents are pitched, no early warning devices are implemented, and no fighting positions are prepared.

3. The PL determines how many members of the patrol will pull security and what the duration and schedule of the guard shifts will be. Typically, hide positions require at least two troops to pull security at a time.

4. Since no anti-personnel mines or trip flares are used, CS canisters or fragmentation grenades are given to the first guard shift and then passed to subsequent guards. If the enemy does walk near the patrol, great discipline must be enforced to allow the enemy to pass by. In the unlikely case that the enemy walks up on the hide position, grenades are used while the patrol makes a quick escape. Direct fire should be avoided at night since

the muzzle flashes from the rifles and machine guns will disclose the patrol's position.

5. Radios are handled in a similar manner—passed from guard shift to guard shift—to be keep in touch with higher command. If the hide position is occupied during nighttime hours, night vision devices are also passed from guard shift to guard shift.

II. Star Method

The star method is used for flat, open terrain such as a desert, high mountain tundra, or grasslands. In this type of terrain, the patrol lies on the ground to lower their profiles.

When the vegetation permits a decent field of observation, the troops lay prone to form a star. Two troops on opposite sides of the formation pull security while others sleep. They kick their buddies to an alert status if a threat approaches. (Photo by Jeong, Hae-jung).

1. The patrol comes shoulder-to-shoulder and the PL instructs them to form into the star. The troops lay in the prone and interlock their ankles. This allows the security team to kick the man to their left and right to an alert status without making noise.

2. Half of the patrol members are designated to ready their sleeping bags and mats while the other half pull security. This responsibility changes hands while the other half of the patrol readies their sleeping bags and mats. Again, no tents are pitched due to being easily visible; however, rain tarps may be used to cover the patrol's sleeping bags.

3. The PL determines how many members of the patrol will pull security and what the duration and schedule of the guard shifts will be. Typically, hide positions require at least two troops to pull security at a time. Also, the two-man watch team will not be positioned right next to each other, but on opposite sides of the formation.

4. It might prove to be a daunting task to find terrain that is difficult to traverse in middle of the grassland prairie. The best a patrol could do would be to place a far distance between it and the enemy position and blend into the vastness of the countryside. Also, due to the ease of enemy movement in the open terrain, the PL may opt to use early warning devices or anti-personnel mines to slow an enemy attack.

On Point

Hide positions provide reasonable security to rest a small or medium-sized patrol. However, when patrols become tired, undisciplined or lazy, hidings are often overused due to the low requirement of security.

The type of hide or surveillance site employed depends on METT-TC. Improvement of camouflage, at a minimum, must be continuous while occupying the site.

Hidings have their purposes, but attempting to use a hide position as a replacement for the patrol base places the patrol in great danger. These two security positions have completely different functions. If a larger patrol needs rest, the PL must establish a patrol base or a tactical assembly area and implement a sleep plan.

Site Selection

The team leader initially selects the tentative sites during the planning phase. He selects the sites by physical reconnaissance (stay-behind), aerial observation, photographs, line-of-site data, soil and drainage data, or map reconnaissance. At a minimum, the team leader selects primary and alternate hide sites, and primary and alternate surveillance sites. Before the team occupies the sites, the team leader conducts a physical reconnaissance of the tentative site chosen during planning. If necessary, the team leader moves the site to a better location.

When selecting a site, the leader should consider the following aspects:

- Line of sight to target
- Within a range that can be supported by available observation equipment to meet the reporting requirements
- Overhead concealment and cover
- Away from natural lines of drift
- Away from roads, trails, railroad tracks, and major waterways
- Defendable for a short time
- Primary and alternate hasty exits
- Concealed serviceable entrance; little noise getting into and out of the hide site
- METT-TC in relation to other site positions (hide, surveillance, communication sites)
- Not near man-made objects
- Downwind of inhabited areas
- Not dominated by high ground, but takes advantage of high ground

Site Sterilization

Before departing hide and surveillance locations, team members must ensure sites and routes have been sterilized.

- Personnel carry out all foreign debris
- If possible, they do not bury waste or trash. Animals will uncover trash and expose it to enemy patrols. If trash is buried, the team buries it 18 to 24 inches deep in sealed containers or covers the scent by using CS or lime.
- The team sterilizes the sites using displaced earth. They use the site to bury overhead material, which contrasts with the surrounding area
- The team camouflages the area by blending the site with local surroundings
- As team members withdraw from the site, they ensure routes are camouflaged to prevent detection

VI. Establishing a Patrol Base

Ref: FM 3-21.8 (FM 7-8) The Infantry Rifle Platoon and Squad, chap. 9 and The Ranger Handbook, pp. 5- 19 to 5-22.

A patrol base is a position set up when the patrol unit halts for an extended period. When the unit must halt for a long time in a place not protected by friendly troops, it takes active and passive security measures. The time the patrol base may be occupied depends on the need for secrecy. It should be occupied only as long as necessary, but not for more than 24 hours--except in an emergency. The unit should not use the same patrol base more than once.

The patrol base is a temporary, forward, static position out of which a patrol conducts a series of missions. It offers cover and concealment from enemy observation. Security is maintained at 360° inside the patrol base, and while there is no requirement to maintain 100 percent security at all times, the percentage of troops maintaining security is kept at a level that work priorities will allow.

Patrol bases are typically used--
- To avoid detection by eliminating movement
- To hide a unit during a long detailed reconnaissance
- To perform maintenance on weapons, equipment, eat and rest
- To plan and issue orders
- To reorganize after infiltrating on an enemy area
- To establish a base from which to execute several consecutive or concurrent operations

The goal is to go undetected by the enemy; the patrol base is never used for more than 24 hours. The patrol base is a temporary position that uses concealment as its primary defense. As such, there is no need to develop fighting positions, bunkers, or trench systems. However, to provide a minimal amount of cover, some build up of the patrol base defenses should be tolerated. Entrenching tools and machetes make a good deal of noise. Barricades are preferable to digging, but hasty fighting positions or 'shell scrapes' are generally permitted.

Site Selection

The leader selects the tentative site from a map or by aerial reconnaissance. The site's suitability must be confirmed and secured before the unit moves into it. Plans to establish a patrol base must include selecting an alternate patrol base site. The alternate site is used if the first site is unsuitable or if the patrol must unexpectedly evacuate the first patrol base.

The rule of thumb on where to place a patrol base dictates no closer than 500 meters from the enemy force, or better yet, to maintain a major terrain feature between the enemy and the patrol base. To further conceal the position and the number of foot trails leading back to the patrol base, all subsequent patrols depart and re-enter the patrol base at the 6 o'clock position.

Establishing a Patrol Base - The Triangle Method

The triangle method is excellent for patrols with three elements (i.e. three fireteams, three squads, three platoons). A crew-served weapon is placed at each of the three apexes of the triangle—6, 10 and 2 o'clock. In this manner, no matter which direction the enemy approaches the patrol base, at least two crew-served weapons are brought to bear on the attacking force.

1. The patrol leader (PL) establishes an objective rally point (ORP) within 300 meters of the anticipated patrol base site. The PL conducts a leader's recon of the patrol base site, leaving a contingency plan with the assistant patrol leader (APL) back in the ORP. The leader's recon includes the PL, compass man, and a six-man security team.

3. They rendezvous with the rest of the patrol in the ORP and inform them of any change of plan. The compass man is now familiar enough with the terrain to lead the patrol forward to the patrol base. If this is done at night, the security team member facing back towards the patrol must have a visual reference for the point man— a chemical light stick, flashlight, or illuminated compass lens. The compass man will link the security team with the rest of the patrol at the 6 o'clock position of the patrol base.

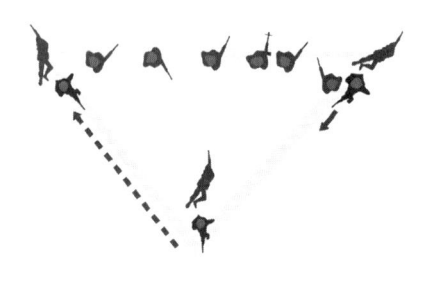

Phase Two: The patrol leader walks the first squad from 6 o'clock to 10 o'clock, and then places them in a line from the 10 o'clock to the 2 o'clock position. The troops face out toward the enemy threat. (Ref: FM 7-8, chap 3, section V, fig. 3-22).

The perimeter is formed in four phases.
Phase One: The patrol leader determines the direction of greatest threat as 12 o'clock, then sets security teams at the 6, 10, and 2 o'clock positions. (Ref: FM 7-8, chap 3, section V, fig. 3-22).

2. Once the PL has inspected the site and is satisfied with the patrol base site (or has chosen a suitable alternative) the PL leaves a two-man security team at the 2, 6, and 10 o'clock positions of the patrol base with a contingency plan. Each security team is placed back to back. The PL and compass man move back to the ORP.

4. The PL then leads the first element from the 6 o'clock position, up to the 10 o'clock position. He then walks that first element from the 10 o'clock to 2 o'clock positions and physically places each member of the element in a straight line between the two security teams. The PL then returns to the 6 o'clock position.

5. Waiting at 6 o'clock is the second element of the patrol. The PL walk the second element from the 6 o'clock to the 10 o'clock positions and physically places each member of the element in a straight line between these two security teams. The PL returns again to the 6 o'clock position.

6. The third element waits at the 6 o'clock security team position. The PL links up with the third element and walks them in a straight line between the 6 o'clock and the 2 o'clock position, physically placing each member of the element.

Phase Three: *The patrol leader returns to the 6 o'clock position to walk the second squad in a line between the 6 o'clock on the 10 o'clock positions. Troops must link up with the apex to their left and right. (Ref: FM 7-8, chap 3, section V, fig. 3-22).*

7. The PL moves to the center of the patrol base to establish the command post (CP). The PL coordinates his subordinate leaders to be sure they are all aware of each other's location and tied into each other's left and right line.

Phase Four: *The PL meets the last squad at the 6 o'clock position and walks them in a line between the 6 o'clock and 2 o'clock positions. Each apex is assigned a crew-served weapon, and the command post is center. (Ref: FM 7-8, chap 3, section V, fig. 3-22).*

Planning Considerations

Leaders planning for a patrol base must consider the mission and passive and active security measures. A patrol base must be located so it allows the unit to accomplish its mission.

- Observation posts and communication with observation posts
- Patrol or platoon fire plan
- Alert plan
- Withdrawal plan from the patrol base to include withdrawal routes and a rally point, rendezvous point, or alternate patrol base
- A security system to make sure that specific Soldiers are awake at all times
- Enforcement of camouflage, noise, and light discipline
- The conduct of required activities with minimum movement and noise
- Priorities of work

Security Measures

- Select terrain the enemy would probably consider of little tactical value
- Select terrain that is off main lines of drift
- Select difficult terrain that would impede foot movement, such as an area of dense vegetation, preferably bushes and trees that spread close to the ground
- Select terrain near a source of water
- Select terrain that can be defended for a short period and that offers good cover and concealment
- Avoid known or suspected enemy positions
- Avoid built-up areas
- Avoid ridges and hilltops, except as needed for maintaining communications
- Avoid small valleys
- Avoid roads and trails

(Patrolling) VI. Establishing a Patrol Base 8-39

Priorities of Work - Patrol Base

Once the PL is briefed by the R&S teams and determines the area is suitable for a patrol base, the leader establishes or modifies defensive work priorities in order to establish the defense for the patrol base. Priorities of work are not a laundry list of tasks to be completed; to be effective, priorities of work must consist of a task, a given time, and a measurable performance standard. For each priority of work, a clear standard must be issued to guide the element in the successful accomplishment of each task. It must also be designated whether the work will be controlled in a centralized or decentralized manner. Priorities of work are determined IAW METT-TC. Priorities of work may include, but are not limited to the following tasks:

1. Security is always the first priority. The patrol base is maintained at a level of security appropriate to the situation. As a rule of thumb, the patrol base does not fall below 33 percent security. That means one out of three troops are diligently watching their sectors of fire.

2. An alternate defensive position is designated. Typically, the PL informs the subordinate leaders that the ORP will serve as a fallback position in the event the patrol base is over-run. This information is disseminated to all of the patrol members.

3. An ambush team covers the trail into the patrol base. A small force backtracks approximately 100 meters from the 6 o'clock position and then steps off of the trail. This ambush team observes the trail for a half hour or so to be certain no enemy force has followed the patrol into the patrol base. This must be done immediately after the patrol base has been secured.

4. Communication is established between all key positions. Field phones or radios are positioned with the CP and each apex at the 2, 6, and 10 o'clock positions.

5. An R&S team conducts a recon of the immediate area. After communication is established, the PL dispatches a recon & security (R&S) team to skirt the area just outside the visible sectors of fire for the patrol base. Everyone must be informed. Otherwise, patrol members may fire upon the R&S team.

6. Mines and flares are implemented. After the R&S team confirms that the area immediately around the patrol base is secure, those positions designated to employ mines or flares carefully place them at the far end of their visible sectors of fire—no more than 35 meters out. These anti-personnel mines and early warning devices must be kept within viewable distance of the patrol base.

7. Hasty fighting positions are constructed. Barricades are the preferred method as digging and cutting can be too loud and may disclose the position. Fighting positions make use of available micro-terrain. If a hasty fighting position is necessary, care is taken to camouflage the exposed earth.

8. Plans are finalized or altered. The patrol's missions may be altered slightly or significantly in time. The PL makes these adjustments and every member of the patrol base is informed. If at all possible, shoulder-to-shoulder rehearsals are carried out in the center of the patrol base, prior to conducting missions.

9. Weapons are cleaned. This is particularly true if the patrol made contact during a mission or if the movement to the patrol base took involved moving through a particularly filthy environment—such as fording a river or being inserted onto a sandy beach. Still, no less than 33 percent of the patrol members maintain security.

10. Sleep and eating plans are initiated. If the situation dictates, the patrol base implements an eating and sleeping schedule, while maintaining security.

(SUTS2) Index

5 and 20 Meter Checks, 8-18

A

Actions by Friendly Forces, 1-13
Actions on the Objective, 6-21
Administrative Movement, 5-29
After Action Review (AAR), 1-55
Alternate Positions, 3-12
Ambush, 6-3
Approach March, 5-29S
Approach-March Technique, 2-8
Area Ambush, 6-16
Area Defense, 3-11
Area of Influence, 1-18
Area of Interest, 1-18
Area of Operations, 1-18
Area Reconnaissance, 5-16
Area Security, 5-6
Arms Control, 4-5
Army Core Competencies, 1-10
Army Operational Concept, 1-5
Art of Tactics, 1-1
Assault a Building, 7-10
Assault Element, 8-6
Assault Position, 2-5a, 2-16
Assault Team, 6-18
Assault Time, 2-5a
Assess, 1-17
Attack, 2-13
Attack by Fire, 1-13, 2-34
Attack Formations, 8-11
Attack Position, 2-5a
Avenues of Approach, 1-33

B

Backbrief, 1-54
Back-to-Back Method, 8-34
Battles, Engagements and Small-Unit Actions, 1-2
Battle Handover Line, 2-5a
Battle Position, 3-3a, 3-24
Bounding Overwatch, 8-9
Block, 1-12
Box Method, 8-25
Branch, 2-22
Breach, 1-13
Bunkers, 7-19
Bypass, 1-13
Bypass Method, 8-24

C

Canalize, 1-12
Checkpoints, 4-12
Cigar-Shaped Method, 8-29
Clear, 1-13, 2-32
Clearing Rooms, 7-13
Close Operations, 1-19
Cold Regions, 7-31
Cold Region Operations, 7-31
Combat Orders, 1-37
Combat Outposts, 5-7
Combat Patrols, 8-1
Combat Power, 1-22
Combating Terrorism, 4-5
Combination Pursuit, 2-25
Combined Arms Maneuver, 1-5
Commander's Critical Information Requirements (CCIR), 1-20
Commander's Intent, 1-20
Contain, 1-12
Control, 1-13
Control Crowds, 4-13
Control Measures,
  Defensive, 3-3a
  Offensive, 2-5a
Convoy, 4-13
Cordon, 7-10
Core Competencies, 1-5, 1-10
Counterattack, 1-11
Counterdrug Operations, 4-5
Counterinsurgency (COIN) Operations, 4-17
Counterrecon, 1-13
Course of Action Development, 1-30
Cover, 1-11, 5-5
Crossing a Danger Area, 8-19
Crossing Large Open Areas, 8-25

D

Danger Areas, 8-19
Decisive Action, 1-6
Decisive Operation, 1-5
Decisive-Shaping-Sustaining, 1-19
Deep Operations, 1-19
Deep-Close-Support Areas, 1-19
Defeat, 1-12
Defending Encircled, 5-27
Defense Support of Civil Authority Tasks, 1-9
Defense, 3-1
Defensive Control Measures, 3-3a
Defensive Tasks, 1-8, 3-2
Delay, 1-11, 3-20
Deliberate Operations, 1-4
Describe, 1-16
Desert Environments, 7-23
Desert Operations, 7-23
Destroy, 1-12

Index-1

Index

Deterrence, 4-5
Diamond Attack Formation, 8-17
Direct, 1-17
Direction of Attack, 2-5a
Disengage, 1-13
Disengagement Line, 3-3b
Dissemination of Information, 5-17
Disrupt, 1-12
Domain Knowledge, 1-3

E

Effects on Enemy Forces, 1-12
Elements of Combat Power, 1-22
En Route Rally Point (ERP), 8-32
Encirclement Operations, 5-27
Enemy Contact, 8-26
Engagement Areas, 3-16
Entering a Building, 7-11
Envelopment, 2-4
Escort a Convoy, 4-12
Essential Elements of Friendly Information (EEFI), 1-21
Exfiltration, 1-13
Exploitation, 2-19

F

Far Ambush, 6-4
File Attack Formation, 8-14
Final Coordination Line, 2-5b
Find, 7-3, 7-18
Finish, 7-18
Fire Support Coordination Measures, 3-3b
Fire Team Formations, 8-11
Fix, 1-12, 7-18
Fixing Force, 3-6
Follow and Assume, 1-13
Follow and Support, 1-13
Force Tailoring, 1-22
Foreign Internal Defense (FID), 4-4
Forms of Maneuver, 2-4
Fortified Areas, 7-17

Forward Edge of the Battle Area, 3-3b
Forward Line of Troops (FLOT), 5-24, 5-25
Forward Operating Base (FOB), 4-11
Foundations of Unified Land Operations, 1-5
Fragmentary Order (FRAGO), 1-40
Frontal Attack, 2-5
Frontal Attack, 2-5
Frontal Pursuit, 2-24

G

Guard, 1-11, 5-4
Guerrillas, 7-9

H

Hasty Operations, 1-4
Heart-Shaped Method, 8-22
Hide Position, 8-33
Homeland Defense, 10d
Human Dimension, 1-3
Humanitarian Assistance, 4-4

I

Individuals, Crews, and Small Units, 1-1
Infiltration, 2-5
Insurgents, 7-9
Interdict, 1-12
Isolate, 1-12
Isolate the Building, 7-3

J

Jungle Environments, 7-47
Jungle Operations, 7-47

K

Key Terrain, 1-33

L

Lead, 1-17
Level of Force, 4-16
Limit of Advance, 2-5b
Line Attack Formation, 8-13
Line of Departure, 2-5b
Local Security, 5-7
Lodgment Area, 4-11

M

Main and Supporting Efforts, 1-19
Main Battle Area, 3-3b
Meeting Engagement, 2-7
METT-TC (Mission Variables), 1-31
Military Aspects of the Terrain (OCOKA), 1-32
Military Engagement, 4-5
Mission Analysis, 1-30
Mission Command, 1-10
Mission Symbols, 1-11
Mission Variables (METT-TC), 1-31
Mobile Defense, 3-5
Monitor Compliance with an Agreement, 4-7
Mountain Environments, 7-39
Mountain Operations, 7-39
Movement Techniques, 5-30
Movement to Contact, 2-7
Moving in the Building, 7-14
Mutual Support, 1-22

N

Near Ambush, 6-4
Negotiations, 4-8
Neutralize, 1-12
Noncombatant Evacuation Operations (NEO), 4-5

O

Objective, 2-5b
Objective Rally Point (ORP), 8-31, 8-32
Observation Posts, 4-12
Occupy, 1-13
OCOKA - Military Aspects of the Terrain, 1-32
Offense, 2-1
Offensive Control Measures, 2-5a
Offensive Tactical Tasks, 2-29
Offensive Tasks, 1-8, 2-2
Open and Secure Routes, 4-13
Operational Framework, 1-18

Operations Order (OPORD), 1-39, 1-42
Operations Process, 1-24

P
Passage of Lines, 5-23
Patch-to-the-Road Method, 8-20
Patrol Base, 8-37
Patrol, 4-13, 8-1
Patrolling, 8-1
Peace Operations, 4-15
Penetrate, 1-11
Penetration, 2-5
Perimeter, 3-27
Planning Guidance, 1-20
Point Ambush, 6-16
Point of Departure, 2-5b
Pre-Combat Inspection (PCI), 1-45
Preparation, 1-45
Primary Positions, 3-12
Priorities of Work in the Defense, 3-18
Probable Line of Deployment, 2-5b
Pursuit, 2-23

R
Raid, 6-17
Rally Point, 2-5b
Range Card, 3-14
Recon Pull, 5-13
Recon Push, 5-13
Reconnaissance, 5-9
  Fundamentals, 5-10
  Operations, 5-9
  Patrols, 8-1
Reconnaissance in Force (RIF), 5-18
Reconnaissance Objective, 5-9
Reconstitution, 3-22
Reduce, 1-13
Regional Environments, 7-1
Rehearsals, 1-51
Relief in Place, 1-11, 5-19
Reserve Operations, 4-13
Retain, 1-13

Retirement, 1-11, 3-21
Retrograde, 3-19
Reverse Slope, 3-27
Risk Management (RM), 1-36
Risk Reduction, 1-4
Route Reconnaissance, 5-12

S
Science of Tactics, 1-2
Screen, 1-11, 5-1
Search and Attack, 2-8
Search, 4-13
Sector Sketch, 3-14
Sectors of Fire, 3-15
Secure, 1-13
Security Assistance, 4-4
Security Checks, 8-18
Security Cooperation, 4-5
Security Element, 8-6
Security Force Assistance (SFA), 402
Security Fundamentals, 5-2
Security Halt, 8-27
Security Measures, 8-39
Security Operations, 1-19, 5-1
Security Team, 6-18
Seize, 1-13, 2-30
Seize, Retain and Exploit the Initiative, 1-6
Sequel, 2-22
Shaping Operation, 1-19
Show of Force, 4-5
Site Sterilization, 8-36
Special Purpose Attacks, 6-1
Stability Operations, 4-1
Stability Tasks, 1-9, 4-11
Staggered Column Attack Formation, 8-15
Star Method, 8-35
Striking Force, 3-6
Strongpoint, 3-25
Subsequent Positions, 3-12
Supplementary Positions, 3-12

Support Area, 1-19
Support by Fire, 1-13, 2-31
Support by Fire Position, 2-5b
Support Efforts, 1-19
Support Element, 8-6
Support Team, 6-18
Support to Insurgency, 4-5
Suppress, 2-30
Sustaining Operation, 1-19

T
Tactics, 1-1
Tactical Doctrinal Taxonomy, 1-14
Tactical Enabling Tasks, 5-1
Tactical Level of War, 1-1
Tactical Mission Fundamentals, 1-1
Tactical Mission Tasks, 1-11
Tactical Problems, 1-3a
Tactical Road March, 5-29
Tactical Victory, 1-3b
Task-Organizing, 1-22, 1-47
Threat, 7-8
Time of Attack, 25b
Train, Advise & Assist, 4-1
Train to Win in a Complex World, 1-10b
Transition, 2-5c, 3-4
Traveling Overwatch, 8-8
Traveling Techniques, 8-7
Traveling, 8-8
Trench Systems, 7-20
Triangle Method, 8-38
Troop Leading Procedures, 1-25
Troop Movement, 5-29
Turn, 1-12
Turning Movement, 2-4

U
Uncertainty, 1-3
Understand, 1-15
Unified Action, 1-10
Unified Land Operations, 1-5
Unity of Effort, 4-16
Urban Environments, 7-1, 7-6
Urban Operations, 7-3, 7-4

Index-3

V
Visualize, 1-16

W
Wagon Wheel Method, 8-30
Warfighting Functions, 1-23
Warning Order (WARNO), 1-38
Wedge Attack Formation, 8-16
Wide Area Security, 1-5
Withdrawal, 1-11, 3-21

Z
Zone Reconnaissance, 5-14

SMARTbooks
INTELLECTUAL FUEL FOR THE MILITARY

Recognized as a "**whole of government**" doctrinal reference standard by military, national security and government professionals around the world, SMARTbooks comprise a **comprehensive professional library** designed with all levels of Soldiers, Sailors, Airmen, Marines and Civilians in mind.

The SMARTbook reference series is used by **military, national security, and government professionals** around the world at the organizational/institutional level; operational units and agencies across the full range of operations and activities; military/government education and professional development courses; combatant command and joint force headquarters; and allied, coalition and multinational partner support and training.

View, download FREE samples and purchase online:
www.TheLightningPress.com

The Lightning Press is a **service-disabled, veteran-owned small business,** DOD-approved vendor and federally registered — to include the SAM, WAWF, FBO, and FEDPAY.

SMARTbooks
INTELLECTUAL FUEL FOR THE MILITARY

MILITARY REFERENCE: JOINT & SERVICE-LEVEL

Recognized as a "whole of government" doctrinal reference standard by military professionals around the world, SMARTbooks comprise a comprehensive professional library.

MILITARY REFERENCE: MULTI-SERVICE & SPECIALTY

SMARTbooks can be used as quick reference guides during operations, as study guides at professional development courses, and as checklists in support of training.

HOMELAND DEFENSE, DSCA, & DISASTER RESPONSE

Disaster can strike anytime, anywhere. It takes many forms—a hurricane, an earthquake, a tornado, a flood, a fire, a hazardous spill, or an act of terrorism.

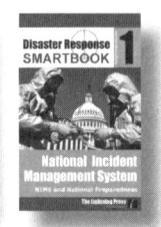

The Lightning Press is a **service-disabled, veteran-owned small business,** DOD-approved vendor and federally registered — to include the SAM, WAWF, FBO, and FEDPAY.

RECOGNIZED AS THE DOCTRINAL REFERENCE STANDARD BY MILITARY PROFESSIONALS AROUND THE WORLD.

JOINT STRATEGIC, INTERAGENCY, & NATIONAL SECURITY

The 21st century presents a global environment characterized by regional instability, failed states, weapons proliferation, global terrorism and unconventional threats.

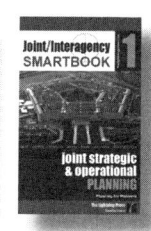

THREAT, OPFOR, REGIONAL & CULTURAL

In today's complicated and uncertain world, the military must be ready to meet the challenges of any type of conflict, in all kinds of places, and against all kinds of threats.

DIGITAL SMARTBOOKS (eBooks)

Our eBooks are a true "A–B" solution! Solution A is that our digital SMARTbooks are available and authorized to a user's Adobe ID and can be transferred to up to six computers and devices via Adobe Digital Editions, with free software available for **85+ devices and platforms—including PC and MAC, iPad, Android Tablets and Phones, and more.** Solution B is that you can also use our digital SMARTbooks through our dedicated SMARTbooks iPad App!

View, download FREE samples and purchase online:
www.TheLightningPress.com

Purchase/Order

SMARTsavings on SMARTbooks! Save big when you order our titles together in a SMARTset bundle. It's the most popular & least expensive way to buy, and a great way to build your professional library. If you need a quote or have special requests, please contact us by one of the methods below!

View, download FREE samples and purchase online:
www.TheLightningPress.com

Order SECURE Online
Web: www.TheLightningPress.com
Email: SMARTbooks@TheLightningPress.com

Phone Orders, Customer Service & Quotes
Live customer service and phone orders available Mon - Fri 0900-1800 EST at (863) 409-8084

24-hour Voicemail/Fax/Order
Record or fax your order (or request a call back) by voicemail at 1-800-997-8827

Mail, Check & Money Order
2227 Arrowhead Blvd., Lakeland, FL 33813

Government/Unit/Bulk Sales

The Lightning Press is a **service-disabled, veteran-owned small business**, DOD-approved vendor and federally registered—to include the SAM, WAWF, FBO, and FEDPAY.

We accept and process both **Government Purchase Cards** (GCPC/GPC) and **Purchase Orders** (PO/PR&Cs).

*The Lightning Press offers design, composition, printing and production services for units, schools and organizations wishing their own **tactical SOP, handbooks, and other doctrinal support materials**. We can start a project from scratch, or our SMARTbooks can be edited, custom-tailored and reproduced with unit-specific material for any unit, school or organization.*